COAL
AS AN
ENERGY RESOURCE
Conflict and Consensus

ACADEMY FORUM

NATIONAL ACADEMY OF SCIENCES
Washington, D.C. 1977

This Academy Forum received support from

American Can Company Foundation
American Electric Power Company
Ashland Oil, Inc.
Atlantic Richfield Company
Bethlehem Steel Corporation
Bureau of Mines, U.S. Department of Interior
The Continental Group, Inc.
Energy Research and Development Administration
Environmental Protection Agency
General Dynamics Corporation
General Electric Company
General Motors Corporation
Kerr-McGee Corporation
National Academy of Sciences

LIBRARY
The University of Texas
At San Antonio

Library of Congress Catalog Card Number 77-92117

International Standard Book Number 0-309-02728-4

Available from

Printing and Publishing Office
National Academy of Sciences
2101 Constitution Avenue, N.W.
Washington, D.C. 20418

Printed in the United States of America

CONTENTS

FOREWORD *Robert R. White*   vii

DAY I

INTRODUCTION TO DAY I *Walter R. Hibbard, Jr., Cochairman*   3

COMMENTARY *Arthur M. Bueche*   5

COAL AS AN ENERGY RESOURCE *Lester Lave*   7
    Discussion, 11

HOW FAST SHOULD COAL BE DEVELOPED?   13
    Gerald L. Decker, 13
    Louise Dunlap, 18
    Discussion, 21

WHERE SHOULD COAL BE DEVELOPED?   24
    William C. Wampler, 24
    Frank G. Zarb, 27
    Discussion, 30

ENVIRONMENTAL IMPACT AND COSTS *Steven Reznek*   34

THE ORDER AND INCIDENCE OF CLEAN COAL UTILIZATION
COSTS *Richard T. Newcomb*   39

    Discussion, 44

DAY II

INTRODUCTION TO DAY II *Alvin M. Weinberg, Cochairman*   51

THE OHIO RIVER VALLEY   53

    Orientation *E. R. Heiberg, III, Chairman*, 53
    Boyd Kennan, 58
    Herbert B. Cohn, 66
    Harvey I. Sloane, M.D., 71
    Jackie Swigart, 74
    Vernon K. Morrison, 78
    Ralph Madison, 79
    James B. Hartnett, 81
    Discussion, 83
    Summary Report to the Plenary Session *E. R. Heiberg, III*, 96
    Inquiries from the Panel, 100

KAIPAROWITS   109

    Orientation *Calvin L. Rampton, Chairman*, 109
    Brant Calkin, 113
    James H. Drake, 114
    Daniel A. Dreyfus, 118
    Steven E. Plotkin, 121
    Discussion, 123
    Summary Report to the Plenary Session *Calvin L. Rampton*, 152
    Inquiries from the Panel, 157

NORTHERN GREAT PLAINS   161

    Orientation *L. John Hoover, Chairman*, 161
    Jeannette Studer, 162
    Michael B. Enzi, 170

    Gary Payne, 176

    Daniel B. Sullivan, 182

    Martin A. White, 188

    Theodore H. Clack, Jr., 194

    Angela Russell, 201

    Robert L. Valeu, 205

    Discussion, 210

    Summary Report to the Plenary Session *L. John Hoover*, 214

    Inquiries from the Panel, 215

PLENARY SESSION DISCUSSION   219

DAY III

INTRODUCTION TO DAY III *Walter R. Hibbard, Jr., Cochairman*	227
COAL TECHNOLOGIES *Thomas V. Falkie*	229
THE KEY ROLE OF COAL *George R. Hill*	236
Discussion, 246	
REPORTS FROM THE WORKSHOPS	250
Workshop 1: Environment *Clinton W. Hall*, 250	
Workshop 2: Labor *Thomas N. Bethell*, 252	
Workshop 3: Dispute Resolution/Rules of Reason *Milton R. Wessel*, 258	
Workshop 4: Technology/Research and Development *Thomas V. Falkie*, 263	
Workshop 5: Coal Conversion *Kenneth C. Hoffman*, 266	
Workshop 6: Habitat *Raymond L. Gold*, 270	
Workshop 7: Health and Geochemistry *Bobby G. Wixson and Albert L. Page*, 274	
SOME LONG-RANGE SPECULATIONS ABOUT COAL *Alvin M. Weinberg and Gregg H. Marland*	277
Discussion, 286	

WHAT MUST BE DONE? 295
    John McCormick, 295
    John F. O'Leary, 297
    Harry Perry, 306
    Eric H. Reichl, 312

SUMMARY OF FORUM *Lester Lave* 319

THE ACADEMY FORUM 325
    Panel for Inquiry, 325
    Program Committee, 325
    General Advisory Committee, 326

FOREWORD

Robert R. White
Director, Academy Forum

The echoes of times past are heard throughout the record of this Academy Forum on coal, convened at the National Academy of Sciences on April 4-6, 1977. The discussions of Kaiparowits, the Ohio River Valley, and the strip mining of the northern Great Plains contain both the triumphs and the human prices of the Industrial Revolution fueled by coal.

It is understandable though ironic that as we confront and begin to live with a new era of escalating fuel prices and possible scarcity, we turn again to a uniquely American source of energy--the vast resources of coal under our land--that is accompanied by old and familiar problems of pollution and dislocation.

In all considerations of the future of coal, however, we must keep in mind that its problems and promises are now set in the context of new social values that have yet to evolve into national purpose and resolve. One strong theme repeatedly emerges from this Forum in all the case analyses and workshops: How can we reconcile our local and regional aspirations and way of life with national needs and purposes?

With Kaiparowits, the issue was the supply of energy to California versus the price of environmental and social disturbances in Utah. California lost.

With strip mining in the northern Great Plains, the issue is the supply of coal and energy to Chicago and other metropolitan areas versus the price of social displacement of Indians and ranchers as well as the perplexities of land reclamation and boomtowns. How do we negotiate equity?

The same kinds of questions arise in discussions of the Ohio River Valley: Should Louisville pay the environmental and social price of adjacent coal-fired plants in order to supply energy to Detroit and St. Louis?

The complex social, political, and economic organizations of our industrially developed world seem to need new mechanisms, institutions, and perhaps the social harmony required to negotiate if we are to find ways to realize the values that sustain the highest quality of human life. Somehow we must find better paths toward accommodation. In this respect, the serious and hardworking participants in this coal Forum--committees, principals, and public alike--established a significant example.

DAY 1

INTRODUCTION TO DAY I

Walter R. Hibbard, Jr., *Cochairman*
*University Distinguished Professor
of Engineering, Virginia Polytechnic
Institute and State University;
Former Director, Bureau of Mines*

This Forum is convened to develop dialogues regarding the use of coal as a fuel. One of its objectives is to identify the issues that will emerge with the increased use of coal, as proposed by the President, and to discuss possible alternative actions. Another is to provide some enlightenment as to actions that might prevent or resolve, in a widely acceptable way, problems of concern to all Americans in the areas of economics, politics, the law, technology, and society.

Coal is our most plentiful domestic energy reserve with mineable amounts equivalent to up to 2,000 years' supply at today's requirements. Our use of coal reached a peak in 1947; although it has lost the home-heating, railroad, and industrial-thermal markets to gas and oil, it retains approximately half of the utility market.

Although we know where most of our coal is and how to mine it, there is a growing realization that, despite what we understand about coal mining, we need to learn more if we are to take coal out of the ground in a way that is acceptable to the people who are affected.

An area of concern to this Forum is that of the vocal, media-dominated electorate, which is effectively involved in participatory democracy and which is distressed on the one hand about utility bills and on the other hand about pollution and land preservation.

We need to know more and to talk about coal-mining boomtowns. Having grown too quickly without appropriate infrastructure, certain towns are paying a high price in lowered quality of life in order to elevate the human life-styles in other cities.

Present federal regulations prohibit the burning of over half the coal that is used today by the utilities. Legislation that has pressured a turnaway from coal is now firmly enforced by large investments. Thus, the return to coal is slow. The 1976 consumption was up 7.6

percent from 1975, but this is far from the projected doubling of production which is projected for 1985.

Mine health and safety laws and their administration are under scrutiny. Strip mine and mined land reclamation laws are a continuing uncertainty. Labor unions are in a state of turmoil. We have moratoriums on the leasing of federal coal lands in the northern Great Plains. It is becoming apparent that, if we want to greatly increase the production of coal, we need to reverse many of our restricting laws and enact new laws as protective as possible of the American and his environment.

Coal mines must be planned five to eight years in advance. The utility market is reluctant to undertake such long-range commitments. No utility plants have been retrofitted to coal in the past three years. If all adversaries succeed, there will be no increased use of coal.

Because there are strong adversaries as well as strong advocates for the step-up of coal production, this Forum brings together both viewpoints, and we hope to point out the required trade-offs between them. Good trade-offs require accurate information about what is possible technically, economically, legally, politically, and socially. We need to think about the consequences of each suggested action. It is the hope of this Forum that, through such discussion of both sides and the possible trade-offs and options, a common perception of and penetration into the complex of problems will be made, common goals will be identified, and actions will later be taken which are acceptable and effective.

COMMENTARY

Arthur M. Bueche

Vice-President,
Corporate Research and Development,
General Electric Company

This Academy Forum is addressing one of the most important challenges facing the nation and the world today. It is also an area where great harm can be done if our deliberations are not handled carefully and well. Let me review some of the reasons for my concern.

My associates and I, over the past couple of years, have been trying to devise an energy scenario of our own. Of course, we haven't tried to develop all the numbers ourselves. What we have done is to take conservative estimates of demand for total energy--not really the *most* conservative but the most sensible conservative estimate that various people have suggested--and we have looked out to the year 2000 to ask how much energy we might really need.

We have taken a very optimistic view of conservation. We have looked at each part of energy consumption--transportation, industry, home heating, and so forth--and have tried to be just as optimistic as we could, realizing, however, that the American people won't be willing to change their way of life very readily. Also, we assumed that in the year 2000 there will be just as much petroleum and natural gas available as we have today. We took an optimistic view of solar energy, anticipating that by 2000 A.D. it will be giving us 2.5 quads of energy and that geothermal and hydro sources will be giving us 6 quads.

Then, we said, let us assume that all the nuclear power plants and coal power plants now planned will be completed. After that, what do we have to make up between now and the year 2000? The answer is really quite striking. It turns out to be 26 quads. That is the estimate without a safety factor, and as one of my associates says, if his wife is inviting ten for dinner she always cooks for twelve, just in case. But I haven't put in the safety factor in this case. We are 26 quads short without a safety factor, according to this analysis.

Now, what can we do? We can certainly start getting all the oil and gas that we know about, and hopefully we will be lucky and can offset some of those 26 quads with additional gas and oil finds. But we would have to be mighty lucky, indeed. Let me just point out that if all the North Sea oil were available for use just by the United States, it would carry us only five years. So even big finds, like Alaska and the North Sea, may change the time scale a little bit, but they really don't make much of a dent. We need some monumental discoveries. The other thing we can do, of course, is perform research and development on things such as fusion and solar electricity and all the other schemes we keep hearing proposed. But the fact is that those won't be ready by the year 2000, even if we are lucky today.

Thus, we will have to get 26 quads somehow, and the somehow appears to be mainly from nuclear fuel and coal, or certainly nuclear and fossil fuel. That tells us what we have to do. I don't know what the balance should be between the two, but I assume that a prudent man's balance would be some of each.

Now, how serious is it if we miss this by, oh, a few quads? If we miss it by a few quads gradually, we probably can stand it. But if we miss it by a few quads all of a sudden, it will be disastrous. Let me remind you that the Arab oil embargo created a shortage of about 1 quad. The energy shortage we had last winter amounted to about 1 quad, and some think even a little less than that. In one of those instances we triggered the largest depression this country has had since the 1930s; in the other we threw a million people out of work and worked considerable hardship on many others. So, a 1-quad shortage is already a pretty serious thing. A 10-quad shortage, even if it is gradual, could be catastrophic. And we need 26 quads to just make ends meet, and without having any safety factor! What we do about coal is one of the most important decisions facing this nation today.

COAL AS AN ENERGY RESOURCE

Lester Lave

*Professor and Head,
Department of Economics,
Carnegie-Mellon University*

I have good news and bad news. First of all, the good news. There is lots and lots of coal. Depending on which estimates one believes, there are between 10 and 30 quintillion Btu of coal reserves. That is enough coal to last a long time even if coal were the only energy resource. Furthermore, the cost of coal is relatively inexpensive. It is the lowest cost resource available to us today, about $0.75 per million Btu compared, for example, with about $2.25 per million Btu for OPEC oil. It is cheap, and it is likely to stay cheap in the future because much of the coal could be strip-mined. That is the good news.

The bad news is that sometime we might have to use that coal. This bad news can be appreciated only after a general overview, so I will begin with that.

In 1975, about 21 percent of our energy came from coal (about 15 of 75 quads). When we look forward to the year 2010, optimistic projections indicate we might be able to get about 25 quads from oil and natural gas, perhaps 30 to 40 quads from nuclear sources, and about 10 quads from all other sources (hydro, solar, geothermal, et cetera). Thus, optimistic projections indicate we might get as much as 65 to 75 quads from these sources, although we would have to be lucky to get 40 quads of nuclear energy.

How much energy will we demand? The most convenient way of forecasting is to assume that the future will be like the past: if so, energy would grow at slightly over 4 percent per year, as it did from 1945 to the present. Such a projection indicates we will need 300 quads of energy in 2010, about four times the present demand. But these other domestic resources are likely, if we are very optimistic about them, to give us about 75 quads, only 25 percent of the needed energy.

Thus, 225 quads must come either from imported oil or from coal. Suppose the energy deficit all came from imported oil; that would be roughly 112 million barrels of imported oil per day, which is more than ten times the current production of Saudi Arabia. Need I say that it would be impossible to buy so much oil? Obviously, most of the energy must come from coal. But that means a fifteenfold increase in coal extraction and use.

A fifteenfold increase in coal is probably both impossible to achieve and insane to desire. It is conceivable that a national commitment to produce and use coal would bring about such an increase. By brushing aside licensing, environmental impact statements, and other procedures that delay and by relaxing standards concerning mine safety, restoration, and emissions from combustion, we might be able to get a fifteenfold increase in coal. My question is, who in the world would want that? Certainly not me.

I was not in Pittsburgh in 1945, but I have heard descriptions of what it was like. I imagine that if we mined and burned fifteen times as much coal as we do now, the nation would look like Pittsburgh did in 1945. Our street lights would be on twenty-four hours a day, and the sun would be the subject of stories for our grandchildren, who will have never seen it. I can't believe the nation would embark on such a path. Therefore, we had better look for some better ways of balancing energy supply and demand.

Let me begin this search with a notion that has become very fashionable: conservation. For a number of reasons (such as forecasts of a slower rate of growth in GNP in the future and of a lower fertility rate), energy use is not likely to grow as rapidly in the future as it did in the past. Added to this is the possibility of conservation. For example, Sweden has approximately the same per capita income as we do and uses about half as much energy per capita as we do. That diminished use of energy comes about equally from two sources. First, they have a less-energy-intensive mix of goods and services in their consumption bundle, and second, they produce the goods and services that they do consume in a more energy-efficient way.

Examination either of our own history or of the histories of other nations indicates that conservation is a real possibility for the future. Let me make a leap and suppose conservation could supply 150 of those 300 quads projected for 2010. Perhaps 150 quads could come from domestic fuel sources other than coal. That leaves 75 quads to come either from imported oil or from coal.

Seventy-five quads is five times the amount of oil that we import today. It seems unlikely that we would be able to buy even twice what we import today; thus, again we are sent back to coal. A three- to fivefold expansion in coal is possible and probably could be done so that the environmental consequences would not be unacceptable.

Another possibility for balancing energy supply and demand is to accept either a somewhat diminished growth rate of GNP or a change in our consumption and location patterns--life-style. With either of these possibilities we might be able, conceivably, to use only 100 quads in the year 2010. If so, domestic resources other than coal could provide 75 percent of needed energy, leaving only 25 percent to come from imported oil or coal, an easy accomplishment.

Finally, there is the possibility of both slower GNP growth and lifestyle change that would freeze energy use at its current level of 75 quads. In either a 75- or 100-quad future we could have the luxury of building in a reserve or choosing among the fuels or technologies by using all less intensively or excluding one as socially undesirable.

Now, I come back to the bad news and will specify why I think mining and burning 225 quads of coal per year would be a disaster. This is a liturgy that I am afraid you are going to hear several times over the next three days, but it ought to be said several times.

First, from the deep mining of coal there are accidents that result in about 200 coal miners' dying each year. There are literally thousands of new cases of coal workers' pneumoconiosis (black lung), which disable thousands of miners annually. In addition there are notable environmental problems that stem either from acid mine drainage or from subsidence. Deep mining is inherently dangerous and destructive of the environment. An alternative is surface-mined coal. Certainly, surface mining results in fewer accidents, but there are the problems that come from attempting to restore the strip-mined land. After mining the coal, transportation is the next problem. We currently mine something of the order of 660 million tons of coal. That represents a very sizable proportion of the total freight that is carried on railroads, for example. Associated with transporting that coal by rail are about 100 deaths from transportation accidents each year.

Finally, we get to the most troublesome part of the story, the burning of the coal. When we burn coal, we release large amounts of particulates, sulfur, oxides, carbon monoxide, and carbon dioxide into the atmosphere. Sulfur oxides and the suspended particulates have been shown to have notable health effects. Eugene Seskin and I have estimated that there are tens of thousands of people who die prematurely each year owing to the burning of coal. In addition to those tens of thousands of people who die prematurely, there are millions of people who suffer either from acute or chronic diseases that are caused by or aggravated by this air pollution.

There is more general environmental damage from air pollution caused by acid rain. For example, it is common in New Hampshire to have rain with a pH of less than 4 (sometimes as low as 3). But the United States is at an advantage compared to places such as Scandinavia, where there are lakes so acidic that no fish can live in them. Indeed, some months ago Scandinavia charged that 20 percent of its acid rain came from air pollution emitted in North America; apparently, even the Atlantic Ocean is not wide enough to absorb our air pollution.

The consequences to human beings and to the environment of burning almost 1 billion tons of coal per year are large. Until recently, we just did not recognize these effects. I find it inconceivable that we would attempt to burn as much as 225 quads of coal each year.

Conceivably, these problems could be solved, or at least their effects largely mitigated. There are a number of additional problems that are not likely to be mitigated, which may prove to be of much greater importance in the future.

The first one is a labor problem. Suppose that 75 percent of our energy in the United States came from coal. I conjecture that we would have created a domestic OPEC in the form of the United Mine Workers. Given the extent of wildcat strikes and aspects of the relationship between the union and the companies and between the miners and their union, that prospect is not a very happy one.

A second kind of problem is regional in nature. The West has a very fragile ecosystem. It gets little rainfall, and the ecology is easily disrupted. Suppose we attempted to mine 5 billion tons of coal each year in the West. This would cause problems with boom towns, water allocation, land restoration, coal transportation, and air and water pollution that would make the problems that were raised with respect to the Four Corners area or Kaiparowits seem small. The resulting environment would be unacceptable to most westerners and would lead to clashes among states and jurisdictional issues whose only precedent would be the Civil War.

A possible solution to the environmental problem is to convert coal into either liquids or gases, thereby alleviating some of the environmental problems stemming from combustion. But generating these synthetic liquids and gases also generates carcinogenic hydrocarbons. Many problems would arise that cannot be anticipated at this point. It is easy to predict that a large synfuel industry would generate a host of problems to human health and to the environment.

A final environmental problem is known as the "cross media" problem. For example, to clean sulfur oxides out of flue gases from coal we might scrub the flue gas with a limestone mixture. That produces large quantities of gypsum sludge. Air pollution is being converted to a liquid waste problem. If I were a utility being asked to invest in a scrubber, I would think several times before doing so, not only because of the costs that are involved, but also because of the possibility that sludge disposal would be more of a problem than the original air pollution was.

I have tried to give you a brief introduction to some of the problems. It is clear that we have a great deal of coal, that we can be domestically self-sufficient in energy if we choose, and that the private cost of achieving energy self-sufficiency with coal (the cost of mining, transporting, and then burning the coal) would be small. The social costs due to a host of undesired consequences would be much greater. Many of those consequences cannot be anticipated now. We hope to clarify the major issues so that we can begin intelligent discussion and to resolve some of the uncertainties that are keeping coal development from proceeding at a pace at which it will have to proceed in the future. We need a lively exchange of views--perhaps "a lively exchange of intelligent views" is a better way of putting it--in order to clarify the various sides of the issue.

DISCUSSION

HIBBARD: Is there anyone from the audience who would like to contribute at this stage?

ROSS FORNEY, Director of the Institute of Mining Automation, Colorado School of Mines: Having spent about eighteen years working with the CEGB, having had several discussions with Dr. Lawther of the Air Pollution Unit in London, and having spent some time in Norway looking into the matter of acid rain, I would like to comment very quickly.

The CEGB, after about ten years, and the Air Pollution Unit in London, after twenty years, have not been able to establish any direct relationship between SO_2 and health. In fact, recently, they ran some tests on monkeys, and eventually on technicians in the lab, in which they had the subjects breathe sulfuric acid mist, and they were not able to determine that there were any bad health results from that.

On the question of acid rain, the condition is that fish are not able to reproduce below, I believe, 3.2. That condition does arise in early spring before the turnover in the lakes. There have been one or two fish kills in Norway, but normally what is of concern is the fact that the fish cannot reproduce. Actually, today they are developing a strain of fish able to reproduce below 3.8.

HIBBARD: Thank you very much. It might be that discussion of this sort would be more appropriate when we get to the subject of environmental impact. By the way, rain is acid anyway. It is 5.6 normally.

FRED SINGER, University of Virginia: I would like to put two questions to Dr. Lave. He mentioned the fact that small OPECs might be set up in the coal business, and my first question is whether he would elaborate on the fact that the western states are beginning to raise their severance taxes considerably--I believe 38 percent in the case of Montana, 33 percent in the case of North Dakota, and so on.

My second question has to do with the value of land that serves both the purpose of farming and also has coal under it. Recently, Secretary Andrus of the Department of the Interior has announced, I guess, a preliminary intention of not allowing strip mining on what he calls prime agricultural land. My question is, does the price of land truly reflect its value for all purposes, or is there some additional value to land that is not captured by the price?

LAVE: I would like to comment on all three questions. Concerning health effects, it certainly is true that pure SO_2 does not have health effects in the environmental concentrations that we see now, as long as it is a pure gas. Laboratory experiments have demonstrated that. To the contrary, sulfuric acid mist does produce substantial reactions, even in relatively small concentrations. I refer you to the proceedings of the Fourth Symposium on Statistics and the Environment sponsored by the American Statistical Association, the American Society on Quality Control, and the National Academy of Sciences a year ago. There is an excellent paper by Mary Amdur, which looks at the toxicological information, and one by Bertram Dinman on SO_2.

There have been a large number of epidemiologic studies done in the United States that demonstrate both for occupational groups and for the population in general that these health effects are quite marked. I will pass the question of acid rain for a second.

Now, on to Dr. Singer's questions. It certainly is clear that individual states such as Montana, Wyoming, or North Dakota could exercise significant power. Even a severance tax of one dollar a ton would be hidden in the cost of mining plus transportation. I am surprised that the severance taxes are not larger than they are. One of the reasons why severance taxes have not been raised more is the vast quantities of coal that could be mined in any one of those states: Each state must fear that if it alone raised the severance tax, mining would shift to other states.

With respect to strip mining under farmland, I understand that the land needs to be disrupted for only two to three years and that it can be restored at the end of that period to the original fertility (if the topsoil is removed initially and care is taken in restoration). Those kinds of regulations don't make very much sense to me, if the land is properly restored after mining.

HOW FAST SHOULD COAL BE DEVELOPED?

GERALD L. DECKER

Corporate Energy Manager, Dow Chemical Company

Before addressing the question of how fast our coal resources should be developed, let me say that I realize full well just how much controversy and honest difference of opinion permeates this whole area of energy needs and growth rates.

Having spent most of my manageable time in the last two years agonizing over the future of coal, I have reached the conclusion that coal must be developed at an optimum rate. If coal is developed too fast, there will be unwarranted damage to the environment. If developed too slowly, the nation will not become free of dependence on foreign energy supplies. The optimum must be determined by negotiation between the proenvironment and proenergy adversaries, as I will explain later.

I believe that I have heard most of the arguments on both sides of this question, and there are some very good arguments on both sides. When I cite some numbers, therefore, I hope that you will understand that I am not making a hard and fast forecast but am giving you some figures that I regard as reasonable and as illustrating the urgency and magnitude of the problem as I see it.

The bulk of the energy needs of the United States are satisfied by natural gas, crude oil, and coal as primary fuels, with nuclear power making an ever-increasing contribution. From 1850 to about 1910, coal production in the United States grew at a rate of 6.6 percent per year but, since 1910, has been relatively constant, fluctuating around a figure of between 500 and 600 million tons per year. After 1910, the growth in energy demand was largely filled by oil and natural gas. They were cheaper fuels, easier and cleaner to handle. With the onset of price regulation of natural gas and, later on, environmental pressures, coal became the least desirable of the fossil fuels.

The combination of price regulation and rising costs made exploration for new gas and oil supplies less and less attractive economically. Subsequent to World War II the ratio of proven reserves to production rate dropped steadily until, in the period around 1972, the oil and gas fields were producing at about the maximum economical rate. Since then, the proven reserves have continued to drop and so have the production rates of gas and oil in the United States. Except for the recession period, however, the demand for energy has continued to grow, and the result has been a substantial increase in oil imports, with all of the accompanying current and potential problems.

This situation cannot be allowed to continue, for the path we are on must, it seems to me, lead to social and economic disaster, or at the very least, severe hardship. Part of the problem, of course, is that some segment of our society finds objectionable features in just about every alternative proposal that has been put forth to alleviate the energy crisis. Nevertheless, I believe that it is becoming increasingly realized that some positive action is needed and needed quickly. I am, therefore, more hopeful now than I have been in some time that action will be taken; but I am not so hopeful that the action taken will really be adequate.

Let's look at a few figures that may help bring into focus the degree of urgency appropriate to the further development of our coal resources. Most energy studies seem to center on the projected situation in 1985 and beyond. For our purpose, I think that it is instructive to consider what might happen during the next five or so years, because it is all too easy to think that there is a lot of time yet to meditate on, argue over, and reconsider conditions that are a decade or more into the future. Five years is a period much closer at hand. It is less than the term of office of a senator and not much more than the term of office of a president.

There is another interesting feature about looking at the next five years. Namely, there isn't very much that we can do to change it very significantly. This is a fact that many of those proposing pat solutions to the energy crisis tend to overlook. Any new nuclear or other power plants that are going to become operational during the next five years are already under construction. None of the intriguing new energy sources that media writers seem to be so fond of--such as solar power, windmills, tidal power, methanol from agricultural wastes, and the like--none of them are going to amount to a hill of beans in the next five years. There are probably not going to be any sweeping new conservation programs to substantially reduce energy demand during the next five years, either. Of course, there will be additional conservation to some extent--after all, industry has been reducing its unit consumption of energy steadily for some years--but my point is that it will be pretty much an extension of what has already been going on. It is not very likely, for example, that *all* of the homes in the United States will have additional insulation installed during the next five years.

Perhaps most important of all, our domestic supply of natural gas and crude oil is not going to go up much, if any, during the next five years--indeed, it will almost certainly continue on its past downward trend for most of that time. I believe that this would be true even if the Congress acted tomorrow to remove all controls from the prices of gas and oil. That would indeed be a powerful motivation to increase our oil and gas supplies, and I think that it is something that should be done in some form or other. But the hard fact must be faced that it takes a period of some years to do the necessary exploration, test drilling, and development of new fields and pipe lines. So again, our course is pretty well set for a time.

To get down to some numbers, let's consider energy demand first. The FEA, in their "1976 National Energy Outlook," projected a demand of around 98 quads in 1985. Some people, who I think are terribly optimistic, tend to regard that as a little high. I, personally, believe that we will find that 98 quads is a bit constraining on our economy and employment. I hasten to add that this FEA figure presupposes a fairly effective national energy conservation program. Using the FEA number for 1985, and assuming a constant rate of growth in demand after 1976, we would expect an energy demand in 1981 of 86 quads.

The pattern of recent years, in terms of new discoveries, suggests that the 1981 production of oil might amount to 14.9 quads, and of natural gas, 19.4 quads. Extension of the trend since 1970 of the generation of electricity from nonfossil sources would give the fuel equivalent in 1981 of 8.5 quads, which seems fairly realistic. Assuming that there may be another quad from miscellaneous sources, this leaves 42.4 quads to be supplied by coal and imports of energy, principally oil.

Let's assume that our present coal mining capacity is about 600 million tons per year. Then, if we wanted to hold our oil imports to 8 million barrels per day, we would have to add coal mining capacity to produce an additional 120 million tons during 1978; then 135 million more tons in 1979; and 126 million additional tons in 1980; and, finally, another 118 million tons in 1981. The total capacity in 1981 would be 1,098 million tons, which is about the best that some knowledgeable people think we can do by 1985.

If we are willing to let oil imports rise to 10 million barrels per day, we would be somewhat better off. We would need another 71 million tons in 1979, which is fortunate, since it does take some years to open a new coal mine. In 1980, we would require 126 million more tons, and another 117 million tons in 1981--for a total 1981 capacity of 914 million tons. Ask any coal mine operator about the problems involved in increasing the total industry capacity by 50 percent in five years!

If we look at the problem the other way around, and assume that coal mining capacity remains at 600 million tons per year, then oil imports could be expected to rise to 9.3 million barrels per day in 1978; to 10.8 million barrels per day in 1979; 12.1 in 1980; and to 13.4 million barrels per day in 1981. With even a modest rate of inflation in the price of imported oil, that amounts to a bill in 1981 of around 85 billion dollars.

If you don't like my numbers, I invite you to work out some of your own, but I would strongly urge that you take a realistic view of how long it takes in the real world to accomplish some of the changes that we all hope will be forthcoming as well as some of the other realities involved in making substantial changes in U.S. society or in our standard of living and life-style. If you do, I suspect that you will come to the same conclusion that I have reached--that is, things are going to get tougher, as far as energy is concerned, before they begin to get better, and that will be true even if we make some difficult decisions in the very near future. And if we don't make those decisions--well, things will get tougher for a lot longer than just the next five years.

Perhaps you will also come to another conclusion that I have reached --that the answer to the question, "How fast should coal be developed?" is really, "Just as fast as we can." Of course, that brings us naturally to the next question, which is, "How fast is it possible for us to develop coal?" and that brings in a whole host of other problems, some of which I will touch on briefly.

A large part of the root of the difficulty is that coal is essentially a dirty fuel by present-day standards, and it is very easy to adversely affect the environment in mining it and in burning it. It is easy, but I really don't think that it is absolutely necessary, to have more than a minimal adverse effect. Because coal is a "dirty" fuel, the list of issues related to its use is long and challenging. Some of the major problems which have been raised are the following: sulfur in the air; acid in the air; particulates in the air; haze formation in mountain air; scarring of the earth by strip mining; polluted water in mine drainage; dust blowing from coal; boom town problems, followed by ghost town problems; safety in deep mines; black lung disease in deep mines; lack of water for pipeline transport; reliability of SO_2 scrubbing equipment, and disposal of sludge from scrubbing.

All of these issues, along with others, have significant problems associated with them. It is my considered opinion, however, that technology now exists to handle many of these problems and in the long run all of them can be resolved satisfactorily. In the shorter term, some reasonable temporary solutions must be negotiated between environmental and industrial advocates in order to allow coal to develop at the pace needed.

I must acknowledge, however, that my opinion is not shared by all of those on either side of the argument over coal. It tends to get confused by emotional and political considerations, economic factors, and other things that even color the interpretation of the status of the technology. Underlying it all is a nagging uncertainty and sense of mistrust over what kind of legislation can be expected and what kind of standards of performance must be met in the future. Opinion has tended to become rather polarized, with industry and the environmentalists being the principal antagonists and other groups taking up pieces of the argument.

Fortunately, an effort is now under way that may help to shed some light on this problem--a problem that has up to this point generated more heat than light. Believing that some sort of meeting of minds will be necessary before any really lasting progress can be made, we have initiated a program which is labeled the "National Coal Policy Project." A number of leaders of the environmental movement and of those industries that consume large quantities of energy are meeting together as task forces and as one large group to address those issues that seem to be most restrictive to the development of our coal resources. The shared objective is to arrive at some sort of middle ground, wherever this is possible, that is tolerable to both parties and to formulate this into a series of policy and action recommendations that can then be publicized to all interested parties.

The first series of meetings of the National Coal Policy Project was held early last July, under the general chairmanship of John Dunlop, former Secretary of Labor, and was considered by both sides to have been highly productive and very encouraging in terms of continuing the project. The organization and funding of the program proceeded under the sponsorship of the Center for Strategic and International Studies of Georgetown University and the administrative guidance of Francis X. Murray, director of national energy programs for CSIS. I am serving as chairman of the industrial caucus, and Laurence Moss, past president of the Sierra Club, is chairman of the environmental caucus. Our general chairman is now the Reverend Francis X. Quinn, S.J., of Temple University.

The first plenary session of the project was held on January 18 and 19, 1977, at Gaithersburg, Maryland. Task forces were set up as the primary working groups of the program, to deal with the following areas: mining, air pollution, energy pricing, fuel utilization and conservation, and transportation. The scope and topics to be considered by each task force were defined, and operating procedures and financial guidelines were established. Since then, the task forces have each had an initial meeting which has gone quite well, on the whole. Funding is proceeding and at present is approaching about 50 percent of our projected budget. Money is coming from a number of foundations, from some governmental agencies, and from industrial concerns which have a strong interest in the future energy situation in the United States.

All in all, the National Coal Policy Project is coming along very much as I had hoped it would. We hope to have our work completed by the end of this year. It is perhaps too early to say just what the results may be, but the people participating are working well together and I am very optimistic.

Just as I said earlier that the supply of relatively problem-free fuels cannot expand to meet the demand by 1981, so I have concluded that the supply of knowledge about coal cannot be built fast enough by 1981 or even 1985 to resolve the problems connected with coal. If the gains we have made on air pollution and other environmental conditions are to be maintained and if the United States is to regain its freedom from dependence on foreign supplies of a resource as basic as energy, there will have to be some trade-offs made concerning coal. The doing of this we think is the dominant need for the next few years.

I feel reasonably sure that we will find enough areas of accommodation and agreement in the National Coal Policy Project to permit our country to develop and use our greatest fuel resource in a way that will not only help supply the energy needed to preserve our American way of life but will also help to preserve our great American environment, not only for our own benefit, but for all those who come after us.

We hope that you participants in this forum will reach the same conclusion. Frankly, we don't see how you can avoid it. And we count on your support for the conclusions of the substantive negotiations of the National Coal Policy Project when they are available at the end of next year.

I have worked with coal for a great many years, when I was directly involved with the power business, and I know what it can do for us-- and I also know what it can do *to* us, if we don't use it wisely. I have been close to the developing energy crisis in our country for a number of years, and I know that we *must* use our coal and we *must* use it wisely. Coal may not be a very glamorous fuel, but just remember the old saying--"There's no *fuel* like an old *fuel*!"

LOUISE DUNLAP
―――

President, Environmental Policy Institute

By way of introduction I will just say that my involvement in coal policy issues has been connected with the debate over a federal strip mine bill that has been before Congress for six years, and I have in that time been working with citizens' groups, agricultural groups, and miners in Appalachia to try to get a decent federal strip mine bill passed in the hopes that the coal industry can get itself more stabilized and that we can have greater coal production.

There is little question that increased coal production is the most logical domestic fossil fuel option available to the United States during the next century. Looming over this apparent inevitability, however, are a series of public policy conflicts that, until resolved, will serve to contribute continuing uncertainties in capital formation within the coal industry, particularly in the midwestern and eastern coal fields close to the major existing demand centers. These continuing uncertainties will serve to jeopardize the realization of specific deadlines projected for greater reliance on coal. Ironically, the coal industry, through its continued resistance to resolution of several of the key policy conflicts, has been a primary factor in the nation's slow shift to increased coal utilization.

The question of how fast coal should be developed cannot be answered until there is a resolution of public policy conflicts in at least six key areas. The vast abundance of our nation's coal resources, 1.5 trillion tons, of which only approximately 10 percent are strippable according to U.S. Geological Survey figures, can be translated into a responsible national coal policy only after there is assurance of protection of other economic interests, specifically, agricultural interests, protection of surface owners and water users in the areas affected, and assurance that the environmental, occupational, and public health and safety hazards of extracting, combusting, and converting coal have been resolved through further clarification of public policy. The six key areas of public policy that need resolution include the following:

First is federal coal leasing, where combined with the need for effective implementation of the recently enacted Federal Coal Leasing Act Amendments of 1976, which were enacted into law over industry opposition and a Presidential veto, a thorough review of the Ford administration's federal coal-leasing policy is essential and will serve to clarify the extent to which the federal government chooses to encourage the development of western, lower-Btu coal, far removed from the major existing demand centers at the expense of stimulating capital formation in the midwestern and eastern coalfields.

Second is federal surface coal-mining legislation. There is certainly a need for clarification of nationwide minimum standards in that part of the coal industry which represents now more than 50 percent of current coal production. The areas that need clarification are requirements for reclamation; protection of surface owners and water users in the affected areas; and placement of prime agricultural lands, national forests, and other specifically fragile or historically important areas off limits to future strip mining. These are all areas that need to be clarified in the federal legislation so that the uncertainties that have plagued the industry can be cleared up and the industry can expand production of this small fraction of the total coal resources of our nation. Through the Abandoned Mine Reclamation Fund, which would be created in this legislation, the new federal strip mine law would also provide the coal industry with a mechanism to abate continuing pollution from abandoned surface and underground coal mines, thereby providing the industry with an opportunity to stimulate jobs and improve the public's perception of the impacts of surface coal mining.

A third area of policy conflict is emerging with the pending Federal Coal Mine Health and Safety Act Amendments of the 1969 Health and Safety Act. Through amendments to improve the equity and the effectiveness of enforcement of health and safety standards, the aggravation of continuing controversies surrounding coal mine health and safety can be diminished, thereby improving work conditions and production stability, particularly in the deep mine sector of the coal industry. Industry resistance to more effective health and safety enforcement will serve as a long-term, continuing disincentive to stimulate increased underground production where the overwhelming amount of our nation's coal resources lie.

The fourth area involves the continuing conflict over the Clean Air Act Amendments. Clarification by Congress of the meaning and intent of the prohibition of deterioration of clean air regions will serve to resolve the long-standing conflict that has impeded capital formation for power plant development. Effective air quality standards need not function as a deterrent to increased coal utilization but may serve to stimulate development of new combustion technologies, for example, the fluidized bed combustion system, which may make the greater reliance on the medium- and high-sulfur coal reserves in the Midwest more realistic. In that area there are 200 billion tons of coal in Ohio, Illinois, and Indiana.

Fifth, there is a need for new priorities in the development of coal R & D. A new emphasis is clearly needed for an accelerated R & D effort to improve efficiency and safety in the underground mining industry, as opposed to an overconcentration on the conversion technologies. The new administration must also review ERDA's coal R & D program, which has been focused on technologies that require the low-Btu, western coals as feedstocks. The critical shortage of water and the difficulties of reclaiming strip-mined lands in that region will make it difficult, if not impossible, to develop a major coal conversion industry there. The new emphasis in coal R & D must be shifted toward conversion of the medium- and high-sulfur coal seams of the Midwest, which coincidentally are located close to the major existing demand centers.

The sixth area that needs to be resolved in public policy conflicts is the need to develop regional coal demands that complement the regional coal supplies. The Carter administration in advancing a public policy of greater coal utilization should, and probably must, encourage the fulfillment of regional coal demands through the production of coal resources of that region or adjacent regions. Promotion of such a policy would reverse the trend of adding exorbitant transportation costs to the price of delivered coal while boosting the local and regional economies of the supply regions without placing unreasonable burdens on specific regions, such as the northern Great Plains coalfields, targeted by the energy industry to become the storage battery of the nation.

I do not have all the answers this evening, but I would suggest that rather than discussing the viability of specific deadlines for meeting certain levels of increased coal production, the Forum, I would hope, in the next two days will discuss in greater depth the public policies before the Congress and before the new administration, which, until resolved, will probably serve as the greatest impediments to increasing our coal production.

DISCUSSION

HIBBARD: Would anyone on the Panel for Inquiry like to raise an issue?

HENDRIK HOUTHAKKER, Harvard University: I was a little puzzled by the focus of Ms. Dunlap's remarks. We all agree that there are very serious environmental problems in coal and that these environmental problems are in the nature of what economists call externalities. I must say, however, that some of the things that she talked about seemed to have very little to do with these externalities. It is not clear to me why agriculture, for instance, has to be protected from coal development. This is a question of balancing different kinds of land use. It is not at all obvious to me that land use in agriculture is necessarily superior to land use in energy production. This leads me to the question of what, other than perhaps a certain degree of attachment to the status quo, the purpose is of favoring apparently agricultural land use over energy land use.

DUNLAP: As the debate over the strip mine legislation has developed in the past six years, it has become evident that there are short-term mining operations and there are long-term mining operations. Lester Lave's comment that a strip mine is only a two- to three-year disturbance of the land may be true for many small operators in the state of Pennsylvania and other eastern states, but it certainly is not true for the large operations in the West, where a twenty- to thirty-year life of an operation is typical, and even in the Midwest, where the life of the mine may be more than a decade.

So one of the questions concerning the conflict of agriculture versus mining is that of interruption of the use of the land. The duration of a mining operation, however, is less the issue than the central point, which is the differing potential levels of productivity for agriculture following mining. Secretary Andrus and the new administration in their recommendation for a moratorium on mining of prime agricultural lands, as defined by the Soil Conservation Service state by state, are responding to some of the results that are coming in from the Midwest. An example is Illinois, where the topsoil is among the best in the nation, where the rainfall is very high, and yet where the early results are showing that while the land is being returned to a state of productivity, it is being returned to a state of considerably lower agricultural productivity after surface coal mining. In cases where some very intense croplands preceded, the mining is being followed by mixtures of pastureland and row cropland, primarily because the land will not support the premining levels of agricultural productivity unless very heavily fertilized and irrigated.

So interruption of the use of the land is more of a factor in the West, where you have long-term mines, than in the East, but the central question for prime agricultural lands--which, true, represent a very tiny portion of total lands in the country but are very important to agricultural economies of certain counties in the Midwest --is, if the land cannot be reclaimed to that premining level of agricultural productivity, is it necessary, given the vast abundance of coal and given the tiny percentage of strippable coal, to strip-mine those prime agricultural lands?

SINGER: To add to this very important discussion, it is amusing to me to see how Ms. Dunlap and Dr. Houthakker talk past each other. I don't think you have answered his question at all, neglecting the fact that, first of all, Mr. Andrus' suggestion amounts to confiscation of property without due process by preventing a farmer to do as he wishes with his land. Let me point out that the prime agricultural land, let us say, in North Dakota right now is around $250 an acre--the very best land. Such land may be underlain by something like 20,000 tons of coal, which means that, roughly, if you can get one penny per ton you have recovered the value of the land. Now, a farmer who owns the land should be allowed to do what he wants with it, one would think, unless there are some substantial externalities, unless, in other words, there is some benefit to the nation that is somehow not included in the price. I think that was Dr. Houthakker's question, and my question too. I would like to see how that is answered.

DUNLAP: I have not seen the actual language that the administration is proposing, but it is my understanding that the proposal is for a moratorium of five years on new mining starts with a grandfather clause for mines where contracts have been committed. So I don't think there is a question of taking here. There certainly is not in the moratorium issue on prime agricultural lands, and the attempt of the proposal as I understand it is to examine in more detail the post-mining levels of agricultural productivity.

DAVID ROSE, Massachusetts Institute of Technology: I don't know what this Panel for Inquiry is supposed to be other than a Greek chorus that gets up every half hour and says "Oh Woe, Oh Woe" and at the end is supposed to produce a *deus ex machina*, but we will leave that to Lester Lave.

Now, regarding that particular question, it is a hoary argument, and there are three general numbers in his example. The cost of the land as bought from a real estate operator is, say, $200 an acre. The cost of reclaiming the land is, let us say, $5,000 an acre, and so the argument is made that it should not be reclaimed, et cetera, and every economist knows that. The value of the coal is, let us say, $100,000 an acre, making it doubly sure that the land should be mined and not reclaimed. But this is purely a social decision and not arguable on economic grounds, really, because the cost of reclaiming the land is small compared to the cost of coal. The cost of reclaiming the land is large compared to the real estate value.

The question you should ask is, what is the true cost of land if it is removed more or less permanently?

For example, if you took the entire United States and valued it at $250 an acre, you would end up with half of the annual GNP. Surely, the United States is worth something like the GNP times the lifetime of the inhabitants. Otherwise, you run into some very severe illogicalities.

So you should ask yourself then something different about the usufructuary price of land versus the actual using of it. For example, would we sell it to, say, the Saudi Arabians at $250 an acre and let them put their flag up there? Perhaps not. Perhaps so. Some might. That is the question.

WHERE SHOULD COAL BE DEVELOPED?

William C. Wampler

Representative from the Ninth District (Southwestern) of Virginia to the U.S. House of Representatives

I would first like to thank Drs. Weinberg and Hibbard, cochairmen of the Academy Forum on coal, for inviting me to participate in what I consider to be one of the most important inquiries and discussions taking place in our country. My only regret is that the Congress of the United States, which should be debating this matter as a matter of highest priority, is not but instead has been considering whether or not to keep employed a committee lawyer to the Kennedy-King assassination review, and how much money a member of Congress should or should not make, and how to improve our image.

It has been said that it is the squeaking wheel that gets the grease, and, apparently in the case of energy and specifically coal, the wheels are just not squeaking loudly enough to get attention.

The topic that has been assigned me is, where should coal be developed? Since I come from the mountainous terrain of southwestern Virginia where coal has been deep-mined and recently strip-mined, my presumption is that I should address myself to this problem.

Before I get very deep into the subject, let me raise a few thoughts for your consideration. The United States Bureau of Mines recently stated, as reported in the *Washington Star*, that "energy use increased 4.8 percent last year, reversing a two-year decline in United States energy consumption, and petroleum accounted for nearly half of all United States power needs, a slight expansion of the role it played the year before, and hydroelectric power production was down because of drought in the West." And in a separate report in the *Washington Post*, our new energy czar, Dr. Schlesinger, intends to downgrade nuclear power to fulfill President Carter's campaign description of the use of the fast breeder reactor as only "a last resort" measure. The new Secretary of Interior has recently stated he intends to review all of the decisions his predecessor made involving leases for oil and gas drilling off the Atlantic coast. Several large hydroelectric power projects have been halted by court order or by the government for reconsideration because of threatened extinction of snail darters or wild plants; and a strong play is afoot to stop foreign oil tankers from entering American coastal waters unless they are rebuilt with double bottoms. Moreover, a revised version of a vetoed coal strip mine bill is in the legislative process, which could for all practical purposes halt strip mining in my congressional area and in parts of Kentucky and Alabama and substantially raise the cost of strip mining of all coal across the land. Meanwhile, unemployment remains a constant problem, and programs for conservation of energy outputs threaten to change life-styles, lower our standard of living, and in many instances, cause more unemployment. Now, having painted a confusing and dismal picture, let me proceed with the question.

For a number of reasons, presuming that we will be permitted to burn coal, I happen to believe that our energy situation is so serious that we ought to expand production at every underground mine, at every strip mine currently in operation. I also believe that we should open as many new mines as we can as fast as we can. In my part of the country our strip mining is mainly conducted on rather steep slopes, above 20 degrees, generally following a contour on a particular mountain. The coal lies in a rather shallow strip, usually several feet thick, running into the mountain for several hundred feet. The coal obtained by this method is high in Btu's and low in sulfur content. Most of it is high-value, high-grade steam and metallurgical coal. Much of this low-sulfur-content, stripped coal is mixed with higher-sulfur, deep-mined coal obtained nearby to form a blend which meets current air quality standards. To stop the stripping of coal in this area and force the use of low-sulfur strip coal of lower Btu value from further-removed sites would, I think, unquestionably raise the price of coal rather substantially and produce a less energy-efficient product. Moreover, the strip coal of Alabama, Kentucky, Ohio, and Virginia, and to a lesser extent West Virginia since it strips less, is more readily accessible to the greatest markets and has a transportation system already established. These two factors should, it seems to me, dictate that these coal-producing areas remain at the top priority in any plan to increase coal production. On top of this, West Virginia, Pennsylvania, Kentucky, and Virginia are our top producers of deep-mined coal. Unfortunately, it takes longer and costs more to open new deep mines. And, additionally, we have a shortage of deep miners today. On further consideration, we should want to expand production in the Appalachian area, inasmuch as this area has been rather economically depressed for some time, although the recent spurt in coal use has improved economic and social conditions in this area.

Now, when I examine the time frame for bringing on line more exotic energy sources, such as the sun or fusion, and at the same time I see the dwindling figures for oil and gas production on our continent, I can't help but say, let us get at the task of digging more coal.

This does not say that I would unleash the coal industry without any restraint. I favor legislation that would adequately and fairly protect the environment. But to those who say stop surface mining coal and to those who would place such restrictions on surface mining of coal to make it economically unfeasible, I say, can our people afford their solution?

A case in point exists in southwestern Virginia. The strip mine bill passed by the last Congress had a provision that would require the land mined to be brought back to its approximate original contour. An exception was made in West Virginia permitting hilltop or mountaintop mining, and another in Ohio, which has little, if any, mountains or slopes over 20 degrees. Neither came under the original contour provision. However, in Virginia, especially in the mountains of the six counties in which coal is located, there are no rolling plains or wide valleys or deep veins of coal. The valleys just don't exist, and level land or gentle slopes are at a premium. A piece of flat land permits one to make a garden, raise a cow, or build a school. My people and the people of Virginia feel if we can put the land to a better use than before, then that makes sense, and today Virginia Polytechnic Institute and State University is engaged in experiments financed by federal, state, and private funds to find out what is needed in reclamation, to allow us to use these areas under less restrictive requirements than are proposed in current strip mine laws--for small crops, fruit orchards, and grazing lands. I am happy to report that these experiments look most promising.

Now, we have recently been encouraged by studies that tell us we may be able to develop mine mouth, steam-generating units and coal gasification plants that would be much more energy efficient than the systems in current use. Development of such systems would be much more efficient than long hauls to the east of the western coal of less Btu value. So, regardless of which system we adopt, we will need all the eastern and western coal we can get before we can come to grips with the looming energy deficit period we face.

In closing, we need to face our problem more realistically and more prudently. It will make little difference to our children if we bequeath them beautiful mountains, clean streams, and clean air and jobs, heat, food, or clothes become much harder to come by than they are today. So, I respectfully suggest that we must seek and solve the problem of balancing our urgent needs for energy and the need to improve our environment.

Frank G. Zarb

Executive Vice-President and Director, Shearson, Hayden, and Stone

Noting the unanimity of opinion here this evening reminds me of my last two years and more specifically of the first time that I delivered a major address after being confirmed by the Senate. A fellow got to a microphone after I had delivered a speech, and he said, "Mr. Zarb, do you understand that with your policies you are raising the price of electricity, that I have to keep my lights off? In addition, you are raising the price of gasoline and I can't drive quite as much. And do you smart fellows in Washington further realize that with those kinds of policies, all you are doing is increasing the size of the United States population?"

Well, at that point in my federal career I was rather timid. The audience was mixed, and I thought I had better not fuss with that one. So I got the microphone, and I said, "Sir, I am going to answer your two-part question [which he ultimately got to] but with respect to your initial comment I think I will pass." And just as I said that, a lady got up in the back of the room and said, "You're going to miss an awful lot of fun."

Well, I must say, I was interested and honored, and I confess a small bit amused when I was asked to speak tonight on the subject of where coal should be developed. I further limited that topic to say where coal should be developed over the next ten years. I too often sat in a hearing room to hear projections over the next forty years, knowing that the projector wasn't going to be around to give an accounting forty years later, and the projection normally reflected that lack of discipline. So ten years is what I used, and I looked at the real world and the constraints that are in place and not likely to be eliminated in time to affect the next ten years.

So I came to the conclusion that it would be safe to tell you tonight that in answer to the question of where coal should be developed, I can say wherever it can, as much as it can, and as fast as it can.

I can't talk about coal without briefly going back over a general energy formula that can answer this nation's problems. Some years ago, a number of us who were thinking about the problem concluded that if we did some very moderate and responsible things over a ten-year period, we could reduce our imports to about 6 million barrels a day. We further thought that if we increased our storage capacity, we could provide at least a one-year protection against embargo, and to get from here to there, we said the nation should reduce its rate of growth and energy consumption from 3.5 percent a year to 2.5 percent a year. We had to double our coal consumption, not necessarily our coal production, but our coal consumption. We needed to double the electricity generated from nuclear power, oil had to go from 8 million domestically produced barrels a day to 12, and natural gas from 20 trillion cubic feet a year to 23 trillion cubic feet. And if we did those things modestly well, in ten years we would still be importing 6 million barrels of oil a day.

Then there were those who said, but wait a minute, we have another option. We can increase our imports. I won't argue the economics and national security issues associated with that position. You have heard my response to those issues before. But I would argue that it is by no means clear that our friends in the Mideast are prepared to agree to that notion. They haven't said that by 1985-90 they will supply us with more than 6 million barrels a day of oil, and, indeed, there are some indications that under certain world scenarios we will have a diet for some long period of time at less than that level. So those of you who would opt for higher imports, please keep in mind that the exporters have something to say, and thus far the things they have said have not been too encouraging.

I would be negligent in my responsibilities as a former federal official not to give one commercial before I get back to coal. All of the goals I described for 1985--the 1 percent reduction in energy consumption and doubling of coal, nuclear energy, oil, and gas levels--all of them now are in serious jeopardy, including the conservation target. It is my view that the likelihood of achieving them in the real world is now fairly minimal, regardless of what this administration proposes and what this Congress determines.

Now, let me get back to coal. What should be done is an interesting question, and I must say that I used to think that questions such as what should be done were highly theoretical. I always felt that if I stayed downtown, I got too much theory, and if I went up to Capitol Hill, I got too much demagoguery. So I am going to try to stay away from questions such as what should be done and talk a little bit about what is being done, because those people out in the real world who are currently making investments or not making investments for the next ten years in corporate America are for the most part going to determine what happens. The decisions they make today and tomorrow in their board rooms are going to produce results five and ten years from now.

In any case, there are some studies that have been made, and those studies have shown that there is a possibility that, on the basis of current plans, by 1985 we can increase steam coal by 598 million tons. Of that, on the basis of real investments and corporate plans, surface mining would supply 65 percent of the increase, and the West would supply 70 percent of that, with Wyoming being the biggest factor.

Now, there are long lead times in coal, two to four to five years, depending upon surface mining or deep mining, location, and infrastructure requirements. As a result I feel more confident with estimates of the first five years than with those of the second five years, and I would say, on the basis of my review and a little bit of red penciling, I would not look for more than 250 million more tons by 1980 of U.S. steam coal, regardless of what other determinations are made by this Forum, and almost regardless of what else the administration says and what this particular Congress does.

As you look at all projections, things can go wrong. We had a poet, Robert Burns, and a fellow by the name of Murphy, who had a saying associated with that. When something can go wrong it usually does, and I would say that we should discount 15 percent of all corporate projections at this particular moment to take care of Murphy's law.

All projections that I have seen have not included mine closings, and these pose an interesting question, because those who are laying out the new additions to our mine capacity are really not overly concerned with how many mines are closing. It seems to me that we can easily count on an average of 20 million tons a year lost over the next ten years in depleted and closed mines.

That brings me to a real-world calculation, based upon dollars invested, lead times, dollars planned, that over the next ten years it is almost impossible to see an average rate of growth of more than 5 to 6 percent in the consumption of domestic coal. Now, to add to that, there are some uncertainties, some good and some bad. Coal mine productivity is down to something close to 9 tons per man as compared to 16 some seven or eight years ago. I assume that has bottomed out, but what if it hasn't? Surface mine productivity is down 20 percent in a much shorter period of time. Again, I assume that is the bottom, but what if it is not? The new Surface Mining Act might be a good, tight piece of legislation providing ample flexibilities to recognize regional requirements and provide lots of certainty, and, as such, the net effect could be to increase coal production. But what if it were like some other versions I have seen in the past that more resembled an emergency employment act for the legal profession, which would undoubtedly lead to even more uncertainty than anything we have seen today.

Further, the Clean Air Act creates an uncertainty. In current years about one half of total steam coal burned actually met compliance standards. And that has kept me wondering. There are now some new amendments that are going to tighten those standards, and what does all that mean anyway? The railroads are an uncertainty, and if you talk about slurry pipeline we are going to get ourselves into a water discussion, which is another uncertainty.

As to labor, will the coming election solidify the labor unit, and will that allow things to proceed so that productivity for the benefit of all is improved? I don't know.

Will stack scrubbers get to be reliable, and acceptable? Will the economics of coal *vis-à-vis* all other alternatives be such that that investment will make coal a better attraction? I must point out that unless we get to real replacement values of oil and gas in our economy, we are not going to get any of the goals that I described earlier--conservation or nuclear energy, coal, oil, or gas increases--and we are certainly not going to get our coal objectives accomplished.

Now, on the good-news part of the uncertainties: In my new private role I have set out to study more carefully some of the technologies that appear to be coming down the road and to select carefully those to whom I listen. I am pleased to report to you that, at least from my perspective, there are some very interesting and some very exciting breakthroughs occurring in the coal gasification liquefaction arena, with low-Btu coal gasification being the first to make a real penetration, followed by high-Btu gas with the elimination of many of the environmental difficulties that we have perceived up until now. These seem to be a lot closer than many other advancing technologies, at least up until now. So it seems to me that good old American technology may yet bail us out if we here in Washington begin to create an environment in which it can get done. And that leads me to my summary, which says simply this:

The next ten years really are almost too late to affect the numbers I gave you by very much. There is a reasonable set of standards that will permit this nation long-term development and burning of coal. Let us get on with the job of determining what those standards are in a way that the costs will equal the benefits, and once we have set out those standards, let us get technology to work so that we can use coal in an acceptable way. It seems to me to be eminently logical and achievable.

As I said before, theory downtown, demagoguery uptown will not get the job done. I am convinced that it is doable. In the meantime, if you are going to limit your observations to the next ten years and want to know where coal should be developed, take my word for it, wherever you can, as much as you can, and as fast as you can, and you won't go far wrong.

DISCUSSION

HIBBARD: Would any of the Panel like to raise an issue or ask a question of our two previous speakers?

DONALD ALLEN, New England Electric System: Speaking from a utility point of view, I would like to pick up David Rose's remark that we may have a Greek chorus on our hands, but I think the words being said are not "Oh Woe" but "Uncertainty, uncertainty, uncertainty." This brings me to the point of my question. Mr. Decker, Ms. Dunlap, and Mr. Zarb have all made quite a bit of uncertainty. Each of them, I think, has had the vision that if we only get to be sensible in a very short time we could put all these uncertainties to rest, and the private enterprise system, the market, Wall Street, the coal industry, whoever, will get on with the job. I would like to know one of two things. How soon can we get that certainty, or much more to the point, as long as we have uncertainty, who is the guy that picks up the tab on the uncertainties?

ZARB: That is a very good question, and the answer is rather uncertain. You know, we are dealing with major public policy change in a democracy. As such, we are going to have periods of debate when major changes need to be accomplished. What we haven't had until now are those responsible for public policy prepared to stand up and be counted, because your current costs are being picked up at the cost of lower capacity than some of you might need in your particular sector. When the brownouts occur in 1981 and 1982, that is going to be a cost that somebody is paying. In the meantime your consumers are paying for it in high-priced residual oils that the producing countries are delighted to have. Long-term decision makers in this instance are public policymakers, and they must either stand up and be counted by saying there is no answer and we are going to pay the desperate price of putting ourselves at the mercy of world oil suppliers or gas suppliers, or indeed we do have an answer. It is going to be more costly, it is going to take some sacrifice, it is surely going to take ten to fifteen years, because we are not going to get it done overnight, but we are here today as your elected officials to vote on these very, very important issues. I predict that is the next step, but when is uncertain.

DUNLAP: I know my place. I always like to follow Frank Zarb. The real question surrounding a lot of these public policy debates is a question of, to what extent the cost will be externalized or internalized? Someone is going to be paying the costs, the taxpayers or consumers, and while a lot of utilities say, oh dear, don't pass a strip mine bill or another piece of legislation because it will increase utility bills, many times they aren't saying that the price of coal does not reflect the cost of production but rather the current price of oil.

But the real issue is that somebody is paying for it now anyway. Many citizens and even consumers would prefer that the costs of production be more honest and be more internalized, because it will either be the Corps of Engineers using the taxpayers' money to clean up reservoirs, rivers, and streams in Appalachia or increased costs of public health care if we don't have good enough Clean Air Act standards. So someone is going to pay. We would prefer that the costs of production be internalized as much as possible.

ALLEN: Could I make a brief response to that? I would like to say to Ms. Dunlap that I am not talking about internalizing or externalizing costs. I want to know whether I can internalize or externalize an uncertainty. The very point you made is that we are having trouble getting investments here as long as there is uncertainty. I don't think you have answered the question.

DUNLAP: At least in the area on which I work, on the strip mine bill, I would be very glad to talk to you about the areas in which we could agree on internalizing the costs.

ALLEN: Fair enough.

DECKER: I am very glad you asked that question, Don, because one of the things we are trying to resolve in the National Coal Policy Project, as I mentioned earlier, is who is going to pay. I know from some very vivid experience in Michigan, the Michigan people are going to pay. I would like to ask Louise the question, what is the difference between the people, the taxpayer, and the citizen? I think they are the ones who are going to pay, and I think they are all the same ones.

HIBBARD: It depends on who sends them a bill.

DUNLAP: I do not think that legislation that requires the reclamation of land to premining levels of productivity or other forms of water and land resource protection are contradictory to internalizing costs of production. In fact, that is what internalizing costs of production is. I am a taxpayer. I am a consumer. The arguments made by industry in opposing meaningful legislation, saying, oh, you can't do that, the consumers are going to pay, are based on the fact that they don't recognize that if the public doesn't pay as consumers, they will pay as taxpayers.

HIBBARD: Isn't it really a question of how the public is going to pay, whether they pay as consumers or taxpayers? It is not a question of whether they are going to pay, it is a question of how it is going to be distributed. Are there some questions from the audience?

BRADLEY VANDERMARK, Chemist, Chandler, Arizona: I am curious, Mr. Wampler, about responsibility. One of the issues involved is whether to let the free market system operate. If people are strip mining, you are going to have to reclaim the land. In southwestern Virginia, there is strip mining with a little bit of acid runoff. How was responsibility shown in West Virginia with the Buffalo Creek Dam, which broke? No one is responsible. I would hope that the people in Virginia or West Virginia, et cetera, who I think are fiscal conservatives, would like to have the free enterprise system operate. This also is true of utility companies. At least by saying this land has to be reclaimed they say, okay, we will put the price on the line. Can coal power then sustain itself or will alternative energy sources take over, such as decentralized power sources? I am just curious if that will ever occur, Mr. Wampler.

WAMPLER: Let me respond by saying that being a southerner in general and a Virginian in particular I believe in states' rights. The General Assembly of Virginia passed a rather meaningful piece of legislation in 1966 in which it was declared that it was proper that the State of Virginia regulate surface mining within its boundaries. It has been twice amended since that time, and, given the opportunity, some even more stringent regulations will go into effect in July of this year. So far as I am aware, every state in the union that has surface mining has a state law regulating it, some more effective than others. But to suggest that the Congress of the United States should pass federal regulatory legislation, I think, certainly is not the answer. There are many reasons why I believe that--for one thing, varying conditions within counties, within regions of a state.

I have a chart that was prepared by the Kentucky Independent Coal Producers to show what would be required to get a permit to surface mine coal under the proposed federal law. I think any fair interpretation of the federal statute would be that it would require a minimum of two years' time in which to secure a permit, up to as much as five years. Now, this means, obviously, that many small, independent operators simply would not have the resources that would be necessary for this burdensome and unnecessary process.

I believe very strongly that we can and must have reclamation. I believe that the states can do it. I think they are doing it. There is much that remains to be done to improve it, but to answer your question, yes, we can have meaningful reclamation, but I believe the several states can do it better than the federal government.

HIBBARD: We have time for a one-minute question.

CAROLYN ALDERSON, Bones Brothers Ranch, Birney, Montana: Congressman Wampler's answer to that last question leads me to a question that I was going to ask Mr. Zarb. I was interested to hear you say that in the West the largest amount of coal production will come from the state of Wyoming. I come from the state of Montana, where we have a considerably more stringent state strip mine regulation law and, to some people in some utilities, a rather stiff state severance tax on the coal. I am curious as to why you think the larger amount of coal production will come from the state of Wyoming--is it the stiffer Montana reclamation requirements and its severance tax or other factors?

ZARB: I can't report on the specific factors. It is clear that Texas is a primary market for a large amount of that coal, and while I said Wyoming was the largest, I also said that the West was going to have the largest percentage of growth. So when you consider Texas as the primary market and also consider that there are many folks in Texas who have been looking toward the eventuality of a slurry pipeline to reduce the long-term cost of moving that coal, that could be a factor. But indeed, if the economics are different and all other things are equal, you can bet that the market is going to the area where the cost is substantially less. So undoubtedly economics does play a role.

HIBBARD: I might comment that under the present circumstances neither Wyoming nor Montana is producing as much coal as Virginia. In fact, 80 percent of the coal comes today from east of the Mississippi.

ENVIRONMENTAL IMPACT
AND COSTS

Steven Reznek

Associate Deputy Assistant Administrator, Office of Energy, Minerals, and Industry, Environmental Protection Agency

Although coal presently provides less than 20 percent of our national energy needs, it is one of the nation's most abundant fuel resources. Production of coal is beginning to grow again and reached 640 million tons in 1975, exceeding the 400 to 600 million tons per year range that had persisted since World War II. Most experts predict that the use of coal will double or possibly even triple before the turn of the century. There are environmental problems associated with all phases of the coal cycle: mining, transportation, direct combustion, and production of synthetic fuels.

RECLAMATION

Surface coal mining, probably more than any other activity, raises the central issue of environmental aesthetics. Adverse effects on the environment caused by surface coal mining are of national concern. Americans are aware that quality of life is directly related to quality of the environment. Ever-increasing demands for coal, coupled with technological advancement in extraction techniques, have increased surface mining of coal. Mining and processing coal for energy utilization have often resulted in what is now perceived to be unacceptable degradation of the environment. Consequences of surface coal mining have been most noticeable on the land. Vegetation has been destroyed, soils turned upside down, and large areas left as bare, unsightly spoil banks. Natural beauty and topography have been greatly altered. Some areas have lost their productivity; only a small percentage of strip-mined lands has been restored. In those parts of the country where high-sulfur coal is combined with humid conditions, surface mining of coal has caused environmental degradation extending far beyond the mine site by pollution of surface and groundwater resources.

The northern Great Plains and Rocky Mountain coal provinces contain approximately 50 to 70 percent of the U.S. coal resources. The two largest regions, Fort Union and Powder River, contain almost 1.5 trillion tons of coal, much of which is owned by the federal government. Although seam depth and thickness vary considerably, some beds are quite thick (50 to 100 feet) and sufficiently near the surface to allow surface mining. Water supplies are not abundant. The average annual runoff amount ranges from less than 1 inch to 10 inches.

Although coal has been produced in a few of the western states for over fifty years, production levels and hence the amount of land disturbed is expected to rise greatly over the next thirty years. For the eight-state western coal province, from 200,000 to 400,000 acres of land may be disturbed, assuming that all of the coal is surface mined. Considering that 43 percent of western coal is within 225 feet of the surface and the existing economies of surface production compared to underground mining, it is reasonable to accept the high end of the projection.

Surface mining destroys the existing natural communities completely and dramatically. Indeed, restoration of a landscape disturbed by surface mining, in the sense of recreating the former conditions, is not possible. It is generally conceded that the original contour, productivity, native vegetation, and utility to wildlife of a strip-mined area cannot be fully restored. Perhaps more importantly, reclamation in terms of re-establishing and sustaining a vegetative cover, is often substantially more difficult in the West and is intimately related to other issues, most specifically, water availability.

The most substantial problems include the natural constraints of climate, soil, overburden, vegetation ecology, and the lack of available moisture. Arid ecosystems simply inhibit natural revegetation except over very long time frames. Several western states fall below the 10-inch minimal average annual precipitation required for successful revegetation. In other states, seasonal distribution of rainfall means that severe constraints are imposed on revegetation for several months of the year. While irrigation can be used to mitigate these problems, water used for irrigation takes water away from other uses and thus worsens existing water availability problems. This emphasizes that water management will be a critical policy component in the reclamation process. Because of the close relationship between moisture and growth, there may be a significant number of cases where reclamation efforts will simply fail or be only marginally successful.

COAL UTILIZATION

Most of the coal used in the United States during the next two decades will be burned in steam boilers. Coal-fired steam electric boilers are primarily located east of the Mississippi, with the heaviest concentration in the industrial upper Midwest. In addition, many new coal-fired power plants are being constructed in the West.

It has been shown that, in combination, sulfur oxides and particulate matter from power plant stack gases can increase the death rate due to chronic disease and can aggravate these diseases. In addition, they can be a causal factor in the occurrence of chronic bronchitis in children and adults.

The oxides of nitrogen lead to photochemical smog reactions. These reactions require hydrocarbon vapors, oxides of nitrogen, and sunlight. Adverse health and environmental effects include eye and respiratory irritation, production of ozone in the atmosphere, interference with visibility, and characteristic forms of vegetation damage. In concentrations of several parts per million, nitrogen dioxide can lead to a condition in experimental animals that resembles pulmonary emphysema in man.

Numerous studies have documented the public health hazards of emissions from coal combustion. Federal agencies engaged in energy-related health research are already directing significant resources toward defining and resolving this problem. However, several previously unrecognized problems have recently come to light.

As examples, measurement of sulfur dioxide concentrations in the working areas of coal-fired power plants has revealed concentrations of about 10 parts per million, a level five times above the currently recommended standard for SO_2 in the work place.

Another potential problem recently recognized involves the potential formation of secondary pollutants in the ambient environment from reaction of nitrogen oxides with amines. One such class of secondary pollutants includes nitrosamines, which are potent carcinogens. In this regard, the long-term, low-level health effects of nitrogen-containing organic compounds derived from photochemical reactions in the atmosphere are not well understood, and this problem could be aggravated by direct combustion of fossil fuels. It appears, however, that photodecomposition may also occur very rapidly, so the problem may be serious only near concentrated sources.

A promising technology for using coal in an environmentally acceptable manner is fluidized bed combustion--a technology under development by ERDA and private industry. Fluidized bed combustors (FBC) burn coal under closely controlled conditions, thus reducing particulates and nitrogen oxides formation and allowing for the capture of the sulfur oxides by sulfur-scavenging materials such as limestone.

SYNTHETIC FUELS FROM COAL

The development of synthetic fuels from coal also is not without environmental problems. Several processes are now being developed to convert coal to clean-burning, synthetic natural gas and lower-heat-content power gas. Other processes are being developed for conversion of coal to low-sulfur and low-ash liquids or solids for nonpolluting fuels.

The conversion processes in such a plant, however, include various operations that can release particulates and hydrocarbons into the atmosphere and potentially hazardous chemicals into water supplies. It is expected that the water effluents from a gasification plant, before treatment, will contain suspended solids, phenols, thiocynates, cyanides, ammonia, dissolved solids such as chlorides, carbonates and bicarbonates, many sulfur compounds, trace elements and tars, and oil and light hydrocarbons. Potential discharge of many other pollutants and the environmental effects these discharges may cause are currently undergoing study.

The hazards to human health associated with large-scale coal conversion can be divided into three areas: (1) hazards to workers in the conversion plant itself; (2) hazards to people living in the vicinity of the plant due to effluent discharges or spills resulting in atmospheric exposure and contamination of drinking water and/or food; and (3) hazards to workers in downstream industries, to distributors, and to the public utilizing coal conversion products or by-products. Perhaps the most serious environmental threat to workers and the general public is from chemical carcinogens. For example, during the operation of a 300-ton-per-day coal hydrogenation plant, the strongly carcinogenic compound benzo(a)pyrene was detected at concentrations as high as 19,000 ng/m^3 near blowdown and steam-cleaning operations. More recently, carcinogenic compounds such as benzo(a)anthracene, benzo(c)phenanthrene, and unspecified mono- and dibenzopyrenes have been reported in the hydrocarbon streams of two coal conversion processes currently under development.

The problems associated with organic pollutants in coal conversion processes are basically different from those associated with other fuel production processes such as petroleum refining. The most prominent difference is the greater concentration of chemical carcinogens in the heavier and more aromatic fraction of coal-derived synthetic fuels. For example, polycyclic aromatic compounds are present in coal pitch in concentrations several orders of magnitude higher than those found in similar petroleum products such as asphalt.

One of the most difficult aspects of development to cope with is the effect upon the people who live in impacted areas. Many mining activities have imposed huge social costs on the public at large. These costs will last for years and are in the form of stream pollution; floods; landslides; sedimentation; loss of fish and wildlife habitats; nonproductive, unreclaimed land; and impairment of natural beauty.

As a result of the Arab oil embargo we find ourselves racing into the development of domestic energy resources and advanced energy technologies, which under more favorable circumstances would have been developed at a more leisurely pace.

The potential environmental insults associated with development of new energy sources are many, particularly for the relatively pristine West, where energy development activities will consume enormous quantities of already scarce water. I have not even mentioned the secondary impacts ensuing from increased population pressures, the boomtown syndrome.

But we should not lose sight of the fact that the ultimate success or failure of the race for energy independence will be measured not only in terms of the commercialization of energy technologies drawing on indigenous fuel sources, but also in terms of minimizing the long-term total costs to society as well. This presents a significant challenge to the biomedical and environmental community--to identify and eliminate health- and ecosystem-damaging pollutants before they enter the environment.

We in EPA do not believe that this is an impossible task. To be sure, it is a complicated set of problems. We must recognize that timing is critical in the overall relationship between energy and health and environmental research and development. We must discover problems as early as possible. Retrofitting existing facilities is generally wasteful and is seldom as satisfactory as designing in all the desired features. As new energy technologies are conceived and explored, we must simultaneously pursue the health and environmental consequences attendant with such activities.

Pockets of ignorance abound in virtually all of the areas in which decisions must be made. For this reason, it seems to me that money and time spent for research is an investment and an insurance policy against excessive future costs for the same environmental and health benefits. If the potential environmental and health insults from increased energy development are to be held in check, then sound research into these effects must proceed in tandem with energy development activities. An anticipatory approach to environmental research and development can avoid the excessive costs of playing catch-up with environmental and technological therapy.

THE ORDER AND INCIDENCE OF
CLEAN COAL UTILIZATION COSTS

Richard T. Newcomb

Professor and Chairman, Department of Minearal Economics, College of Mineral and Energy Resources, University of West Virginia

Like all fuels, coal in the twentieth century has been incredibly cheap. Coal's real price fell by more than a third in the thirty years following World War II, but with rare exception (in 1918-23 and in 1946-47) the industry has faced buyers' markets and conditions of chronic excess supply. Coal costs, of course, reflected the cheap energy regime obtaining for all fuels. Public policy fostered this regime by placing premiums on aggressive market competition among fuels and by encouraging or stimulating supply. While the policy was a success, as with all successful policy, there was a casualty ward. In this case it was conservation and the environment.

As Irving Fisher and other economists pointed out in the 1920s, the full social costs of energy are the sum of market costs plus the cost of waste, environmental damage, and other costs external to the market. The latter costs were so-called because they were not easily collected by producers or assigned to consumers. However, they were always recognized. On the producers' side there were thirty bills introduced into Congress from 1900 to 1940 attempting to control the external cost problems associated with excess supply conditions in the U.S. bituminous coal fields, none of them successful. On the consumers' side a rising tide of complaints about coal dust arose from the mayors of large cities. However, these concerns faded after World War II as coal utilization yielded to cheaper and superior oil and gas. Inexpensive nuclear power was to have been the final solution.

The prospect of much higher priced energy regimes and the recognition that nuclear power, when it arrives, will not be inexpensive, has revived forecasts of extensive increases in coal utilization. With these, arise forecasts of extensive external costs. I wish simply to sketch briefly my views of (1) the magnitude of full social costs such an expansion will evoke in the future and (2) some idea of their incidence, i.e., who will have to pay them.

THE ORDER OF COAL UTILIZATION COSTS

Market Costs in Production

The market costs, while not uncomplicated, are the easiest to assess. If we ignore for accounting purposes the costs of current regulations on mining and the control of emissions of sulfur dioxide inspired by the concerns over externalities, there is little doubt that coal is the least-cost solution to present boiler fuel scarcities. Unconstrained by environmental concerns, coal costs about $0.60 per million Btu (MBtu) compared to $2.50 per MBtu for incremental oil and gas or the conventional nuclear fuel cycle without subsidies or an acceptable breeder. Clearly, the capital costs for coal expansion as a boiler fuel are less than for any other energy form. In addition to direct use of coal under boilers, some conversion systems for indirect coal use, such as fluidized bed combustion, may become commercially competitive with natural gas or oil in meeting final demands in some regions.

External Costs in Production

External costs, however, cannot be ignored. When these costs are added to estimate the full social costs of coal utilization, different patterns emerge. I shall assess these briefly for production first and then consumption.

Health and safety costs affect underground mining chiefly. However, in the major producing centers, competition equates the marginal cost of surface mining with the cost of underground output. Thus, the rise in the average cost of deep mining affects complementary surface mine costs similarly. These costs, and the associated output effects on unit cost identified with the labor and management response to coal mine health and safety, add $0.30 to $0.40 per MBtu underground. This higher underground average cost vents more aggressive stripping, driving down the output to overburden productivity ratios in surface mining and driving up surface coal prices. It also induces production on the surface from more distant basins, adding to the delivered cost of coal to the eastern power supply districts where oil and natural gas are scarcest. Thus, the gap between delivered surface and underground coals tends to close under competitive mine expansion.

Reclamation costs for new underground mines can be controlled generally within the cost of new mine designs. On the surface, however, costs rise as high-wall heights become more extreme in the East. Expansion of western mining brings the special problems of reclamation in fragile, water-scarce regions. The expense of solid waste and water pollution control in the West, as has safety in the East, will almost double the cost of mining.

In any event, therefore, the cost of delivered coal to buyers can be expected to rise from $0.90 to $1.00 per MBtu as a result of the enforcement of health and safety regulations and restrictions on surface mining. This estimate includes depletion effects. The cost of meeting air pollution standards is an external cost best related to consumption.

Market Costs in Consumption

Assuming that the primary fuels are deregulated so that the field price of natural gas and oil rises above $2.00 per MBtu. Clearly, the conventional utilization of low-sulfur coals that meet environmental regulations is the cheapest solution in all forms of energy use in which solid fuels are currently competitive. To the extent that variances are given to high-sulfur coal users, all future power generation will be coal based. Alternatively, if western surface coals are used and the environmental costs to the producing region are ignored, some of the synthetic processes appear competitive in final markets (e.g., high-Btu gas, but not as boiler fuels) with the deregulated price of natural gas or oil distributed in these markets. Thus, there may be competitive gas and liquids from coal selling at $5.00 per MBtu in open markets where distribution of the synthetics is a minor factor of cost and the distribution of natural gas or oil alternatives is a major cost factor.

External Costs in Consumption

Although the epidemiology of sulfur and other trace elements from coal is not well enough understood to assess precisely the danger to health and safety of consumers, I accept here both the reality of concern over sulfur oxides and the choice of measures EPA has selected for their control. This leaves me only the question of measuring the damage to health if variances are given at the same time that coal utilization is expanded. In assessing epidemiology the economist is on much vaguer grounds than he is in assessing the costs of new technology. He must resort to multiple-regression techniques of the most general sort to estimate coal's potential damage functions. Unfortunately, while multiple regressions show that coal utilization is closely correlated with premature mortality, they also show that SO_x emissions add little explanatory power to the other variables, such as income used, as arguments in such regressions. These more customary arguments relate high mortality rates closely with general urban population statistics. Even without taking coal use into account, the higher-age and lower-income characteristics of urban consumers invariably account for higher mortality rates in the city than in the country. Therefore, we have only dubious damage functions to deal with if we wish to claim that SO_x emissions cause premature mortality.

Fortunately, we can ignore these problems, in my opinion. First, the cost of damage prevention via the use of coal cleaning and stack gas scrubbers is only from $0.25 to $0.55 per MBtu and will be much lower than the costs for alleged damages. Where they are effective alone, the cost of mineral preparation systems is even cheaper. By employing mineral beneficiation prior to scrubbing of stack gases, a great deal of the sulfur can be removed, as well as many of the trace elements in coal. These elements are present primarily in the clays surrounding coals.[4] Even with both preparation and cleaning, the ultimate cost of most clean coals will be between $1.15 to $1.55 per MBtu, well below the comparable cost of deregulated oil or natural gas. Consequently, clean coal can be expected to drive the other fossil fuels out of boiler markets under current EPA restrictions.

THE INCIDENCE OF HIGHER COAL COSTS

Table 1, summarizing the above analysis, shows that concern over external effects has raised the social costs of coal utilization from $0.55 to $0.95 per MBtu above market costs and to $1.55 at the margin. This is a substantial increase. Who will pay it? In the past, because policy recognized only the market prices of fuels, the burden fell largely on the producing regions. To date, mouth-of-mine plants have been the only expanding sector for bituminous coal, so few urban consumers were damaged. The excess reserve capacity in the coal industry produced the result that coal sold at the producers' cost. No producer felt he could "afford" controls. The ravages in the coal fields have been well documented in Appalachia and other regions. The poverty of the industry, combined with the environmental concern over emissions in distant urban centers, is indeed responsible for the notion that "the best thing coming out of Appalachia is an empty bus." Within producing regions, these attitudes toward coal have become typified at best by

TABLE 1 Estimated Costs of Coal Fuels[a]

Market Production Costs	
Underground mining or surface mining, unconstrained	$0.60
External Costs	
Mine health and safety	
Reclamation and control	$0.30-$0.40
Total production costs	$0.90-$1.00
Market Consumption Costs	
Incremental costs of meeting admissions standards	$0.25-$0.55
Total utilization costs	$1.15-$1.55

[a]1975 dollars per MBtu.

the stance of regional commissions, which have largely ignored coal mining, and of environmentally conscious groups within these states, which actively have sought to diminish or abolish mining. In the mining states, the public response to coal activity has become "Who needs it?" and "Who wants it?" The present prospects for coal, however, make it improbable to maintain this stance for the future. However, the social cost of mining will have to be considered.

The implication of most economic analyses I have reviewed is that the energy consumer will pay these higher costs of coal if oil and gas prices are deregulated. If, alternatively, the environmental or safety regulations are relaxed, the producing region will pay instead. If transfers are made from general revenues to producing states; e.g., for black lung and reclamation, then indirectly all consumers will pay. However, this sort of arithmetic is too simplistic, for it ignores the role of user costs in minerals, sometimes called "economic rents."

User costs in the coal-producing regions can be guessed roughly. My own estimate is that these will range from $0.00 to $0.50 per MBtu and may average $0.25 per MBtu. The distribution of these rents is unknowable, simply because the data to estimate them for a supply region are lacking and the variance in production conditions within seams and among mines will be too great to predict them. However, their existence in a market regime in which energy is chronically in excess demand rather than supply will transform the industry from a declining and ineffectual one into an expanding and profitable one. Thus, the means will be at hand for the first time on a prolonged basis for the producing regions to require good behavior on the part of the industry, and the incentive will exist for the industry to comply. The industry can generally pay for increased control. As for the consumer, although the cost of implementation will be high, this will encourage his habits of fuel conservation. Moreover, it appears that this will still leave coal lower in cost than any other viable alternative for boiler fuel consumers.

Synthetics, unfortunately, will not be used as boiler fuels when the costs of environmental protection are added, nor will, in my view, the costs of *in situ* coal conversion be favorable enough to interest the power industry. This means that synfuels and exotic underground conversion methods will have to seek the much higher prices available in future deregulated final markets for liquids or gas. These may reach peak levels approximating $6.00 per MBtu for residential users. At present levels, approximating $3.00 per MBtu for residential users, domestic natural fuels or imported oil will remain cheaper than any direct synfuels.

DISCUSSION

ROSE: I have four questions, the answers to which would help to put some of the problem in perspective.

Coal is not looked upon as an issue in itself. For years the nuclear argument, although argued as if nuclear energy were in a vacuum and compared to nothing else, was really not a single option. I would like somebody to summarize the relative epidemiological and environmental damage of nuclear power per MBtu under some reasonable circumstances. To put it another way, how far away, for example, are the fossil fuel emission standards from observable, clear views of health damage? On what basis were they set, and how sure are we of them?

Also, how much is spent on doing research on environmental and health consequences of coal compared to those of nuclear power. That is a number that I think would be interesting to know.

Third--the question has been raised by many, including many in this audience--can we decouple the question of rapidly increased use of coal from the question of rather large environmental hazards, the problem of the world's CO_2? This hasn't been raised this evening but will be raised in other days. What is our attitude there *vis-à-vis* the rest of the world, because we don't live alone.

Fourth, do we actually look on coal as a long-term reserve or as, at most, a medium-term reserve, especially as moderated by these other thoughts? The thousand-year supply that people talk about, with a growth rate of 2 percent per year, means that the coal would last 100 years. Now, that is a long time according to the rate of return on investment, but I am glad the Romans didn't think that way when they built the Roman Empire. At least they didn't seem to. But perhaps they did in some respects. They cut down all their trees, and as a result they ran out of energy and made the Mediterranean littoral what it is now. So is it such a long-term reserve, and should it be looked upon the way it seems to be these days?

NEWCOMB: I can't attempt to answer all of these provocative questions, but I can give a few comments on perhaps some of them.

For one thing, I seem to feel that you are asking what is going to restore competitive balance if everything except coal is objected to and then one looks at the true concerns over coal. I recall, although it is not mentioned much, that EPA's drive for elimination of sulfur is really that. The idea is to go from 1.2 pounds to 0.6 pounds, and then to 0.2 pounds. Even then you can make 1.2 with some commercial viability, maybe at $1.50 at the outside for troubled coal. You can't go much below that. You are in real trouble to scrub down below that, and that is really where EPA wanted to go.

But it is clear that for many, many years coal will compete with coal. In other words, there is such a massive amount of supply relative to the current needs, that you will have a restoration of market competition. So, given time, I think, one could answer satisfactorily that part of the equation for you.

With respect to mind-boggling hazards from either coal or nuclear power and with respect to the true length of time that one can buy with a fossil fuel technology: I think the embarrassing thing to an economist is that any time you give us a twenty-year supply of anything, we don't worry about it after that. We are like W. C. Fields with a twenty-year supply of whiskey. You could always get a free drink at our house. A further problem is that we have been asked to go to Congress and defend nuclear technology or necessary technologies on a budget basis rather than on a scientific basis, so you have to bring a cost-benefit analysis in, and any benefit received twenty years from now at current interest rates is zero. So Congress only faces the costs.

I think we ought to engage in scientific activity for the sake of the knowledge and the necessary knowledge that there will come a time beyond discounted present value of things, of projects, when that technology is vitally needed. I think we should be frank with our congressmen and tell them to make an investment where the payback period is longer than we had hoped.

DECKER: I would like to take a shot at that first question relating to coal versus some of the other alternative fuels that we might use. I feel I am qualified to comment because I have spent seventeen years burning a million tons of coal a year, at least in the plant that I was responsible for operating, and I have been trained as a nuclear physicist and have at least some association with nuclear projects. Concerning the relative environmental aspects and safety aspects of the nuclear and the coal plants--I am just giving my own personal judgment now--I would prefer to work in a nuclear plant rather than a coal plant, from the standpoint of all of the aspects involved.

I think the nuclear situation is one that has gotten badly fouled up. The nest is so badly fouled from the standpoint of the emotional aspects. I am not saying that there aren't a lot of aspects that need to have very serious consideration for the long term, as there are in coal as well. I think the CO_2 problem in the upper atmosphere, which you brought up, is one that needs to be looked at in the same context as some of the long-term problems relative to nuclear power. But the one aspect of coal that I feel very strongly about is that we know how to burn coal and we know how to mine it, and, in my opinion, we know how to do this in a way that is technically feasible from the standpoint of producing only a very minimal amount of damage to the environment.

I would like to correct one thing that Steve Reznek from the EPA mentioned. There are some chemical engineers who have been looking at the scrubber problem. I know of at least one utility that has had several chemical engineers working on the scrubber problem and has been able to make a scrubber operate satisfactorily and even reliably. There is still, I know, the problem of sludge. What do you do with the sludge after you make the scrubber work satisfactorily? I mentioned that there are chemical companies that are willing to work with utilities and other industrial firms in trying to use chemical technology and experience in solving the problems, and there have been discussions between some of the chemical companies concerning that. But in relation to natural gas, we would all love to have some magic happen so that we could all continue to use this nice, cheap natural gas or even petroleum; but, in my opinion, it looks like we are running out of the oil and gas at a very significant rate, considering what we have left, and I think we are going to have to go to coal and nuclear power. I think that conservation is going to be a must, but that is not going to solve the problem by itself.

HIBBARD: Are there questions from the audience?

AL SCHULER, U.S. Geological Survey: I would like to comment on Mr. Newcomb's remarks about the cleanability of coal. A great deal of information has been developed recently on the mineralogical form or the association of many of the potentially hazardous elements in coal, and very little of this information bears out his assertion that coal can be cleaned largely by removal of the clay. Such elements as sulfur itself, antimony, arsenic, zinc, cadmium, and mercury are associated with the sulfide minerals. The element beryllium is associated mostly with the combustible coal material. These may be problems, and although they will not be universal problems in all coals, they may be problems in many individual basins. They will not be removed, these problems, or lessened, by the removal of clay.

HIBBARD: I think Mr. Newcomb meant that the clay was associated with the particulate.

SCHULER: The reason I commented was that the term "dirty coal" is used, usually, to embrace all of the various environmental and chemical hazards, and I was afraid that his remarks therefore might be reflective of the conceptions people have of these other elements.

NEWCOMB: Of course, those are just a few of the trace elements that are found in coal, and, of course, the mercury and cadmium are the main villains, as you mentioned. The trace elements of these in most of the coals that are combusted are so terribly minor, there are many other sources of these that would be of concern before coal. The trace element problem with coal consists largely of those metals involved in the synergistic development of SO_3 and SO_4 from SO_2. These are the ones that wash out, and they are the large ones. For instance, 30 percent of your ash will be alumina. These are the things that are causing meteorologists and others problems in looking at the synergism of acid rain.

The U.S. Geological Survey is to be admired for getting into the trace element debate, but I don't think that it has shown at all that the dangerous trace elements in coal cannot be removed by mineral preparation. In fact, if given more time I think I could establish for you that the opposite is true.

JAMES BOYD, Materials Associates: I do have one general comment, Walt. As I sit back here and listen to all of the confusion--reiteration, discussion, and concern--I think we need to go back and trace somewhat the cause of that confusion--something resulting from our abandoning the basic philosophy of our form of government.

We have a tendency to start today from the assumption that the way to solve a problem is to enforce what-have-you's on somebody else, even though we don't at this particular time know what our views are based upon. We have a great deal of doubt--scientific doubt, doubt of knowledge, even doubt of economic theory.

We have given up our commitment to the solution through our basic principles of government, and we are turning toward regulation. I would like to take one example--something I have been close to, for a number of years: the safety example in the mines. There, we had a tremendous difficulty back in the early part of the century. And then came federal research, education, and demonstration, and the accident rate in the mines dropped precipitously, for about fifty years, almost sixty years. Then it seemed that there wasn't so much urgency about it, so the budgets were continually decreased, causing a decline in mine education. The way to answer that was to go into regulation. Now we regulate every movement in the coal mines and in the metal mines to the point that the people who are enforcing the regulation are finding out that it is counterproductive. So, they are going back toward the old principle of research, determining the cause of accidents, and then taking it back and demonstrating it in the mines. The result is that today we find young people coming into the mines uneducated and untrained, and, therefore, the accident rate is not as low as it should be. That to me is a particular example of overregulation, what is causing much of our economic concern and what will give us difficulty in solving this energy problem.

DAY II

INTRODUCTION TO DAY II

Alvin M. Weinberg, *Cochairman*
Director, Institute for Energy Analysis, Oak Ridge

In this session we will follow a case analysis method and deal with three specific situations in which the utilization of coal poses not general or abstract dilemmas, but very specific and concrete ones. The three cases that we have chosen to depict the benefits and the problems in using coal are the Ohio River Valley, which is largely dominated by coal; Kaiparowits, which incorporated many plans for the use of coal but was never actualized; and the northern Great Plains, with particular emphasis on Colstrip and Gillette as two towns that have been impacted by the use of coal.

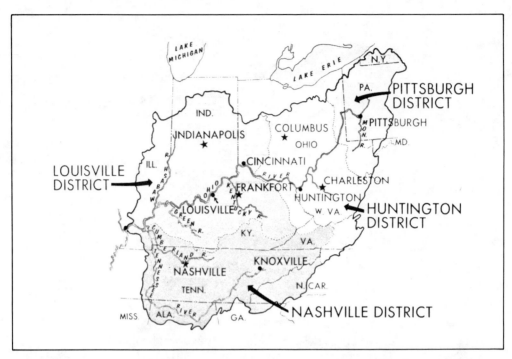
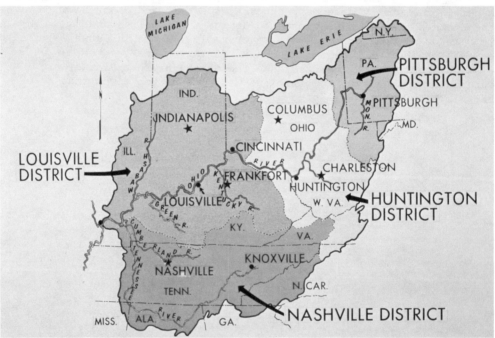

OHIO RIVER BASIN: U.S. Army Corps of Engineers district organization.

THE OHIO RIVER VALLEY

ORIENTATION

E. R. Heiberg, III, *Chairman*

Brigadier General, Division Engineer, U.S. Army Engineer Division, Ohio River

The Ohio River Valley presents most of the issues and problems--and opportunities--of coal use in industrial and populated areas of the continental United States. Coal is the most important Ohio Valley energy source, and coal has been the source for both growth and problems for many years. Pittsburgh and steel are synonymous through the collocation of Pennsylvania coal, Ohio basin water, and economic opportunity (labor, capital, and markets). Coal is by far the largest volume commodity moved on the inland waterways. A hugh slice of the valley's economy is founded on mining, transporting, and using coal. Pittsburgh and Cincinnati evoke images of baseball and football for some. Huntington and Charleston, West Virginia, suggest chemicals and the doorway to Appalachia. Louisville means horses to many, and the Wabash suggests farming. Nashville means music, Ashland means oil, and Indianapolis might suggest pistons and wheels. Yet, tying together every corner of the Ohio Valley are two common themes: availability of coal and plentiful water.

Today the issues important to coal as an energy resource can be discussed under two major headings: jobs (or economy) and environment (or quality of life). There are five interrelated areas: mining, conversion, transportation, water supply, and governmental regulations and investment.

People settled into the Ohio Valley because of the opportunities, which included the cheap energy available from the coal resources. There are jobs involved in mining, transporting, and converting coal to energy and jobs in other coal-dependent industries because that energy source is available at reasonable cost. Anyone who appreciates the energy-related industries in this region fully understands why I title this point "jobs." Perhaps it is more sophisticated to say "economy." Lest you not appreciate this point, tune into a local community when a neighborhood plant announces a layoff! The environment is important, because people's health, their quality of life, and the future of their children all depend on how activities associated with the use of coal as an energy resource interact with the environment. Environmental problems that have already developed in the Ohio Valley include the following:

- *Air quality*. The valley air quality has improved since the deaths from smog at Donora, Pennsylvania. Now there are new standards for alleviating air pollution, but these also make use of coal more difficult. There are several disturbing themes here that we will wrestle with in our panel discussion, relating not only to the use of coal but to the specific location of that use, the nuclear power issues (if not one type of plant, will the other come?), and so forth.
- *Water quality*. Efforts to avoid thermal pollution have caused new worries about consumptive use of water. The acid mine drainage legacy of previous coal mining has caught up with us.
- *Sedimentation*. This is a pervasive problem whose seriousness is enhanced if erosion from strip mines and spoil banks is not controlled.

Whatever is said about jobs and environment must relate to physical and political conditions and scientific and technical realities. During the discussions that follow we hope to touch on some of the important points that will affect coal as an energy resource in five separate technical areas, including the following:

- *Mining*. This familiar activity must occur where the coal is, with a tremendous range of effects on jobs and the environment depending on the surface terrain, the coal depth, and the governmental regulations which apply.
- *Conversion of coal to energy*. This can occur underground, at the mine mouth, at some central location, or at the point of use, with widely varying impacts on jobs and the environment.
- *Transportation*. Bringing coal to the point of conversion and energy to the point of use involves conveyors, barges, trucks, railroads, pipelines, electricity, transmission lines, and all the varieties of transfer and storage facilities.
- *Water supply*. The supply of water does limit the location of conversion facilities and may also limit the growth of energy demand.
- *Governmental regulatory and investment matters*. One example is today's discharge or construction permit requirements, which were

developed in response to past and present problems and will affect future development of coal as an energy resource. Another is the limiting aspect of many of the transportation elements in the valley, including the bottleneck and aging locks on the waterways and the geographical or perhaps fiscal constraints on the railroads in the area.

Central to the movement of coal in the valley is the inland waterway system, which is not confined to the Ohio River with its 981 miles of locks, dams, and navigation pools. The system in the valley includes huge tonnage movements on the Tennessee, Cumberland, Green, Kanawha, Allegheny, and Monongahela rivers. Let us examine this coal movement. First, all types of freight movement can be approximated by looking at the ten Bureau of Economic Analysis economic areas considered to be the Ohio River trade area. The volume of freight moved in that intercity commerce in 1974 totaled 245 million tons. Of that, 60 percent, 149 million tons (or well over half), was coal. Additional coal transported short distances to power plants on privately owned systems is not counted in those totals.

The Corps of Engineers is responsible for maintaining or improving the inland waterways for transportation. We are, therefore, responsible for the Ohio Valley waterway system. There are 2,200 miles of developed waterways that are used for commercial freight traffic. In 1974, those waterways carried 100 million tons of coal, about two thirds of the coal moved any distance from mine to users in the Ohio River trade area.

The 981-mile main stem Ohio River carried 70 million tons of coal in 1974, almost half of the total for the trade area. This transportation is accomplished without recognition by more than a few river-oriented people. Frankly, only occasionally do we even notice the passing "tows" (as towboats pushing barges are called). They are quiet, generally accident free, and away from much of valley life--with apologies to the river traffic fan in Louisville, Cincinnati, or Pittsburgh. The development of coal transportation growth has differed from the patterns of growth of several other commodity groups. Those other goods are also vital to the industries of the Ohio Valley, but coal tonnage moved by water has been doubling every twenty years.

With a few specific exceptions, the existing waterways are generally adequate for additional growth to the year 2000. The waterways, in brief, are not much limited by the physical nature of the river today. The Ohio is generally deep enough, wide enough, and passes enough water. The limits occur at the specific points where we have placed the dams and locks to provide navigable streams. On the Ohio River, these points occur roughly every fifty or sixty miles. An old-fashioned bucket brigade, trying to pass buckets to the fire, could pass water no faster than the slowest brigade member. Similarly, the passage of tows can move up- or downriver no faster than the constricting set of locks allow. That does not imply, however, that the waterways can operate without problems apart from restrictions posed by the smaller or less efficient locks.

Ice was a problem during the winter of 1976-77 because of cold weather and associated low flows. In most of the valley, this was the coldest winter recorded over the past 177 years or so for which records have been kept. Not only did nearly all the Ohio freeze over, as did a neighboring section of the Mississippi, but the basin-wide freeze reduced the flow to amounts too small to allow traffic past some points. Either warmer weather or higher water runoff would have reduced these problems. This winter reminded many people that projects and plans must contemplate contingencies that will not occur most of the time. That may mean additional emergency storage for energy or coal near the point of use; it may mean planned outages or phased shutdowns; or it may mean relaxation of some environmental standards when conditions reach such an extreme that we must do so under explicit and carefully controlled parameters. The major effect of this severe winter on the Ohio and upper Mississippi River systems, frankly, was to show the river-dependent customers the importance of the river to their needs. While I don't want to minimize the effects of the severe weather, it did not last long enough to strangle any industries. A few more weeks would have meant a drastically different story. But, the severe weather of the past winter is not a normal occurrence.

The Corps of Engineers does have a federal role in water supply. This aspect is not prominent in the Ohio Valley, although there are danger signs for the future. Most areas have sufficient water for most uses. However, water for use in conversion of coal to energy is generally not available in the headwater areas of the basin. The Ohio in the 1970s has had a plentiful supply most of the year. Water storage for power plant cooling or pumping water to a power plant in a headwater area for cooling may be possible but may not be economical. On larger streams, water is available, although concentration of usage could reduce streamflows below volumes desirable to maintain water quality. Water supply is one of the factors that will enter into the evaluation of every specific coal use project.

Federal and local regulations are another factor that will impinge on every project. As an example, the Corps of Engineers is responsible for permits for construction of facilities in waterways. Environmental assessments, and sometimes environmental impact statements, are required on every permit. These regulations are intended to assure that all factors are considered before action and that public values will be protected. The difficulty with today's regulations is that we are not always successful in preventing one kind of problem without creating other problems. We share this regulatory responsibility with a number of other federal agencies. As one of those "feds" to whom the Congress has delegated regulatory powers, I quickly admit that the weight of the requirements on a coal, nuclear, or water industry user can be stifling-- and I am using a kind word. But every one of the regulations a federal agency exercises is a direct response to and result of a law on the federal books and, therefore, represents a "political" conclusion on what needs to be done. I know the feds appear to the permit applicant as a many-headed hydra, and, seemingly, the heads spit and bite at each other. However, I have no quick solution. President Carter has stated that he is looking at the organizational streamlining options which may be available. I must also add that each state government has laws and regulations of its own, and often the permitting requirements add another layer to the stifling aspects of the federal regulatory process.

Energy is a current subject in need of tidying up. The pending national examination offers challenge and change to the coal and coal-related industry. I know there is an expectation through the executive, legislative, and public policy discussions ahead that national directions in the energy arena will soon be agreed to, and the safeguards and limits discussed and decided. I confess, though, that, to my mind, transportation and water are two areas which will be tough to reorganize in a manner that will make federal regulations easier to administer. But I suggest, in conclusion, that both transportation and water use are intimately related to coal and energy. We must wrestle these bears to the ground, or at least into the ring, if we are to hope to find the answers to the coal and energy issues.

The need for a national water transportation policy has been illustrated by the current hiatus in Corps of Engineers navigation structures planning and construction. Our old methods of projecting tomorrow's lock needs, so that needed rehabilitation can be performed in a timely fashion, have been successfully challenged in court. A decade of new environmental laws has necessitated some planning changes. It took the Corps of Engineers (and many others) a few years and several court cases to learn how to do a complete job in planning a project. The testing point today on an existing waterway is on the upper Mississippi, at Alton, Illinois. Our old locks and dam 26 must be replaced, but it seems everyone has a different idea of how, where, what size, and for how much. That controversy has affected in some manner virtually every planning effort underway throughout the inland waterway system, including several in the Ohio Valley system. I am concerned, because the relatively efficient water mode will be restricted soon on the Ohio, the Kanawha, and the Monongahela. Perhaps the most dramatic are old locks and dam 3 on the "Mon" upstream of Pittsburgh, which are today seventy years old and seriously deteriorated. Yet it will take us several years to replace them, and we don't know today how to "prove" what *size* the replacements should be.

Perhaps you now have heard more about the water side of the Ohio Valley coal issue than you needed to hear, but I trust I have underlined the indivisibility of water and coal in the valley. My fellow panel members will bring out other vital facets of the issue.

Boyd Keenan

Professor of Political Science, Institute of Government and Public Affairs, University of Illinois; Department of Political Science, University of Illinois at Chicago Circle

General Heiberg has provided you with the basic physical characteristics of the Ohio River Valley. He has asked me to present the broad framework in which various "publics" are raising their voices in the valley either to support the emerging coal and public utility system or to demand modifications in that system. I take this to be an invitation to discuss the politics of coal in the Ohio River Valley.

I need not tell this group that very few--if any--political data can be handled in a scientific or even a systematic manner. Here, we are dealing almost entirely with intuition and impressions. In this hallowed hall of science and precision it seems almost sacrilegious to make this admission, but I fear it is a reality with which we must begin. Yet we must have some political impressions--primitive as they might be--to place openly upon the table if this session is to be productive.

The Ohio River itself flows southwest 981 miles from Pittsburgh, Pennsylvania, to Cairo, Illinois. And its basin--or valley--contains parts of eleven states. But the publics who are straining to be heard are centered in that six-state region through which the river flows. These six states are Pennsylvania, West Virginia, Ohio, Indiana, Kentucky, and Illinois. Latest figures released by the National Coal Association show five of these states to be the nation's largest coal producers, in this order: Kentucky, 140 million tons per year; West Virginia, 108 tons; Pennsylvania, 83 tons; Illinois, 58 tons; and Ohio, 47 tons. The sixth state, Indiana, ranks ninth in coal production, behind Virginia, Wyoming, and Montana.

It became evident following the Arab embargo of 1973-74 and the accompanying quadrupling of oil prices that public utilities in the Midwest must increasingly be dependent upon coal. It was also realized that a greater percentage of our energy needs must be derived from electrical power as opposed to oil and natural gas.

Many factors pointed to the Ohio River Valley as the most likely Midwest location for the required new power plants. The river itself was a major factor, both as a supplier of water for cooling and processing purposes and as a means of transporting coal as well as heavy equipment and plant elements. Also, as noted above, the river itself has a close proximity to coal fields in the nation's five largest coal-producing states.

Finally, environmental regulations precluded construction of either coal-fired plants or nuclear facilities in or near urban communities. And many stretches of the Ohio River flow through relatively rural sections. Thus, there has been a virtual stampede on the part of public utilities in the Midwest to acquire land along the Ohio River for use in future years as sites of coal-fired power plants. As this probability of a concentration of power plants in the valley has become more evident to the general public, a complex set of political forces has been formed that almost defies either description or comprehension.

If we could somehow devise a political Geiger counter to pass across maps of the area, which "publics" would be measured as being most significant in the valley?

"GEIGER" MEASUREMENT OF POLITICAL FORCES

Starting at mile 0 in Pittsburgh, where the Allegheny and Monongahela rivers form to join the Ohio, we would soon notice political action on the Geiger counter. It would initially record minimal conflict on the northern reaches of the Ohio between industrialists and developers who wish to see more power plants, on the one hand, and environmentalists, on the other. But loud "clicking" on the counter would not be noticed until we had moved a considerable distance down the stream.

From about mile 400 (measuring from mile 0 at Pittsburgh), in the vicinity of Maysville, Kentucky, to about mile 600, near Louisville, the counter would shift into a virtual spasm of activity. On this 200-mile stretch of the river we would find utilities, industrialists, and developers in conflict with environmentalists and agriculturalists at a political level that is beginning to match that in any other section of the country.

Certain of these environmental coalitions are opposing plans for construction of both coal-fired power plants and nuclear facilities. In other cases, the groups have centered their attacks on one fuel. In a few instances, the environmentalists have been joined by state, county, and local governments, and we shall hear later this morning from representatives of certain of these governmental units.

FOCUS ON MADISON, INDIANA

Of the host of environmental groups operating in this 200-mile stretch of the river, the most prominent is probably an organization known as Save the Valley (STV), centered at Madison, Indiana--about mile 570. STV is an alliance of environmentalists, farmers, leaders of small businesses, and others. I shall note its major role in efforts to stimulate study of the impact of a probable concentration of energy facilities on the river later in this presentation.

Increasingly, however, over the past year or so, an informal coalition of leaders among coal producers, public utilities, industry, and labor organizations has argued that the actual numbers of people involved in the environmental movement have been exaggerated by the press and other media. These spokesmen have particularly been critical of the reporting and editorial policy of Louisville's *Courier-Journal*, undoubtedly one of the most influential newspapers in the valley and indeed the entire Midwest. This development coalition also argues that the environmental-agricultural movement to stall the planning for additional power plants in the region has been blunted by the severe 1976-77 winter.

There are many other ways of slicing through these six states to identify the most significant publics and devices utilized in seeking political action. It is true that environmental and agricultural opposition to power plant construction transcend state boundaries, particularly in the 200-mile stretch of the river from Maysville to Louisville (particularly in Kentucky and Indiana). But preoccupation with the environmental-developer cleavage overlooks the fact that the states themselves, as units with distinct power structures, represent a broad set of actors in a considerable measure independent of the environmental-developer cleavage.

The states *do* have energy and coal personalities of their own. These personalities are fluid and are constantly being modified, but they must be recognized. Of course, it is hazardous to attempt to characterize any of these state personalities. Yet, failure to notice that both executive and legislative leaderships in these states are developing some tentative and distinct personalities would be to ignore the reality of coal politics in the valley.

For example, for a number of complex reasons, the state of West Virginia, particularly through its legislative branch, appears to be a political entity which is *seeking* those power plants that environmentalists in Indiana and Kentucky don't want. Thus, it appears that environmentalists in Indiana and Kentucky may be aiding the cause of development-minded West Virginians who are eager to move proposed facilities upstream, one of the most interesting interstate developments in the valley.

It was probably no accident that President Carter chose Charleston, West Virginia, as the site of one of his energy seminars. The broad national political dynamics of interstate relationships in the valley are considerably affected by leadership changes in the U.S. Senate which give West Virginia interests high visibility there and will inevitably bring greater attention regionally and nationally to that state's concern. These changes relate chiefly to the elevation of West Virginia U.S. Senator Robert Byrd to the position of majority leader and the enlarged role of Senator Jennings Randolph in broad matters relating to energy and the environment.

RESEARCH ACTIVITIES IN THE VALLEY

Central to the future of the coal and public utility system in the Ohio River Valley are a number of research activities being sponsored by state, regional, and national entities. Prominent among these is a complex study of the "lower" Ohio River basin by the U.S. Environmental Protection Agency (EPA). In part, this study was initiated as a result of efforts by STV to demonstrate the validity of their efforts to U.S. Senator Birch Bayh of Indiana. After STV leaders visited Senator Bayh late in 1974, he persuaded his colleagues on the U.S. Senate Appropriations Committee to direct EPA to produce a study of the environmental, social, and economic impacts of the probable concentration of power plants in the lower Ohio River basin.

In a highly unusual move, the U.S. Congress, through the Senate Appropriations Committee, specifically mandated this study and directed that the study area include portions of the four states in the lower valley (Illinois, Indiana, Ohio, and Kentucky). In turn, EPA itself took an unusual step by awarding grants to researchers at six universities in the four states to conduct the actual research. The universities are (1) Ohio State University, (2) Purdue University, (3) Indiana University, (4) the University of Kentucky, (5) the University of Louisville, and (6) the University of Illinois (both Chicago Circle and Urbana-Champaign campuses). Grants for an initial one-year study effort were awarded in August of 1976. EPA is considering an extension of the project for two additional years.

This particular research activity is known as the EPA Ohio River Basin Energy Study (ORBES). I am associated with the ORBES project as codirector with responsibility for coordinating the institutional and governmental aspects of the undertaking. Also present here today as a resource person for this panel is Professor James Hartnett, director of the Energy Resources Center at the University of Illinois and coordinator of research activities being conducted by the Chicago Circle Campus.

Under the EPA "work plan," to which the various universities responded with their research proposals, a public advisory committee was established. Every major identifiable sector has been given representation on this committee. Among those serving on the body are General Heiberg, whom you have already heard, and two other members of our panel. They are Dr. Harvey I. Sloane, mayor of the city of Louisville, and Mrs. Jackie Swigart, chairman of the Kentucky Environmental Quality Commission. Still another member of the advisory committee is Ralph Madison, president of the Kentucky Audubon Council, who is present today as a resource person to the panel.

Also represented on the advisory committee are coal producers, public utility corporations, environmentalists from a broad spectrum of organizations, and public officials from all levels of government.

THE POLITICAL INTERFACE OF COAL AND NUCLEAR ENERGY

This audience should not overlook the political interface of coal and nuclear energy in the Ohio River Valley. If we should use our imaginary political Geiger counter once again to scan that 200-mile stretch of the river between Maysville and Louisville, we would identify considerable but very little subtle activity joining the two fuel sources in political controversy.

With our preoccupation at this Forum with coal, it is tempting to dismiss the significance of nuclear fuel cycle activities in the Ohio River Valley. A prominent historian of the Midwest, commenting on the Ohio River Valley, recently made this observation: "In no other district in the world are so many men and machines at work with the atom."[*]

For this discussion, the most significant work of "men and machines" with the atom takes place at two mammoth gaseous diffusion uranium-enrichment plants constructed in the valley in the fifties by the U.S. Atomic Energy Commission. These facilities are still operating--and indeed may be expanded--at Paducah (Kentucky) and near Portsmouth (Ohio). (Of course the Atomic Energy Commission was abolished in 1975, with the new Energy Research and Development Administration [ERDA] now operating the plants.)

The Portsmouth plant is particularly significant in understanding why Madison, Indiana, has become the eye of the environmental "hurricane" in the 200-mile stretch of the valley between Maysville and Louisville. As is well known, uranium enrichment plants require huge amounts of electrical energy. One of the factors in selecting Portsmouth as the site for such a facility was the availability of electrical power from coal-fired power plants. A group of Midwestern public utilities formed a consortium known as the Ohio Valley Energy Corporation (OVEC), which built two large coal facilities--one of them near Madison--for the purpose of supplying the Portsmouth enrichment facility with power. The Madison power plant was built 200 miles downstream from the Portsmouth uranium facility for national defense reasons.

At the Madison plant--labeled "Clifty Creek" by an OVEC subsidiary--the world's tallest stacks were built to disperse the coal polluting emissions. But community resentment grew from the midfifties onward, and by 1974 some Madison residents had become aroused sufficiently to provide leadership in the creation of STV. The point here is that a coal-fired plant built to provide energy for a federal nuclear facility 200 miles distant in another state had sensitized a community to a level at which it is now providing leadership for opposition to construction of coal plants on a major stretch of the river.

POLITICS OF FUEL COMPETITION

Earlier, I noted that each of the six states bordering the Ohio River were developing distinct "energy" personalities. It was noted that most of West Virginia's leaders are unabashedly proclaiming their willingness to offer power plant sites to any public utilities that may wish to locate them in their state. And, as second only to Kentucky in coal production, West Virginia producers have coal for new plants. (In a related controversy, which is far too complex to discuss here, West Virginia spokesmen argue that much of their coal is of a

---

[*]Walter Havighurst. *The Heartland: Ohio, Indiana, Illinois.* rev. ed. (New York: Harper & Row Publishers, 1974), p. 358.

low-sulfur variety and could be burned in compliance with federal EPA regulations.)

Thus, West Virginia's present energy personality is apparently built on a desire to attract power plants, which would burn indigenous coal within the state. Kentucky's energy personality seems less easily evaluated. While its leaders appear to be ready to push even further ahead as the nation's leading coal producer, there is a preference among many of these leaders to ship that coal out of state for burning in electric generating plants beyond its borders. Needless to point out, Kentucky businessmen, public officials, and environmentalists all agree that the state should be well paid for this coal when it is transported to other areas.

The matter of transportation within the broad Ohio River Valley political arena is of much importance but little understood even by those most involved. Hauling of coal became critical in the early seventies when federal and state legislation forced the public utilities to reduce their burning of high-sulfur coal, particularly in urban areas. As is well known to this audience, most Illinois, Indiana, Ohio, and western Kentucky coal is of the high-sulfur variety. Relatively small quantities of eastern Kentucky coal and a great deal of West Virginia coal, however, is rated as low-sulfur coal.

Faced with the prospects of converting more and more from oil and gas burners to coal facilities, many midwestern utilities sought low-sulfur coal supplies in Wyoming, Montana, and other western states. Now, these midwestern utilities are even transporting some western coal to their plants in the Ohio River Valley, much to the chagrin of coal producers in the region. The question of whether such long-distance hauling is sound public policy for the nation as a whole is likely to be bitterly debated during the coming years. There are many dimensions to this controversy and, as noted above, few students of the problem fully understand it.

Aside from the strictly environmental aspects, there are broad institutional questions now being debated. Certain West Virginia spokesmen lean to a "conspiratorial" theory. They argue that their low-sulfur coal, which is also rated as having desirable volatility properties when compared with western coal, is being ignored by the larger utilities for nonenvironmental reasons. A host of charges have been made, including the contention that the utilities have moved into the business of owning and operating their own mining and transportation systems in the West. Also, fuel-adjustment clauses in some states are said to allow the utilities to pass the cost of buying and shipping coal on to consumers without approval from their state regulatory commissions.

Much of the frustration in West Virginia is exacerbated by the fact that the Tennessee Valley Authority (TVA), the country's largest coal user, has apparently never purchased low-sulfur West Virginia coal, while it has imported much western coal. It seems inevitable that TVA will increasingly become linked to the complex coal politics of the Ohio River Valley.

COAL-NUCLEAR FUEL RIVALRIES

It was noted above that the uranium-enrichment phase of the nation's nuclear fuel cycle process was centered in the Ohio River Valley in plants at Paducah and Portsmouth. (The only other enrichment facility for preparing uranium for commercial nuclear plants is located on the edge of the Ohio River basin in Oak Ridge, Tennessee.)

Probably in large measure because of the availability of coal in the valley, public utilities have planned very few nuclear plants in the region. Only two nuclear facilities are certainties at this point. Pittsburgh's Duquesne Light Company has built a nuclear facility thirty-five miles downstream from the city. And Cincinnati Gas and Electric Company is constructing a nuclear plant at Moscow, midway between Cincinnati and Maysville (to the southeast).

Our political Geiger counter would probably identify the most dramatic political confrontation on the river at the present time as surrounding plans by Indiana Public Service (PSI) to construct a nuclear facility on the river only ten miles southwest of Madison. STV is leading the opposition in preliminary hearings now being held by the Nuclear Regulatory Commission (NRC). Also opposing plans for the PSI plant--known as the Marble Hill facility--are the governor of Kentucky, the state's attorney general, and a host of other Kentucky governmental units, including the city of Louisville.

Spokesmen for the Kentucky governmental units have been accused by Indiana officials and by the public utility itself of fearing the fuel competition that nuclear energy would bring to Kentucky's coal in that portion of the valley. One of Kentucky's major weapons in attacking the construction of the Marble Hill plant is state's ownership of the Ohio River itself. River water will be used for cooling, and the project opponents fear both radiation and thermal pollution. Legal research by Kentucky officials on this matter suggests that the state may utilize this ownership in future controversies over the siting of power plants.

EDUCATION OF THE PUBLIC

One affirmation may be made with confidence. Unless a broad effort is made to educate the diverse "publics" to the complexity of the politics of coal and public utilities in the valley, the nation will be afflicted with time-consuming and energy-dissipating controversies. But how should the educational process begin?

All "publics" are often categorized as fitting into one of the four following classes: (1) elected officials, (2) government administrators, (3) the informed public, and (4) the uninformed public. For the first two categories, short courses and seminars on energy and environmental affairs are the usual (but uninspiring) answers. Governments at all levels and universities no doubt will be forced to attempt to provide imaginative offerings both for elected and appointed officials as well as for the informed public. And service on advisory committees, such as that described above in connection with the ORBES project, is a "learning" experience.

For the uninformed public, including youth still in their formative years, some existing institutional arrangements might be utilized. The array of organizations which operates under the umbrella of the various land-grant college programs are suggestive. An extension service for energy, utilizing either existing "educators" such as county agents and home demonstration agents or new agents, should be explored. Colleges of agriculture around the country, in which such extension services are located, are eager to move into energy education. But, likely, there will be long debates as to whether these units are prepared to provide help for problems which have such an "urban" orientation.

Other institutions not generally viewed as "educational" in nature should be considered as arenas for such education. Public libraries in towns and cities, large and small, on the Ohio River itself and in basin communities some distance from the river have a ready-made information system that might be utilized effectively. Librarians have often argued that libraries are more than repositories, and they should be given an opportunity to assist in this national need.

In times of need, churches in America have often responded with educational and quasi-educational assistance. They should be encouraged in this instance to study the problems of energy and environmental affairs. But topics relating to fuels, public utilities, and other energy issues are ideologically controversial, and too heavy an emphasis by churches upon these topics could embroil them in controversy with government. And great as the energy challenge might be, we need to be mindful of the danger here of violating the tradition of separation of church and state.

On the 200-mile stretch of the Ohio River between Maysville and Louisville, noted so frequently above, a little-noticed phenomenon is worth noting.

By most definitions, this section qualifies as a part of the "Bible belt." In earlier generations many young people of high school and college age were spending Sunday evenings at worship services and Wednesday evenings at prayer meetings. Now, there is an increasing number who find the challenge of participating in public discussions of environmental matters to be more satisfying than church-related activities. Whereas their fathers and mothers traveled the river to attend revival meetings and other religion-oriented sessions, these young people are pursuing questions that may be equally profound. Observation of these young people in action leads one to wonder if the environmental movement is as dead as "conventional wisdom" would have us believe. If concerns over possible dangers from energy shortages could be linked with environmental fears, the forums offered by these young people could be invaluable to the nation.

Finally, there are the "main line" public interest groups, which have survived over the years despite the fickle and fashionable moods that have afflicted the country. The League of Women Voters immediately comes to mind. It typifies the best of participatory democracy. Yet the League, like so many worthy organizations, has sometimes been constrained by those outdated and outmoded state boundaries that defy reason in a technological era. Perhaps the League or other similar organizations should experiment in developing devices for examining interstate and regional problems. Periodically, in American history, politicians and educators have advocated regional approaches to our most serious problems. The energy question, more than any other issue in our nation's history, cries out for education in regionalism.

Herbert B. Cohn

Vice Chairman, American Electric Power Company

I gather that my primary role on this panel is to express the point of view and concerns of a representative of an electric utility that takes very seriously its legal and moral obligation to provide the electric power requirements of a substantial part of the Ohio River Valley and to summarize the reasons why we believe coal is a vital energy resource and why its use in the valley is of essential importance to the valley and to our nation.

In these introductory remarks, I propose to outline briefly the nature of our energy problem; the importance of energy to our economy and national well-being; the available sources of energy; the role of electric power in helping to provide the required energy; the facilities required to produce and deliver electric power; and the necessity to balance the essential need for electric power with other aspects of the public interest.

THE NATURE OF OUR ENERGY PROBLEM

We do, indeed, face a most serious energy problem, with only the tip of the iceberg having yet been seen by the American people. There are two principal aspects of the energy problem. Although they are related, it helps, I think, to refer to each of them separately.

The first relates to our increasing dependence on imported oil. Such dependence has risen from 23 percent of the oil we consumed in 1970, to some 30 percent in 1973 at the time of the embargo, to over 40 percent in 1976, and it is still increasing. The *New York Times* recently reported oil imports for one week reaching 10 million barrels against a domestic production of only 8 million barrels. The cost of oil imports in 1976 was $35 billion.

This increasing dependence on imported oil--and the very real potential for the imposition of additional restrictions on supply and still higher prices--creates major risks for:

- Our national security--with an obvious relationship to our defense apparatus, which is so heavily dependent on oil, and a very direct effect on the strength of our position in international affairs.
- Our national economy, employment, and standard of living--where even temporary shortages of oil can create chaos in our economy and where a $13-per-barrel price and an annual expenditure of $35 billion for imported oil makes a substantial contribution to our adverse balance of trade and to our inflation.

There is, I think, a pretty clear consensus that we must take steps to reduce this increasing dependence on foreign oil.

The second major aspect of our energy problem derives from the fact that our principal sources of energy at the present time--oil, gas, coal, and uranium--are nonrenewable resources and are finite. At best, oil and gas, on which we now rely most heavily, are likely to be largely exhausted in the next thirty to fifty years. The domestic uranium supply appears to be limited. And, although we have very large coal reserves, they too will eventually run out.

Accordingly, we must find new--and, wherever possible, renewable--sources of energy.

The end

THE IMPORTANCE OF ENERGY TO OUR ECONOMY AND NATIONAL WELL-BEING

There is a high degree of correlation between an adequate and reliable energy supply and our economy, employment, and standard of living. A concise and, I think, most persuasive demonstration of this correlation appears in a study done by the Energy Economics Division of the Chase Manhattan Bank entitled "Energy, the Economy and Jobs" issued in September 1976. And, for a discussion giving particular emphasis to the relationship between energy and jobs, I commend to you a highly perceptive paper entitled "The Threat to Jobs" by Vincent A. O'Reilly, director of utility operations for the International Brotherhood of Electrical Workers.

Abundant energy has been essential to the economic development of the United States. It is even more essential now. Without it our nation cannot sustain the high standard of living that most of our citizens enjoy. We cannot, in good conscience, deny our children and our grandchildren the opportunity to enjoy a similar standard of living. Even more important, an inadequacy of energy supply will impose a major obstacle in the way of improvements in the standard of living of our economically disadvantaged. Bayard Rustin has made the point, in the most forceful terms, that no growth "would measurably worsen the nation's--and the world's--economic plight" and would shatter "the hopes of those who have never had a normal role in the world economy, among whom the darker-skinned people of the world rank most prominently."*

AVAILABLE SOURCES OF ENERGY

What are our available sources of energy at the present time? Our principal sources are gas, oil, coal, uranium, and hydropower. Gas and oil are running out and should be conserved for uses for which we have no substitute. Additional hydroelectric sites and hydro potential are limited. This leaves coal and uranium, with a good deal of uncertainty as to how long our uranium supplies will last and considerable uncertainty as to how extensively we will be permitted to use nuclear fuel. We do have very large supplies of domestic coal that can provide a major part of our energy needs for at least several hundred years, and coal must in fact be the major part of our answer for at least the short and midterm.

Fusion, solar, geothermal, tidal, biomass, hydrogen--and, indeed, any other new energy source which has any promise for ultimate success-- must be the subject of intensive research and development. But it is clear that the so-called "exotic" new sources of energy are not going to be able to make a material contribution to our energy supply before the end of the century.

Conservation to eliminate waste and to improve efficiency is highly desirable and, indeed, essential to help us get through the short and midterm, but conservation, by itself, can do only a part of the job of eliminating the gap between demand and supply.

THE ROLE OF ELECTRIC POWER

It is clear that electric power must play a major part in supplying our energy needs. For the short and midterm, in addition to supplying our conventional needs, electric power, based on coal and uranium, must be utilized to substitute, wherever possible, for gas and oil, so that gas and oil can be conserved for uses where substitution is not possible. This is being done now to a limited extent; such substitution can be-- and, indeed, must be--increased greatly. A major area in which very little is now being done is electrification of surface transportation. The technique is well-known in the railroad field, and it is within sight for the automobile and bus.

Electric power now provides about 27 percent of our total energy requirements. Taking into account the possibilities--and indeed the necessity--for using electric power as a substitute for oil and gas, it is estimated that by the year 2000, electric power should provide about 50 percent of our total energy requirements. And, in the longer

---

*Rustin, B., *New York Times*, May 2, 1976.

term, it seems quite clear that electric power will provide the essential mechanism required to utilize the newly developed sources of energy.

Yet, despite the current and future importance of electric power, the fact is that conflicting governmental policies and requirements, the interminable delays associated with the need to obtain required governmental permits and licenses, the failure to eliminate existing governmentally imposed obstacles, and the imposition of new obstacles have left us with the prospects of power shortages that could have the most adverse effects on our economy and on the "quality of life"--in its broadest sense--for all of our citizens. The Federal Power Commission, in its December 1, 1976, report on "Factors Affecting the Electric Power Supply, 1980-85" concludes that electricity shortages by as early as 1979 are "distinct possibilities."

A study, just published by the Energy Consulting Division of the Chase Manhattan Bank, entitled "Potential Electric Power Shortages" concludes that unless present expansion plans are completed *and increased by at least a third*, "utility customers will be subject to service interruptions or 'brown outs' resulting in serious economic and job-impact implications" by the mid-1980s. And the Chase study indicates that the single most vulnerable region of the country, in this respect, includes the Ohio River Valley.

THE FACILITIES REQUIRED TO PRODUCE AND DELIVER ELECTRIC POWER

Under current and foreseeable technology we cannot have the quantities of electric power we need without large, central station generating facilities and transmission lines to deliver the power to the load centers.

Power plants have certain basic requisites for siting. They should, of course, be properly located in relation to loads. Both fossil fuel and nuclear plants require extensive sites and, under current regulatory requirements, nuclear plants must be located some distance from concentrated areas of population. Both types of power plants must have sufficient cooling water and a means of transportation to permit the delivery to the plant site of large-scale equipment required in the plant's construction, maintenance, and operation and, in the case of fossil fuel plants, the delivery of large amounts of fuel. Where the fuel necessary to conform with environmental requirements can be obtained only outside the area, this latter consideration of water transportation is of particular importance. These factors, based on today's technology, dictate the location of such plants most often on major bodies of water including major streams, such as the Ohio River and its tributaries.

The Ohio River Valley is frequently, and properly, referred to as the heart of industrial America. It includes, and it is contiguous to, large concentrations of heavy industry and of population. The requirements in the area for electric power are large and will be increasing. It is inevitable that there will be increasing use of the Ohio River and its tributaries in the production of such electric power. This must, of course, be done with the most careful consideration and coordination to minimize adverse effects on the environment and on other aspects of the quality of life in the valley.

In this connection, as you may know, the East Central Area Reliability Council, which includes a regional group of electric power systems serving in the Ohio River Valley, is undertaking a study to project the need for electric power and energy in the area; to relate those needs to existing and projected power plants; and to indicate the location of all existing plants and, to the extent possible, future power plants. The study will explore the unique requisites of the Ohio River Valley for power plant siting and the severe--and in most cases impossible-- constraints in locating power plants elsewhere in the river basin area and will relate--and place in perspective--power plant land use as a proportion of overall land along the Ohio River.

BALANCING THE NEED FOR ELECTRIC POWER WITH OTHER ASPECTS OF THE PUBLIC INTEREST

The construction and operation of power plants and transmission facilities will, of necessity, have some adverse effects on the environment. The most obvious relate to the effects on air, water, and land use.

We must, of course, do all that we reasonably can to minimize any such adverse effects. This means the use of precipitators to minimize emissions of particulates, improvements in the quality of the coal used by washing and cleaning, steps to take out the pyritic sulfur from the coal before burning, and new and innovative combustion processes, such as the fluidized bed. It means the use, where it makes sense, of cooling towers or cooling ponds to minimize harmful effects to our lakes and streams. And we must, of course, make every reasonable effort to minimize harmful effects to our land and accommodate to the desirability of esthetically pleasing structures.

The essential fact that has to be understood, however, is that we cannot have pristine air and water, on the one hand, and an industrialized economy and a high standard of living, on the other. There are some inevitable conflicts, and there must necessarily be an accommodation and trade-offs to achieve a compromise that best furthers the overall public interest. This can, in fact, be achieved through such techniques as cost-benefit analyses and a proper balancing of all aspects of the public interest.

There are risks and costs associated with any course of action--or inaction. In evaluating adverse environmental effects associated with the increased production of coal and electric power, and the alternatives of inadequate energy supply and increasing dependence on foreign oil, we should keep in mind the observation of a writer in Denver's *Rocky Mountain News* that "it is far more likely that our children will die in a war for energy than from the pollution that energy causes."

Harvey I. Sloane, M.D.

Mayor of the City of Louisville

My background is medicine, and I have also been an advocate for clean air. I served on the local air pollution control board. Jackie Swigart and I started a citizens' group for cleaner air; we fought the battles. But I also like to read by electric light and not candlelight. I like a fairly warm house; 65 degrees is fine.

With the advent of a real possibility of a nuclear power plant some thirty miles upstream from Louisville, Kentucky's interest in coal is not only traditional, since we are the largest coal-producing state in the country, and not only based on economics, but also stems from the safety factor. We are much more interested in coal now that the controversy of the Marble Hill plant in Indiana is upon us.

Comments have been made about the conflicting agencies and regulations. Let us go over a few of them. We have the Army Corps of Engineers (and they are the least controversial in my opinion), EPA, and the Nuclear Regulatory Commission. We have the Rural Electrification Administration and the Department of the Interior. All this results in enormous problems in getting decisions made.

It reminds me about the story of a fellow named Michael, who had to get up to go to school every morning. His mother would come and wake him up and say, "Michael, it is time to get up to go to school." She then would go down and make breakfast for Michael. One morning she came back up to find that he had not risen. She shook him and said, "Michael, you have got to get up and go to school this moment." Michael opened one eye, and he said, "Mother, why do I have to go to school?" She said, "Well, there are two reasons. First of all, you are forty-two years of age, and second, you are the principal."

I have never met the principal of the development of an energy policy or the regulation of the environment to satisfy that energy policy. I think Dr. Keenan explained the politics up and down the river well.

Let me talk about that 200-mile stretch from Maysville to Louisville, which happens to be mostly in Kentucky. We want to see power developed, Mr. Cohn, and we think it needs to be developed, but let us think about the synergistic action of some 100 plants that are talked about along that area, the synergistic action from an air pollution standpoint of nuclear plants and of coal-generating plants and the synergistic action on water pollution. Let us think about that Ohio Valley region, which has significant problems of inversion and impacting. Let us think of the problems of water pollution and what that does to the quality of streams.

We made, as I think Dr. Keenan has mentioned, a considerable effort to get an environmental impact study headed in the right direction. I think it is going in the right direction, and ORBES is going to address many, many issues. It is going to take three years to complete.

That is my understanding. It will result in a recommendation. It will not involve a final decision. During that time there are going to be power plants built that are going to have an impact on what subsequent development happens. From a purely economic standpoint, as an elected leader of Kentucky, I am concerned not only about generating power for our immediate region. Let us understand that the majority of power generated is going to go out of our region. We have use for the land for industrial development. That brings jobs also. That brings positive aspects to our economy, which need to be balanced against the possible creation of a river valley that is going to supply the midwest grid, eleven states, and perhaps jeopardize the economic development that we desire and wish and that does not speak to the environmental impact on the surrounding area.

If there is a lack of credibility in government, I find it nowhere more pervasive than in the area of environmental planning. Anybody who has been to hearings at which the Nuclear Regulatory Commission is discussed can feel the real hostility and antagonism toward that agency, the lack of credibility that it has. EPA, unfortunately, does not enjoy a great reputation. It was unfortunate that Russell Train, whom I by and large admired quite a bit, made his comments about nuclear power, expressed his concern about nuclear power, after he left office but did not make those comments while in office. Perhaps that is a part of the political process, but it does not help the credibility of governmental agencies ultimately responsible for implementing aspects and developments that are going to have an impact on the lives of many, many people.

Let me focus on some of the problems we have in terms of regulation. Indiana, Illinois, and Ohio are in Region V, which is headquartered in Chicago. Kentucky, right across the river, is in Region IV, headquartered in Atlanta. West Virginia is located in Region III. These are three different regions of EPA. Sometimes they do not agree. Sometimes they have conflicting recommendations, and to develop an overall strategy for the Ohio Valley development with such a fragmented approach leaves the average citizen confused and bewildered and in the end mobilizes that person politically.

Dr. Keenan has mentioned the political developments that relate to the Ohio Valley region. They are an offshoot, in my estimation, of the lack of governmental unity, the lack of governmental direction, and the lack of governmental credibility in this area. Not only in this particular part of the country, but all across the country, we do not have a common, unified thrust that people can have input into and argue about. If there were a unified direction, people might not be satisfied, but at least there would be a process that made sense. The present lack of unity is so bewildering, even the most sophisticated of us are not able to weave our way into it and use our knowledge to get action.

We have a further complication in making specifications, for instance, for proposed plants in the Ohio Valley. Writing the impact statement for the Cincinnati Gas and Electric Company is the Corps of

Engineers. The Environmental Protection Agnecy is doing the same for our Louisville Gas and Electric Company. The Rural Electrification Administration, under the Department of the Interior, is writing specifications for the Spurlock plant in Mason County, which is in Kentucky. The lack of coordinated planning polarizes people into political action.

Some days ago in Louisville, one of the major consumer advocates in this country made a speech, and after the speech I talked to him about the problems of the Marble Hill nuclear plant. How did he believe that the community should approach it? He said, if you are against it, create a hostile atmosphere in that particular community against that plant, and you will delay it; therefore, for economic reasons it may not be built. I do not think that is the right way to approach our power problem, our environmental problem.

The fragmented bits and pieces that we have experienced and the lack of unanimity or lack of a uniform plan create such frustration, in the end we are going to see, I think, a very poor plant development that may have serious impact, not only on the energy generation for our country, but on the environment that is being affected by that particular area. Understand that most of the power that is going to be developed is not going to be developed for Kentucky; it is not going to be developed for that basin in Indiana, for Ohio; it is going to serve a midwestern grid. That is important. I do not think we should just serve ourselves. We can export the coal. Our coal will be used no matter in what area the plants are. But we do not want to develop a civil war between Indiana, the Public Service Corporation of Indiana, and Kentucky, a development which is happening now. You have the major political figures in Kentucky opposing this nuclear plant in Indiana because the Indiana Public Service Commission did not really consult Kentucky citizens. They went ahead and did it on their own.

The Nuclear Regulatory Commission, which, in my opinion, is the most unresponsive agency in the federal government, perhaps following the CIA, is creating such a hostile environment, it is going to develop the confrontation that none of us wish to see and that will hinder a positive development of our whole country and our whole area. We need a professional analysis that is able to be incorporated into a decision-making process. The citizens have to be involved. The developers have to be involved. The environmentalists need to be involved. The people who are desperately concerned about the energy crisis in our country need to be involved. We need a direction; we need to go forward and to make the Ohio Valley region an asset, not only to that area, but to the rest of the country. We also need to make the Ohio Valley a livable community, a community that does not have such environmental hazards that people cannot live there healthfully or the total development of other industry is obliterated.

So that is my message. If I had a few people from EPA and the Nuclear Regulatory Commission and a couple of other agencies to collar personally, I would, but I hope this Forum will be able to convey some of these feelings to the national authorities.

Jackie Swigart

Chairperson, Environmental Quality Commission for the State of Kentucky; Cochairperson, Ohio River Basin Citizens' Advisory Council

Kentucky is an excellent example of what this discussion is all about. There is probably no other state in the basin that really brings into focus all the discussions, all the concerns--social, environmental, economic, legal, you name it, they are there--and they have existed for a long time. It is some of those issues that I have dealt with over the past twenty years upon which I base my remarks.

There are those of us in Kentucky who would agree with Representative Wampler about states' rights, and there are also quite a few who recognize that when you talk about a resource used all across the world, we definitely need some federal intervention. What has been lacking is an energy policy coming out of the federal government, and I really am pleased to see the Carter administration attempting to move into this most controversial area, one that I think definitely needs to be resolved before we can come up with a solution.

I would like to emphasize the need to define the various publics; we have touched on this a number of times at this meeting. Who speaks for the public? This is a question that is raised at every single meeting I attend. I am afraid that over the years we have become much too fragmented, and as Dr. Sloane has said, we end up on opposite sides of the hearing room. This does not solve problems. It only creates more. To begin with, we probably need a little understanding, a little agreement that I appreciate where you are coming from and you appreciate where I am coming from. This country is made up of a variety of viewpoints, and that is where the conflict is. To start out with the feeling that I respect your point of view and you respect mine is a good starting point for discussion.

We dig coal in Kentucky and have for more than 100 years. Kentuckians are proud of the fact that we are the number one coal producer in the nation, mining 146 million tons in 1976, which was 23 percent of the nation's total production. In January of 1975 it was estimated by the U.S. Bureau of Mines that approximately 9 billion tons of coal lies under Kentucky soil, with 3 billion tons to be surface mined in western Kentucky. This, combined with 8 billion tons lying underground in the eastern part of the state (and 3 billion tons of that that can be surface mined), adds up to 25 billion tons of a mineral that the Carter administration is now looking favorably upon as an energy source for the next 25 years or more.

It was not until 1964 that the leadership of Kentucky recognized the impact of this powerful industry on the environment and passed Kentucky's strip-mining law. The coal companies have always contributed generously to political campaigns at all levels of government, so it came as a shock to them to realize that not only was regulation of the industry here to stay but would only grow stronger as the number of permits increased and the public's resistance to environmental damage grew.

Government's responsibility in controlling this industry and its powerful ally, the utilities, is tremendous, and at times it seems like an impossible task. I think all of us recognize the very real problems that government at all levels has in trying to hire good technical people, people who can get paid probably three or four times more by working for an industry, but who, through a sense of dedication, have tried to remain in the bureaucracy and see that the job gets done. Added to these problems in the state of Kentucky, and, I am sure, in other coal-producing states, is the expectation that an inspector, being paid $10,000 a year at the most, will stand up to a multimillion-dollar corporation offering him a bribe and look the other way. These are some of the realities that government has to face in regulating a very powerful industry. The job of expressing environmental concerns, which are understood as being meaningful, is a difficult task, at best, in Kentucky.

Kentucky's coal is embedded in two distinct coal fields, the eastern and the western. The coal in eastern Kentucky is high in volatile matter; relatively high in heating value, producing 12,000 to 14,000 Btu per pound; and low in sulfur and ash content. Western Kentucky coal produces 11,000 to 12,500 Btu per pound and has a sulfur content of 3 percent or more. Kentucky coal is produced by almost 2,000 companies, some large and some very small. About 96 percent of the western Kentucky coal is consumed by electric utilities. Forty percent of that goes to TVA alone, while about 60 percent of eastern Kentucky coal goes to utilities.

Utilities tend to buy coal on long-term contracts from large producers, placing the small operators in the position of being dependent on the spot market. When you talk about coal in Kentucky, you are not talking about one kind of a problem. You are talking about a lot of different kinds of people, different kinds of coal, and different accessibility because eastern Kentucky, as you know, is very mountainous and western Kentucky is low pastureland type of terrain.

Some of Kentucky's coal is exported—a great deal of it, as a matter of fact, and as Dr. Sloane pointed out, we use less in Kentucky than we ship out.

The lack of assured markets and inadequate rail transportation at this time create some very real problems for the coal industry. If you are a coal operator, the owner of mineral rights (the broad form deed is still valid in Kentucky), a provider of heavy equipment, or a utility, the large amount of coal reserves tends to make one very optimistic, especially in light of the current emphasis at the federal level to insure coal's position in the energy picture.

However, this joy is not shared by all Kentuckians, and I would like to explain why. There is a great deal of concern about the impact of increased production of a finite resource, especially in mountainous areas, and I really am pleased to see the utilities at this meeting refer to coal as a finite resource. In past numerous discussions I have had with utilities people, they were not even willing to admit that, so I was pleased to hear them refer to coal as a finite resource today. It is a step in the right direction.

It is ironic that the most desirable coal from a low-sulfur standpoint is also in a part of the state that is more difficult to strip-mine and where strip mining is environmentally more damaging because of the terrain. Even with a federal strip-mining bill, enforcement will be a difficult task. The lack of enforcement of Kentucky's reclamation laws has a long history and even served as a thesis subject for a Yale graduate student, along with a critique of corruption in the New York City police department. These concerns will not disappear with another layer of government added to the picture, and I think all of us know that the laws are only as good as their enforcement. There is a long, long history in Kentucky of lack of enforcement for a number of reasons.

Another issue expressed by the public before the Environmental Quality Commission, which is the advisory panel that I chair and which advises the governor and the Department for Natural Resources, is that of the combined impact of these coal-fired power plants, planned for the stretch of the Ohio discussed today, and the three nuclear power plants scheduled for the same stretch. Concern has caused citizens up and down the valley to speak out, sometimes in violent opposition, as in the case of the Marble Hill nuclear power plant. No one has been able to assure the public that the overall impact of increased river transportation, industrialization, and deteriorated air quality will be minimal. It is hoped that the ORBES study, which Dr. Keenan and his colleagues are working on, will provide the decision makers with some meaningful information about this impact before it is too late.

Another concern expressed by the citizens in the valley is the possibility of a contaminated drinking water supply from a nuclear power plant. The risk of a nuclear accident, which we are told is minimal; the siting of power plants in areas of prime agricultural land; and the disposal of radioactive waste are other public concerns expressed in Kentucky.

We are fortunate or unfortunate in having the Maxey Flats burial site in Kentucky, which receives 99 percent of its waste from other parts of the country. There has been a whole lot of concern evolving around that site, which is built in a part of the state that happens to receive a great deal of rain. In the past the dumping of waste has not been done very carefully, and there is some concern by geologists whether this waste is leaking out into the rock fissures. That is another issue very definitely related to the impact of planned power plants in Kentucky and one in which the Environmental Quality Commission has also been involved.

With the advent of these proposed coal-fired plants, we must also consider the disposal problems associated with large amounts of sludge, which could be generated if scrubbers were applied. I notice American Electric Power did not say much about this issue. Six hundred tons per ton of high-sulfur coal could be generated, which is a pile of waste. Where do you put it? Nobody is sure yet what the answers are. No one is talking about the overall impacts. Everyone is thinking in terms of that individual plant that is going up. There does not seem to be any discussion about the impact of all of these plants. In the absence of a power plant siting law and with the Ohio River providing an interstate boundary, it is a very helpless feeling that people have about what can be done.

There have been many, many hours spent by citizens preparing for public hearings in each one of the instances in which the public finds out a power plant is going to be built, and finding that out is not an easy process. There has been a lot of time and effort put into preparing testimony, some of it at a highly technical level. There is still no one governmental agency that can give the public an answer. Again, the Environmental Quality Commission expressed this concern to the governor last year and urged him to work with the governors of the surrounding states to seek a solution. He has agreed to do this and, as chairman of the Natural Resources Committee of the National Governors Conference, is in a position now, we hope, to influence some of these decisions.

Now, getting a commitment from the politicians is the easy part of the battle. The tough part is seeing that something gets done, and one of the very important roles that the Environmental Quality Commission plays is to watchdog governmental agencies at the state level, to see that they do what they are supposed to do and follow through on their promises. However, the number one question throughout all of these discussions is, who should make these decisions? Who speaks for the public?

In my opinion, there are four publics. We have the elected officials. We have the bureaucrats within the government. We have the informed citizens, and we have the uninformed citizens; and then there is a question of who educates the uninformed citizens. The public is not just one group. It is a variety of people who make up this country, and I think because of that reason, answers and solutions are made very difficult.

Traditionally, the decisions have been made by the elected officials in cooperation with the vested interests. I think the time is long overdue for the public to discuss some of the trade-offs involved. The public is the consumers and the ones who will ultimately bear the costs. This past winter has really dramatized the fact that our resources are finite, and with skyrocketing utility bills people realized that we have been getting by pretty cheaply in this country.

Given the suggestion that the year 2000 is the time when the supplies of oil and gas will disappear, we have some very difficult decisions to make immediately, and it is reassuring that the Carter administration is attempting to look at these options. But, again, who is going to be making these decisions? How will these decisions be influenced?

In spite of the popularity coal seems to have as a long-range solution, there are many serious implications that this solution would have on the environment. As Professor Rose said, sure, we can go ahead and mine the coal, but we should be thinking in terms of what the impact will be. It is important that we give some thought to these trade-offs. Perhaps it is shortsighted to rush to get this coal out of the ground if we have to give up areas of irreplaceable beauty and destroy streams in the process, things that we cannot manufacture.

In Kentucky we are blessed with more miles of free-flowing streams than any other state and an adequate rainfall. I often wonder whether we would be so careless if we had a shortage of water, as the western states have. In my opinion, and that of others, energy efficiency is the only answer. We must never let ourselves become dependent totally upon foreign resources, although a healthy exchange is certainly necessary. Cutting back in meaningful ways, such as building energy-efficient buildings, manufacturing cars that consume less gasoline, and encouraging through incentives a conservative approach, will allow us to explore other options.

To plunge ahead and do business as usual will only lead to society's grinding to a halt in a very short time. I am encouraged by President Carter's serious approach to considering conservation efforts. It appears he is providing the leadership we have so badly needed for the federal government in defining an energy policy. I think it is important that we talk about goals. I think Louise Dunlap referred to goals. All the organizations struggle and flounder until they have a clearly defined goal. I think you need something to work towards, and I hope that this can be done. From my years of experience in working with the public and with citizen groups, trying to educate people to what these issues are, I would say that Carter is exactly right when he senses the American people lack confidence in their government. I see this everywhere, and more power to him if he can restore that.

It will be extremely important that any direction that the Carter administration provides to the resolution of these issues be meaningful, practical, and honest so that the people can once again have some faith that there is some leadership that can be given to these issues.

We have gone under the assumption for far too long in this country that the air, water, and land can provide free dumping grounds for our waste, and until all the costs from beginning to end are accounted for, we will continue to be wasteful and consumptive.

Vernon K. Morrison

Manager of Coal Marketing, Union-Mechling

My training has to do with river transportation. I am oriented in Pittsburgh to most of the industry that is involved. We have problems like many of you. It is hard to disagree with some of the things that have been said; I do agree with a great many of them. I am available if you have any questions about our business on the inland waterways or anything that has to do with the marketing of coal as it would be related to river transportation.

Ralph Madison
―――

President, Kentucky Audubon Council

The first thing that I would like to do is point out that we do not have enough decisions. I am talking about decisions on the need for power. About a month ago I saw a projection on the need for power from 1952 to the year 2000. It showed a gross rate of approximately 1.9 percent. That projection is not used anymore. There is a new one, which was started in 1975 and now goes to the year 2000. The rate now is between what we call the BOM rate and the Ford Technical Fix. It is about halfway in between.

It just happens that the Ford Technical Fix has about the same rate as the old ORBC projection. Why the difference, and why do we have so many projections? Anybody who can read statistics is entitled to make a projection and the more publicity, the more it is believed, but we do not have any decision upon the need for power. We have certain projections by this group and certain projections by that group, but it does not come out to anything finite. It is infinite.

Second, just to put things in perspective on what we are doing with coal, a few years ago a gas pipeline company applied for a permit from the state of South Dakota to erect seventeen coal gasification plants in that state and use that coal. This permit was never granted. However, upon investigation it was found that if those seventeen plants had been constructed, they would have seriously depleted the water in the Garrison Dam at a time when that water was needed to increase irrigation for agricultural land. To put it another way, if those seventeen plants were located in Kentucky, they would use up all of the water in the middle fork of the Kentucky River, all of the water in the Tradewater River above the confluence just south of Madisonville, and all of the water in Pond River before its confluence with its receiving part.

This is something that we ought to think about. Now, it was estimated that 147 of those plants will be needed to satisfy the gas demand as we approach the year 2000. This is at an approximate rate of 250 million cubic feet per day. Those figures, again, are projections, and they have a certain amount of elasticity always built into them, but you can see the magnitude of the problem.

Talking about power plants, we expect to have 159 more power plants in the so-called ORBES region. Let us consider the amount of coal. Someone made the statement last night that our total output from mining of coal was approximately 655 million tons. Well, of course, all of that is not burned. Much of it is exported, exported clear out of the country, and much of it is used in steel production and other areas, but there is a great deal of it used in combustion.

Let us assume that that might go down to 480 million tons. One-sixteenth of that is going to be SO_2 if you assume a sulfur content of approximately 3 percent. When you take one-sixteenth of 480 million tons, you get 30 million tons of SO_2. Now, 30 million tons is a tremendous resource. Nobody ever thinks of it as a resource, however, because it also is a poison. It is a definite detriment to the air quality in this country and in the ORBES region itself. Apparently, according to the study which has been going on, we are going to have a 41-percent increase in coal usage between the year 1975 and the year 1985. That gives you an idea of the magnitude.

We begin to see that there is a consensus that coal is going to be used more and more, and I believe it is and that increasing our use of coal is what we have got to do. But there is a parallel consensus that this looks like the time when we can bust the EPA standards on air pollution, on the ambient air standards. It looks like the greatest thing that has come along.

Look at some of the things that we have. At one plant, and this is one of the plants that supplies power for the enrichment plant at Portsmouth, 280,000 tons of SO_2 pour out of the stack every year--every day, all night, all day, every year. What if we increase output, and keep on increasing it, and the state of Indiana does not do anything about it, does not live up to the standards, lets the stuff spew into the air? Upon inquiry, their Department of Pollution says there are no laws or regulations for that. Now, something has to be done about such things or, before we realize it, there will be a very serious problem of SO_2. It does not matter where it is coming from. We are going to use coal, and we are going to have SO_2. Somebody has got to address that problem very, very sharply and very, very accurately in order to get the people to understand.

Mr. Cohn made a remark that I thought was very interesting. He covered the idea of coal and other materials as resources for fuel and then said, we must develop alternate sources of energy, and I agree with that. We must. Later on in his remarks he made a statement absolutely demolishing any chance for certain things, such as geothermal energy, solar energy, hydropower, et cetera. Now, speaking directly on solar energy, I believe that the utilities are very much in the background of saying that we do not want to use solar energy. If they would put their minds to it, they know they could do it, but they do not want to do it.

James B. Hartnett

Professor and Director, Energy Resources Center, University of Illinois, Chicago

I am also associated with the Ohio River Basin Energy Study, but many of my remarks this morning I would not ascribe to that group. You will hear something about the study and something about my evaluation of the energy situation in the valley.

Let me mention the two scenarios that are currently being addressed by the technology assessment groups working in the ORBE study. The two scenarios are basically as follows. In the Ohio River basin study region, and that includes all of Kentucky and most of Illinois, Indiana, and Ohio, excepting the northernmost tier of those three states, in 1975 there were installed 58.6 thousands of megawatts--58.6×10^3. In the year 2000 the Bureau of Mines high scenario shows 245.2 megawatts installed--245.2×10^3. It is essentially a quadrupling, then, of the electrical energy installed over what is there today.

Now, the Bureau of Mines high scenario in effect says let demand go where it will. Do not impose any major conservation efforts and run to get the production up to where the demand requires it. That is basically the Bureau of Mines high scenario.

The Tech Fix, on the other hand, says push the supply side down as hard as you can by aggressive conservation, and then you do not have to run as hard on the production side to do the job. It does not say change the quality of life. It does reject the thesis that GNP and energy are 1:1 related. It does accept the fact that they are related. The relationship is not known in detail, but it assumes that we can squeeze much more productivity out of our energy than we have in the past, so that the Tech Fix scenario goes like this: In the year 2000, we will have 94.7 megawatts. The electrical utilities in the Ohio River basin region have already on the books (as you know, they must engage in the planning process; it takes ten years lead time) 88.5 megawatts planned. This is, if I may say, in the works, and it falls somewhere between the Tech Fix scenario and the Bureau of Mines high scenario.

Now, we have taken subsets of these two scenarios. We can go from here to here by combinations of coal and nuclear fuel so that we have one scenario which is 80 percent coal and 20 percent nuclear fuel; in another one, we have 50 percent coal and 50 percent nuclear fuel. When I make those statements, I mean add-ons from here. We know what the plants are going to be out to a point. The add-ons have the two mixes that I just mentioned, and we then look at the relative impacts--economic, social, and environmental.

On the Tech Fix scenario, what we say is, if we were to go that route, if history pushes us or if some other force pushes us in that direction, these plants will not be built by 1985 but rather will be smeared out. They will come on line by 1995, and in that last five years we will add new plants, either all coal or all nuclear plants.

Those are the scenarios that are being addressed. What are some of the outcomes? One outcome is that if you go the high route, you need a lot of capital. In the Ohio River basin region alone, just for electrical power plants, you need $200 billion. We are not talking about coal mines or anything else.

If you look at the concentration of power plants along the Ohio River basin, it really sets you back on your heels. It turns out that the Ohio River basin, the Ohio River, runs in a direction such that the plants are lined up in the direction of prevailing wind, and you begin to see why there need to be some federal regulations above and beyond the state regulations: All that stuff that is being generated in Indiana is blowing into Ohio, and the collection is blown downstream into Pennsylvania and West Virginia, and there is no way in the world you are going to meet current environmental standards, given the maze of power plants that come out of the Ohio River BOM high scenario.

There are several ways to go. One is to relax environmental standards; the other way is to tighten them up so that it becomes very difficult to build the plants. A third one is to try to alter the prevailing winds. That might be the easiest of the three. I am not sure, considering the way things are going.

The Tech Fix scenario clearly has very substantial advantages from a capital availability standpoint, from an environmental standpoint, and it has some drawbacks. I will not say that it is all black or all white. In the study, we are not recommending that you go either route. What we are saying, ultimately, is, here are the implications. Here are your policy options.

I want, before I finish up, to say one thing about nuclear fuel. I come from a state where we generate a good deal of our power by nuclear means. We are by and large not as afraid of or as emotionally involved with nuclear power as some of our friends in Kentucky. I would not reject nuclear power construction as a viable option. Indeed, it is a must. There is no way between now and the turn of the century to satisfy our energy requirements on either scenario without nuclear power. We must have nuclear fuel.

It is true that there are some problems with respect to radioactive waste disposal. There are problems of safeguards, but it strikes me that the problem of safety is rather an interesting one. The analogy between coal and nuclear fuel is something like the analogy between the airplane and the car. We have a long history of what the safety records are in the coal industry. We know we are going to lose people. We know that if we burn the amounts of coal we are talking about here, we are going to lose people as a result of the public health impacts.

We do not have a very complete record of nuclear fuel. We have got twenty-five years of experience, and over those twenty-five years, the record has been a great deal better than the coal record. To turn back completely on nuclear fuel at this juncture would be the result of emotional arguments. It would be a serious mistake in my judgment.

It seems to me that we have some factors at work which tend to allow the use of coal consistent with public health. One is the high cost of oil. Thanks to our friends in the OPEC countries, we will generate a sufficient level of income for the coal industry and for the utility industry so that we can afford scrubbers and other devices. In that sense, OPEC has helped us. The cost of energy is high enough that we can afford to do these things.

Additionally, the reliability and cost of pollution control equipment is coming very close to the reliability and cost of lawyers. When you cross over there, you are home free.

DISCUSSION

HEIBERG: Do the panel members want to speak on any of the remarks that we have discussed?

COHN: I would like to address myself to the last remark made by Professor Hartnett. We are wasting a great deal of money on lawyers today, and we should stop that just as well as we should stop excessive expenditures in any other area, including any pollution control equipment that is unreliable, inefficient, and wasteful. I think I am the only lawyer on the panel, so I had to deal with that one.

ALLEN: I have two related questions, and I think they go pretty much to the Kentuckians but also to the whole panel. I am fascinated by the undertone of the Kentucky position that says we are greatly endowed in both coal and natural resources. We are happy because it is tradition to export our coal, but we do not want to use it here and export electricity. Further, we should fully price those natural endowments and the environmental impact before we do any selling of things from Kentucky to the outer world.

I come from New England, where we are pretty regional, and we realize that we have to export in order to pay for imports. Has Kentucky considered the problem of its favorable or unfavorable trade balance in relation to the rest of the nation if it continues this posture? I have in mind that Kentucky must want to import cars from Detroit, gasoline from Texas, and a certain number of table vegetables from either California or Florida. My second question-- there is a growing transmutation of environmental words into environmental costs, and it begins to get twinned with "you are not paying me enough." It looks like a bargaining posture. Is it?

SWIGART: I wish I could speak for the leadership of Kentucky. Unfortunately, I am just commenting on a posture that has puzzled me, too. In the past sessions of the Congress dealing with federal strip mine legislation, there has been a tremendous resistance to this by Kentucky. What I saw as a need for Kentucky's industry to survive was the regulation of coal in the rest of the country. And I have been reassured by this current session of the Congress, in which Kentucky has been in the forefront. The leadership for the state has been quite vocal in supporting the legislation with, of course, a great number of reservations about the environmental restrictions. But, at least there is a recognition that there is a threat out there in the western part of the world.

I wish that I were in a policy making position on some of these things, and I really hesitate to speak to that because it is not my role.

I am a little puzzled by the second question, and I wonder if you could clarify that, unless Harvey Sloane has a good answer for it.

ALLEN: Let me take it away from Kentucky and take it out to Colorado or Utah. I think I am hearing a great deal of posturing perhaps-- I am not sure how to interpret it--saying do not come into pristine Colorado, where we excluded the Olympics, do not come into Utah, where we have God's gift of scenery, and make us the resource base or the powerhouse or whatever of the nation because these are great values and we put a high value on them.

Increasingly, I think I am hearing a trading posture, and it is a transmutation of the unquantified environmental values to something that is beginning to get quantified. The effort is to say that we would like to put a value on these natural endowments we are asked to give up. You are not paying enough. We want more. It becomes a trading position. Is this a possible development that is going on in the heretofore pristine environmental movement? I think that is what I am hearing from various sources.

SWIGART: Once again, this is the problem in Kentucky. You have the leadership exhibiting one posture that I feel is very strongly motivated by the vested interests, and the citizens, you know, are trying to counteract this. But I think there is a growing recognition, with the importance of coal, that there has got to be a trade-off somewhere, and it is my argument that those are the things that need to be discussed. I do not think we have had that meeting of the minds yet in Kentucky, and maybe that is why I am presenting somewhat of a confused picture, but--

ALLEN: No, you and I are on the same wavelength. Essentially, I am saying that part of that trade-off may be putting the right value on what the environmental intrusion is, and this gets into a very different realm of discourse, not necessarily the right one, on the environmental trade-offs.

HEIBERG: Mayor Sloane?

SLOANE: I do not know if your price could ever be high enough for the environmental trade-off in my estimation, seeing what is proposed along this 200-mile stretch. The professor here mentioned the prevailing wind, the synergistic action of all these plants. My posture is that before we develop these plants, we have got to have some view about what is going to happen to that area.

I am also interested economically in that region and interested in what kind of development could occur other than power generation.

Let us get to the coal situation. Eastern Kentucky has long-term contracts with Japan. A lot of coal, good, low-sulfur coal, goes to Japan. They are pretty bright over there because they let us rip up our land and then they utilize our coal. They generate the steel and then sell it back to us, and we enjoy that technology and also pay some of the price for it. But I do not think that we can embark on this project as I see it. I do not think, sir, there is any price that you can pay Kentucky or Indiana or Ohio for an air quality that is totally unacceptable for the people who live there or a pollution of the river, whatever it may be, heat or minerals, that is unacceptable in terms of any kind of quality of life.

Again, I am not saying that that area should not be developed, and I think it will be developed, but until we know what the final product is, I am not prepared to give my endorsement to the magnitude of the development.

HEIBERG: Professor Hartnett?

HARTNETT: I happen to be a native of Massachusetts, so I can say what I am going to say without hesitating. Massachusetts has also worried about its pristine purity when it has come to energy developments off the coast, so you have trade-offs as well. This is not unique to Massachusetts and to Kentucky. The same thing is true in Colorado and in the West, and I think that this is one factor that is going to assist in holding the energy level down in the sense of getting rid of the waste.

I do not say that it is going to cut the energy growth down to the point where we change our style of life, but it is going to play a role in seeing to it that this country begins to use its energy efficiently because in Massachusetts you are going to have to drill off the coast, whether you like it or not, and in Kentucky you are going to have to mine coal and generate electricity and ship it out of the region whether you like to or not. These decisions are going to push us in this direction, absolutely. I would wager my life on it.

MARK J. BAGDON, New York State Energy Office: We burn some coal in New York, not a lot by your standards, about 7 million tons a year to generate electric power, and we are clearly interested in looking into the possibility of burning more.

The New York Power Pool is projecting about 80 percent of its added capacity over the next fifteen years as being nuclear fuel, and there are a lot of people who are clearly concerned about that.

We have some very lovely mountains in the northern part of New York State, as some of you may know, called the Adirondack Mountains, and recent studies have shown that about 90 percent of the high-level mountain lakes are devoid of fish because of the decreased pH, which is largely due to acid sulfates, a good number of which have come from the Ohio Valley region.

We are presently completely in compliance in terms of our sulfur emissions. We are burning coal, which is maybe a maximum of 3 percent sulfur, and we certainly expect that our future power plants will be within new source performance standards. According to the Federal Power Commission, Form 423, which most of you are probably familiar with--it outlines all the characteristics of all coal shipments--the east, north, and central regions are approximately 70 percent not in compliance in the sulfur content of the coal it is burning.

Indiana is 89 percent not in compliance. It is burning coal which is 5, 6, 7, and 8 percent sulfur, and we are the ones who are bearing the brunt of this. We are not interested in loading up our air with any more of the acid sulfates as long as we are getting this tremendous load from the West.

I would like to ask Mr. Cohn, why is it that, on the one hand, the power companies seem to be very much against scrubbers but, at the same time, are not burning coal that is in compliance with their own standards and so far above them?

COHN: First, I would reject what purports to be your statement of fact as to what is going on. On the question of the fish in the Adirondack lakes, while I do not have any background to be able to suggest the answer, I am reasonably clear that it has not been demonstrated that any loss of fish was due to emissions from the Ohio Valley. But more importantly, to get to the central thrust of the question, in the case of American Electric Power, we have spent a very large amount of money to try to bring the lowest-sulfur coal possible from the West in order to meet the standards that are in effect in the states in which we operate and the standards of EPA.

We are still in the process of working up to meeting all the standards. This is a process that obviously takes a considerable period of time. I am not aware, certainly not at American Electric Power, of any situation in which we have not either taken steps to comply with the applicable requirements or entered into some kind of a compliance program where we will meet the standards within a prescribed period of time, which apparently has been acceptable to the environmental authorities, state and federal.

I had occasion last week to be in a meeting with a group of representatives of power people from Western Europe, including the man who runs the Central Electricity Generating Board of the United Kingdom and a senior official of the power system in Sweden. Perhaps this may be of some help to you; I give it to you for what it is worth. The people in the United Kingdom, you know, think that we are a little bit batty and have gone overboard concerning the effects of SO_2 and the requirements relating to SO_2 emissions. They think the answer is very clear. They believe that ambient standards can and should be set, that they can meet the ambient standards with high stacks, and that is what they are doing. Apparently there does not seem to be too much controversy about this in the United Kingdom.

However, the man from Sweden stated that there are some who believe that the emissions from the United Kingdom are killing fish and causing some other problems in Sweden. The official who runs the Central Electricity Generating Board says that studies have been made that demonstrate that that is not the fact.

All I can suggest to you is that the answer has by no means been demonstrated, but to the best of our ability we are going to meet whatever standards are imposed by the federal and state authorities.

BAGDON: Could I ask more specifically, are you planning to go largely to western coal or are you planning to seek out and try to develop low-sulfur coal sources within the East?

COHN: I am glad you asked that question. Someone--I think it was Mr. Madison--made some references to scrubbers and the position that American Electric Power is taking. I think I earlier pointed out that I am just a broken-down lawyer, but I go to lunch with the engineers, and I listen to what they have to say, and I think we have some pretty able people. They have been in the business a long time.

Our people believe that the scrubber is not proven. It is not technologically sound. They believe it is unreliable, and I should say that there is clearly a difference of view about that. They believe it is highly wasteful of energy, and here I do not think there is any difference of view. It does use a considerable amount of energy. Moreover, if the scrubber is attached to a power plant, and if the scrubber goes out, as things now stand, the power plant is down, and the effects in that connection on energy production can be very severe.

The costs of scrubbers have been over, and over, and over again greatly underestimated. They are very substantial indeed. The latest figures I have heard, and they grow every week, are something like $100 a kilowatt, which is what we used to pay in order to build the entire power plant.

Third, if we go ahead with scrubbers today and on our system that would cost many hundreds of millions of dollars, that will be a sunk cost. At that point, the dollars will have been spent. We have nowhere else to get the money but from our customers. It has got to be put into the rates, and our customers will have to bear the cost.

All kinds of things are coming along day by day. A few of the speakers made reference to the fluidized bed as a form of combustion which may take care of the problem. We have thought from the beginning that either a front end process to clean the coal before it is put into the power plant or this fluidized bed, which we are now spending a good deal of money on in an effort to research and develop it, are much better answers.

At the present time, under the applicable regulations, we have been given a choice. We can either burn low-sulfur coal or put in a scrubber and burn higher-sulfur coal. We believe, for all the reasons I have suggested, the choice of burning the low-sulfur coal--although such coal costs a lot more--is still more economic, imposes less of a burden on our customers, gives us flexibility if new developments come along, and is the preferable choice. That is why, as long as we have that choice, we are rejecting scrubbers and we are choosing the option of burning low-sulfur coal, which will come primarily from the West because that is the only place we can get it in the quantities we need.

HARTNETT: I recently attended a briefing by the research arm of the electric power industry, and one of the major speakers on that occasion said that he was pleased to announce that he was absolutely certain that the day of the scrubber has arrived, that we would see, within the next three to four years, these units coming in across the country. I think that offers a somewhat different perspective on scrubbers than the view we just had.

I do not know whether American Electric Power talks to Electric Power Research Institute, or whether EPRI talks to AEP, but I did think it was worthwhile showing the other side of the coin.

I do understand that there are problems with down time, but I gather that what generally happens is, you put in an extra unit, an extra scrubbing unit, so that when one is down, you switch over without having to turn off your electrical power station. It was this practice that I thought was currently involved and does account for the increasing use of scrubbers across the country. If you look across the country today, you will see a substantial number going into operation.

SLOANE: Let me just be a little chauvinistic about Louisville and our gas and electric company. The Louisville plant has installed scrubbers on all their new units. They are working. The customers are accepting the increased cost. We feel that is a great advancement, and we are very proud of our local utility.

HEIBERG: Let me ask a question, and I think it goes back to you, Mr. Cohn. I think you will want to come back into this one again. I am not clear in my own mind if we brought out this part of the question. What difference does it make whether we are still meeting the air quality standards, either state or federal? Now, we might want to argue about what those standards ought to be, but as long as American Electric Power or any utility is meeting the air quality standards, either by scrubber or by use of the proper mix of coal, I am not clear in my own mind why it makes a difference. Can you cover that, Mr. Cohn?

COHN: Well, I think, General, the point you make is certainly a very valid one, and I attempted to suggest that the course of action we are taking will be taken as long as we have the use of low-sulfur coal as a permissible alternative to comply with the standards that are set by the environmental agencies.

There is some possibility in the Clean Air Act amendments that are now under consideration that that option may be taken away from us. If the option is taken away from us, we will do whatever is required to comply with the law, but we will take the alternative, if alternatives are available, that we believe makes the most sense in terms of compliance and imposes the smallest possible burden on our consumers.

Now, I would like to refer to something that Professor Hartnett said. Number one, I should say that our people, and I think we could pretty well document this, talk to everybody, including EPRI and Louisville Gas and Electric, and read everything that comes out on scrubbers and, indeed, on just about every other technological advance--or technological change, perhaps I should say--that is being made. Our people, who I think are very able engineering people, believe that the scrubber technology has still not been demonstrated.

Now, with respect to the statement that Professor Hartnett referred to, which he ascribed to somebody from EPRI, I am willing to accept for the time being, until I get a chance to check it, that the statement is an accurate one. All I can say about that is that having read the literature over the years, I can remember similar statements having been made at least once every six months. For the last three years our people went out and checked each successive scrubber to which the statement was applicable, and in each case the statement turned out not to be supportable. Generally, the scrubber that was regarded as better than sliced bread and demonstrated that the technology was here broke down within the next month or so, so you have to take all of these things with a grain of salt.

Then I come back to what General Heiberg said, and that is: If we have a choice and we think we can do as good a job environmentally at a lower cost to our consumers and preserve our options for the future, we think it is very much in the best interests of our consumers that we do so. Incidentally, I did not refer to one additional problem that is also associated with the scrubber. I think Ms. Swigart referred to it, and I thought she was very accurate about it, suggesting that there is a new problem and that is how to dispose of the residue that comes out of the scrubber, which is not a small problem by any means.

JOHN H. ANDERSON, University of Pittsburgh: I wanted to say first of all that hearing Mr. Cohn's expression of what has been the usual line of the energy establishment, of concern for what would happen to social welfare if energy growth stops, made me wonder whether American Electric Power or EPRI or any other organization has begun to put serious study into launching space satellites to which we can go when we can no longer continue to increase our energy usage on earth. I have two questions somewhat related to that point.

The first is how was this lower technical fix figure of 94,700 megawatts arrived at? A second question goes with the first and is not meant to impugn that decision at all. Any kind of exponential growth whatsoever, whether it be at 5 percent or your 1 percent per year, eventually arrives at infinity. I sometimes have a feeling that even environmentalists are not leveling with that great part of the American public, that fourth part of the public which is not supposed to be informed, in not being willing to face the fact that someday all these curves have to turn over unless we do go into space satellites on a large scale. I would like to have a comment on that.

HEIBERG: I would suggest, Jim, you take the first one. Professor Hartnett?

HARTNETT: We spent the summer, a subgroup within the Ohio River Basin Energy Study, to decide on which scenarios we would select as scenarios to study. They may not conform to the real world, but they are scenarios of possible growths, and there are a whole array of these as you know. The Bureau of Mines high scenario is essentially a business-as-usual scenario. It gives you about a 5.8 percent electricity growth, which is about what the electrical industry says is going to happen, and the Tech Fix is the one that I described earlier.

What we did basically was to say that the percentage of energy used in the four states in 1975, that same percentage, is going to prevail in the year 2000. You have got the national Bureau of Mines scenario. We take the percentage; it is 16 percent. That gives us what happens in the four states, and then state by state we apportion to the region that proportion which was valid in 1975. So we just take the photograph that we have in 1975 on a ratio and project it ahead to the year 2000.

That is the answer to the question of how we arrived at the numbers. Now, one can criticize that, and we know, ourselves, a number of the weaknesses in that way of projection, but it did give us some scenarios to look at.

You talk about growth and I will address that very briefly because I think that is a question the other panelists would want to look at. Over the years between now and the year 2000 there is going to be some population growth. Within the framework of that population growth there will be industrial growth, and we did not feel that a zero-energy-growth scenario from now to the year 2000 made any sense. We have got a lot of slack to take out of the energy system.

We cannot take it out right away. I turn the question back to Mr. Cohn. That is, even if we get fluidized bed combustion, by and large you are not going to go back and retrofit all of these existing power plants with fluidized bed combusters. And by and large, if you retrofit them, you are going to have to retrofit them on the stack side. So even if we get fluidized bed combustion, I do not see that it is going to solve the problem any more than having thermal codes today is going to solve the problem of heating in the residential community area. You have got to go back and retrofit your existing buildings. That was the tenor of our thinking in looking ahead to the year 2000.

HEIBERG: Do any of the panel members want to add to that?

COHN: I do not want to monopolize this, but I think I am a majority of one here, and perhaps I ought to try to respond.

First, on the question of growth generally, you may recall that a couple of years ago there was a report put out by something called the Club of Rome suggesting that within a reasonably short time the world was going to collapse or the equivalent. Interestingly enough, that group has since changed its mind at least to some extent.

Number two, some years ago there was a fellow named Malthus, who announced that the world was going to collapse about 1930 or whatever, and it developed that he was wrong.

These predictions discount advances by reason of technology, and the fact is that in recorded history, technology has somehow been able to find a way. Sure, we will come to the end of our resources some day, but I was involved in a study conducted by the Edison Electric Institute over a period of a couple of years in which we concluded, after a great deal of thought, that that time was so far into the future, it was not something that had to be regarded as a matter of great concern today.

On the growth of electric power, and this is a point that I want to get across because I think it is a very vital one, Professor Hartnett gave a couple of forecasts. In the most conservative case, there appears to be something like a little bit less than a doubling by the year 2000. Obviously, that means that in order to take care of the needs predicted for the year 2000, plants will have to be built.

I think these exercises in forecasting for the year 2000 are interesting. They ought to be carried out, but I can tell you that we have a pretty good group of planners--maybe fifty people. Every year they revise their predictions for obvious reasons: they learn something in the intervening year. It is true that we have got to plan about ten years ahead, plan in the sense of having to begin to do something today in order to be able to achieve a result ten years hence. But then next year we will have a better notion of what the situation will look like in 1987, 1988, or whenever, so the big problem today is to plan for the mid-1980s.

It is absolutely vital to understand that the risks and costs of an erroneous forecast on the high side and of some degree of overbuilding are infinitesimal in comparison with the risks of a mistake in underestimating and underbuilding because it will then take so long to catch up. All you have to do is take a look at what happened over a very short period of time when there was a deficiency in gas supply in some parts of the country to get some notion of what would happen to the country if we had a shortage of electric power over a period of several years.

If you have overbuilt, it is very simple to catch up. It costs relatively little. You catch up very quickly. You slow down what is under construction. You use the more efficient plants. You do not use the less efficient plants. The costs are really very small in comparison with the very large costs and risks of underbuilding.

ALLEN F. AGNEW, Congressional Research Service: I am senior specialist for environmental policy in the area of mining and mineral resources. I am a geologist. I have worked underground in the coal mines in the Middle West. That is not to threaten. That is to say that I am going to ask a different kind of question.

I have also taught at universities. Here is a public, a series of publics here, addressing ourselves in one particular kind of forum. Early on, Professor Keenan and, later on, Ms. Swigart mentioned that we have an involvement of all kinds of publics, but I did not hear the word "youth" mentioned. So my question is a double-barreled one. How are the youth involved in this particular project that we are discussing here, and as an outgrowth or perhaps an inter-tie with that, how are the church groups involved? The reason I ask the question is that I have come upon two paperbacks in the last three months. One is called *Spoil*, and, of course, it is about strip-mining effects. It is written by Richard Cartwright Austin, who is a Presbyterian minister down in Wise County, Virginia, the county that is totally flooded today. Perhaps, that is the reason he is not here. The other publication is one for youth and is a magazine called *Youth Magazine*, which is sponsored and available to a number of Protestant churches and is also accepted by the Roman Catholic Church as being a proper communication outlet for their kids.

Both of these have what I think is a fully objective view of the strip-mining issue. So with that as background, how are the youth and the religious aspects of this project being handled?

HEIBERG: I had earlier said that my experts on the publics are, by definition, Dr. Boyd Keenan and Jackie Swigart. I would like for both of you to take a crack at that if you would. Boyd?

KEENAN: I particularly appreciate the question because as I was straining to finish up my presentation, I wanted to get into the question of how some of the activities by environmental groups in the valley could be likened to earlier religious "great awakening" kinds of activities in American history. You asked if the project, the ORBES project, has given any consideration to the involvement of these people.

I might mention--it just comes to mind very quickly here--that one of the members of our Advisory Committee is a man named Tom Behan from Cincinnati. I think he is executive officer of the Tri-State Air Committee. He is a young man. He represents, I think, the youth group. I have talked with him recently, and where he tells me that, in Cincinnati anyhow, there is a kind of a neighborhood coalition made up of youth groups who feed into this Tri-State Air Committee, so we think that we are getting some input from such groups.

Another possibility, I would think, is the utilization of the land grant colleges, who have their county agents working in the field throughout these states. Although I am associated with the University of Illinois, I would have to say that Purdue University and the state of Indiana have probably done more relating to this project than those of us in Illinois. At Purdue, the College of Engineering is working with the College of Agriculture to try to get some input from, I suppose, groups like the 4-H clubs, the Future Farmers groups, and others who have long worked with the land grant colleges and the colleges of agriculture.

In fact, early along, when a group of universities were putting this proposal together for EPA, there were many who argued that the extension service approach, that is, the extending of this kind of knowledge through the county agent system and on to the young people on the farms and, increasingly, in the urban areas, might possibly be the greatest contribution that this study could make.

SWIGART: We have had an activity in Kentucky that I think is reassuring along these lines, and that is the development of an environmental education plan for the state of Kentucky. The state of Florida was the leader in this area, and ours was patterned somewhat after theirs. The Department of Education employs a coordinator for this effort and started out by pulling together an advisory group composed of people who had access to youth groups, not only through the school systems, but through some of these things that Boyd mentioned, people who had also over the years been trying through their own efforts to bring about some kind of environmental education in the school system. That group met for a couple of years and evolved a plan that was adopted by the Kentucky legislature. They are hoping to expand on this next time around.

I think their philosophy was a valid one. They did not want to introduce a package that dictated this is environmental education, learn this. Instead they wanted to integrate the concept throughout the entire system from grades kindergarten through twelve. Now they are providing materials, working with the teachers, who are in turn working with some of the outside groups. The Environmental Quality

Commission was very supportive of that effort and provided some funds to establish workshops throughout the state to bring in designated coordinators from the school systems who would then carry the experience back. That is an effort that I personally am very proud of.

KEENAN: After giving the commercial for the agricultural colleges and the extension services in the land grant colleges, I think I should mention that we have had some ideological problems since we have tried to bring in every sector in this study. I think we would be putting our heads in the sand if we did not admit that there are some sectors, and in one of our states it was the agricultural sector--representatives of the agricultural sector--who attacked this study as being an un-American study, who said that planning of this sort should not be carried out by government, certainly not by universities, and that the free enterprise system did not allow for the education of young people in the environmental field.

MIGNON SMITH, *Washington-Alabama Report*: I have a two-part question. Kentucky's governor was on Capitol Hill yesterday and also meeting with Mr. Lance down at OMB, vigorously testifying in favor of the Tennessee Tombigbee waterway, which has become an endangered species under the Carter administration. My question is, how does the Ohio Valley feel about this possibility, particularly since coal is the most commonly transported commodity on the Ohio? This would connect, of course, into part of that system. Also, in the case of the high-sulfur coal being exported to Japan and so forth, is this made possible primarily by the artificially held-down price of gasoline, and if this were removed, could we then afford to keep our high-sulfur coal and import less gas? How would this affect the development of coal in the Appalachian region?

SWIGART: I think it is the quality of coal that is needed. That is why it is exported, because it is high-quality coal. It can be used, you know, in the manufacture of steel. I do not think there is a relationship there, although someone else may have better information on that than I do.

WILLIAM RICHARDSON, Office of Technology Assessment: I think that you will find that the coal being exported to Japan now is of an extremely low sulfur.

SMITH: I meant to say low-sulfur coal.

RICHARDSON: It is a metallurgical quality coal that is being burned in the power plants now, of course. This is another issue, which I think the panel should be addressing itself to.

HEIBERG: Anyone else want to try to take that on? I will not speak for the governor of Kentucky here. Yes, Bill?

RICHARDSON: First of all, may I say that all of my opinions are my own. My remarks are not connected with my agency, of course, but I would like to congratulate the many knowledgeable reports that we are listening to this morning.

There was a question from our man up in New York, who asked whether scrubbers were, in fact, viable. For the past week I have been looking at the latest EPA report on the feasibility of the scrubber system. At the moment the EPA predicts that by the year 1985 there will be roughly 49,000 megawatts of scrubber capacity on the line. At the moment they allege that there are just over 6,900 megawatts of operating capacity, but when you look through their voluminous report, you see that only 3,900 megawatts were, in fact, functional during the November-to-December, biennial period, and out of that 3,900, only 53 percent were operating at all.

I know at the moment that the coal-fired capacity in the nation is roughly 205,700 megawatts and that we have probably got another 105,240 coming on line within the next eight years. Now, if we go the way that the lady on the panel was saying, we know that with the present coals that we are burning--quite a lot of it is high sulfur, well over 2 percent--you need 1 ton of lime for 13 tons of coal to produce between 550 to 600 pounds of sludge for every 1 ton of, say, coal input. You have another onerous thing to look at in the future. One of the utilities--I think it is in Pennsylvania--is alleged to have the most effective scrubber system. It is a 835-megawatt unit. At present, it brings its coal only 80 miles up the Ohio River, but it brings its lime, I think, 300 miles up the Ohio River. This is another problem that you have to look at in the future. There are, I believe, only about four or five possible sources of this very good-quality lime in the country. It is the only kind that will prevent the scaling of gypsum in the scrubbers. We have to look again at the very, very serious transport situation coming up within the next two to three years.

HARTNETT: I just want to mention that we have addressed that problem in the Ohio River Basin Energy Study. That is a serious problem.

HEIBERG: And it is one that is of some interest to the Corps of Engineers, which has to try to project our traffic so that we will know what to build or repair next. Mr. Cohn?

COHN: General, the young woman who asked the last question started by referring to a water project that I gather is being delayed. I do not know that I am familiar with the particular project, but I did want to make a last plea for rationality and pick up on something that Mayor Sloane said in his remarks. That was that he had been told that it was possible to delay practically anything for practically any length of time. I think that was essentially what was communicated to him, and I think whoever said that knew what he was talking about.

I have been making some notes on what has happened in this country in the last several years in terms of delays of energy projects. I think if a man from Mars came down today, he would not believe what is happening in this respect.

There is the saga of Kaiparowits. That was held up some twelve to fifteen years, a project involving $3.5 billion, and finally abandoned because the sponsors just could not afford to sit around and wait any longer for a final decision.

I happen to have been associated with a project that was kicking around for some several years. It involved our application for a governmental license. It was something called the Blue Ridge Project in Virginia and North Carolina. It kicked around for twelve years, and then the Congress, in effect, reversed the decision of the Federal Power Commission and said that it could not be built.

The Con-Edison people have been trying to build a project along the Hudson River for something like fifteen years, I believe it is, and they are still in the courts, up in the air as to whether they can or cannot build it. Within the last year or so an alleged threat to a plant called the Furbish lousewort has been holding up the building of a $700-million project at Dickey Lincoln in Maine. The snail darter has been permitted to hold up a TVA project in which $116 million has already been invested. I have a clipping-- I would be glad to show it to anybody--stating that the Cumberland monkey-faced mussel is holding up a project in the TVA area, called the Columbia Project, involving some $142 million. And we have the spectacle of a possible threat to clam larvae in the Atlantic Ocean holding up the Seabrook nuclear project in New Hampshire, involving $2 billion.

There is, I think, something wrong with the decisional process in this country, and I think we ought to do something about it.

SUMMARY REPORT TO THE PLENARY SESSION

E. R. Heiberg, III

Professor Keenan discussed the publics and the politics that surround the Ohio River Basin Energy Study and the need for us to try to recognize and try to put some order into the--as he called it--fantasy and imprecision. This issue involves the political opinion side of the whole question as it involves the Ohio Valley.

The Ohio Valley includes the five top coal-producing states in the nation, led by the state of Kentucky, with its current rate of 140 million tons per year of coal production. Professor Keenan discussed the 981-mile long Ohio River. He used the term "Geiger counter" with

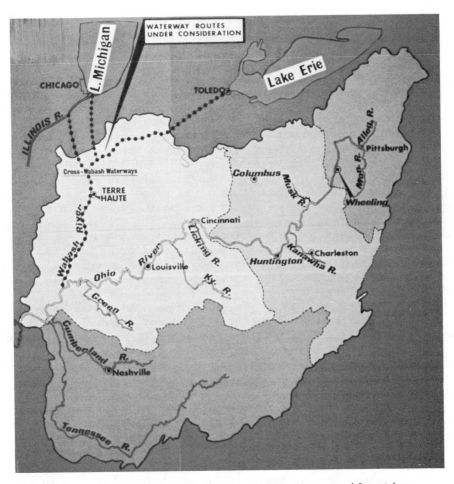

OHIO RIVER BASIN: Waterway routes under consideration.

respect to trying to find out where the political hullabaloo is occurring, where the opinion, still in the business of jelling, is, and where the problems are. He identified mile 400 on the Ohio River, downstream from Pittsburgh at about Maysville, Kentucky, to mile 600, which is about where Louisville is. That area and the area a bit downstream from Louisville on both sides of the river (Ohio, Indiana, and Kentucky particularly) are where the focus of the problems surrounding coal use and energy production is taking place.

For some reason at the upper end of the river, above mile 400, up to and including Pittsburgh, the Monongahela, the Allegheny, and the state of West Virginia, the Geiger counter isn't clicking too much with respect to discussion in public forums of the pros and cons of coal-related development. In fact, there was some discussion later by the panel of the desire and the pursuit by the state of West Virginia for some of those power plants that weren't going into some of the other areas where the Geiger counters were reading high.

On the lower end of the Ohio River, over the past several months, there has also been a growing coalition that might be described as industry and labor, a very loose coalition, which is beginning to provide the other side of the argument. I guess this means the other side of the environmental concerns.

Also, it must be recognized (and I hinted at this by my comment earlier about the state of West Virginia) that these states that we are discussing in the Ohio Valley (Indiana, Illinois, Ohio, Kentucky, West Virginia, and the western side of Pennsylvania) have very distinct and different personalities. This thought was echoed very clearly by Mayor Sloane from Louisville, as he was presenting the views from a politician in the western part of Kentucky.

Boyd Keenan raised the issue regarding public education and the issue of ensuring that we have the right publics conferring, getting their ideas on the table, reaching a consensus, as much as possible, so that we get the issues out and allow the policy makers to make the right decisions for the area.

Herbert Cohn of American Electric Power made one point very clearly and one that I think most of the panel accepted: the dependence on imported oil today, the fact that we exceeded 40 percent dependence on imported oil in 1976 and that this dependence is today increasing. We sometimes forget the twin specters of security and dollars if we have further restrictions in the supply from overseas. It can put us into a situation of dependence.

Further, we were reminded of the fact that we are today so dependent on the nonrenewable resources--oil, coal, gas, and uranium--with maybe thirty or fifty more years of oil and gas ahead of us. It is not a question of whether we are going to run out; it is a question of when. This brings forth the twin issue of sustaining a standard of living. Mr. Cohn suggested that we must not forget that we have an obligation not only to ourselves but also to our children and their children.

Second, Mr. Cohn made a point that we did not come back to. I regret that we didn't because I think it is a very interesting one. He stated that today's economically disadvantaged, and I took this to mean both the disadvantaged within the United States and the disadvantaged of the nations overseas, would probably be the ones who would be hurt the most if we had some problem with respect to progress, if we had some dips in the progress curve. If we have major changes in the rate of growth the first people who would be hurt would be those, perhaps, who could not afford it.

He also made a third point. What we should be focusing on is reserving gas and oil. We should reserve those fossil fuels that are running out earliest for those areas that are nonsubstitutable and increasingly turn our attention to substitutes, the electric highway if you will, and focus on trying to find ways to make substitutions quickly.

I think I can report an agreement at this point among the panel members, if not among the people in the entire room. We all perceived the absolute need to pursue at all speed the so-called far-out solutions to our energy problem: fusion, thermal energy, solar energy, and so forth--intense R & D. But Mr. Cohn did point out that there is wide agreement among the scientific community that, from the standpoint of major numbers of Btu's before the year 2000, there will not be much help from these more esoteric ways of producing power. Conservation will only fill part of that gap; we do need some work with coal.

Mr. Cohn pointed out that we must use bodies of water in order to produce power by current means, that we have to face the trade-offs. I think there was general agreement among the panel members that those trade-offs are essential, that we meet them, face them, decide what they are, and get on with it. I will get back to that in a second. He stated that there are risks and costs whichever way we go, not only risks and costs of whatever action we take but also very high risks and costs of inaction.

Mayor Sloane focused on his area as a case in point--the Marble Hill nuclear plant issue, which brings together several questions surrounding energy issues in the Ohio Valley. He discussed this in terms of resolution of conflict, again coming back to the point that Mr. Cohn had made. He really focused on something that I had referred to earlier from my perspective as a "fed," a permit maker or a policy decision influencer within the federal government. This focus is that we have many, many conflicting agencies and regulations that make the federal government muscle bound, if you will. He brought up several cases in point where not only different offices of the same agency are focusing on the Ohio Valley, but also different agencies are taking on different pieces of the same pie. It is very, very confusing--different policies, different bosses, different regulations--yet we are all focusing on the same area-- the need for energy and water policy, water and air quality issues, in the Ohio Valley. I can only echo that by saying, I agree with it.

He mentioned the 200-mile stretch where the high Geiger counter activity that Dr. Keenan had mentioned is going on and the concern of the people in that area, why they have coalesced in their feelings over the fact that they are now becoming the recipients of power plants, yet much if not most of the generated power is leaving the area. Tomorrow's generating plants seem to be crowding in. He says, "We live there, and we don't really enjoy this prospect of 'Ruhrification' of our part of the Ohio Valley."

He questioned very seriously the credibility of the federal establishment. He gave a specific case of a decision maker who left the government--and we have seen this happening before--the guy who is the boss leaves and then says, now let me tell you how it really was while I was there. This is the kind of thing that caused people at the state and local level, the various publics that we are discussing, to question the direction they are getting from the federal side.

He referred to a very interesting comment that a consumer advocate had made in Louisville. The advocate was quoted as saying, essentially, if you are against a certain project, what you should do is go down to the locality and create a hostile atmosphere to that project locally, and then it won't be built. It is going to go somewhere else.

Mayor Sloane used that incident to say, that is not the answer, and the panel agreed with him on that. It is the worst possible answer, particularly from the standpoint of our responsibilities here, where we are trying to wrestle with national issues. If it goes somewhere else, or the decision is postponed with other costs that we don't anticipate, that is the wrong answer, and I think we had general agreement that what we need is timely decisions, not postponement of decisions. It is better to have perhaps a "yes" or a "no" today rather than a "no answer."

Look at your conflicts, decide on your trade-offs, get the publics involved, and then have somebody in a position of leadership make a decision and move out.

Ms. Swigart, reporting from her view as the consumer commentator of the panel, discussed how it all comes together in Kentucky, the largest coal producer in the nation. All the major issues, one way or another, are played back in Kentucky. And she discussed Congressman Wampler's comment yesterday with regard to states' rights--that we ought to be able to say what direction we should be going. We don't need all this help from the federal establishment. She says that in Kentucky many people would quickly agree with that. However, we do have to have, in the issues that concern energy and water resources, clear federal policy direction and leadership. I think I can report that the panel agrees with that statement.

She noted the optimism in the Kentucky coal business. A lot of people are very happy in Kentucky. They see coal on the ascendancy, and there is a lot of coal to be mined in Kentucky, both in the eastern and western parts of the state. They are number one, and apparently the prospects of staying number one are pretty good. But she pointed out that not all Kentuckians are overjoyed with this prospect. She went back to the conclusion that was one of the major ones that came out clearly from our five panel members and many of the rest in the room: we see a very deep need for decisions on a timely basis; they have to come from the federal level, ultimately, and we have to get on with it. As I said, the postponed decision is perhaps usually the wrong decision.

INQUIRIES FROM THE PANEL

WEINBERG: I would like now to turn to the Panel of Inquiry and invite inquiries.

JAMES BOYD, Materials Associates: Many of the things General Heiberg has talked about, of course, are common concerns of the country. He addressed four publics, the elected officials, the bureaucrats, the informed citizen, and the less-informed citizen.

The problem that worries me is the mechanism for being able to put each of these publics on the same footing, because most of our problems are due to the fact that we have differences of information, or we seem to have differences of opinion based upon incomplete information. Did the panel have any feeling on how you can acquire that information, how you can get it in terms or in language that each of those publics can understand so that they can come to a general conclusion as to what should be the policy?

HEIBERG: Yes, we did address that question. Specifically, Ms. Swigart and Dr. Keenan took that on from their perspective of the Ohio River Basin Commission work that Jackie Swigart has done and of the work that Professor Keenan and Professor Hartnett have done with respect to the Ohio River Basin Energy Study that is now ongoing. This study in particular is making a very specific effort to try to get public participation, and this is somewhat new in the federal government. Let me talk for a minute from a perspective that we did not discuss as a panel but that I have.

The Army Corps of Engineers over the past eight years since the major environmental watershed (that, I would say, would be 1969) has been right in the middle of this, and we have been trying to figure out how to play this business of getting the right publics to make their views known at the right time so that the environmental, the economic, and the other issues involved in my kind of business get brought out at the right time. We have had some agony doing this, because what you run up against immediately is, if you don't do it right, you are ignoring the other side, which is the first public you mentioned, the elected official. In our way of doing business, ultimately it is the elected official who is held accountable every two years, four years, or six years for his action. You have to play very carefully this business of trying to figure out what the public says if you aren't at the same time making sure that the public official who was elected as a public official has his chance to input. This is part of the issue that is going on now in the water resource projects, as the President questions the nation's water projects. The Congress is saying (or many of them are saying), "We have already spoken. We have said what we are approving, and we have given you so much money for this, that, and the other."

I feel I must mention the difficulty a public official like myself has in trying to bring the public in. Now, having said that, I think we have made a lot of inroads into public participation at the right time, the decision-making time. You do it if you first learn that it is important to listen carefully to the public. That is the first and hardest step for a public official to make, not to go in with a stiff neck and say, okay, the Congress has told me to do this, so we are going to go ahead and do this, and that is it. It takes some finesse to be able to play it differently, so you can get the right publics talking at the right times. You get the discussion going in the media, and you get more informed publics than you had when you started.

The ORBE study, the work done under the EPA, the six-university consortium, all have done good work in this area. They have done it through public notices, through a large number of public meetings at various steps along the way, and the publics have had a chance to talk. It is coming out and being played, I think, at least rather accurately in the media.

I don't know if that directly answers your question. My point is that we are learning in the federal establishment how to bring out public views at the right time in a way that is sensible, that doesn't go to the extreme of ignoring the Congress or ignoring the President and playing to the media. That would be just as wrong.

ALLEN: Let me see if I can phrase a rather complicated piece. I attended the Ohio River Valley session, and found it quite fascinating in the analysis of public attitudes. I was disappointed, however, since I went as the utility man from the east coast, wanting to find the experience both of the utilities and of the publics in what certainly is the heartland of coal, and the heartland of the coal burning of the electric utilities.

I heard in the discussion that there was lots of coal; Kentucky loves to mine it. I heard a good deal of discussion that Kentucky also is very anxious to ensure that if the coal is burned in Kentucky, there be full environmental protection. And I heard quite a bit of discussion about scrubbers.

I didn't hear at that point anything that told me whether the public in the form of the electric consumer was happy to pay the extra cost of scrubbers, and I assume that that question has not been focused on except perhaps by Louisville Gas and Electric, which has led the parade.

Beyond that, I didn't hear anything that takes care of the risks of somebody who signs up to burn coal. The experience today has been that the old plants, which were designed when we designed solely for economics, have now had a cross-cutting absolute thrown at us, first in the form of how you mine the coal--strip-mine or no, reclamation-- and second in how you burn it, which is the clean-air side.

Let me concentrate on the clean-air side, which I know a bit about. I think it is quite clear that by now we have found that the scrubber will at least do a temporary job of taking care of sulfur. We have not yet found out how we are going to fare in these regions on particulates, but we assume that precipitators will do the job.

There are two things in the wings, and they are the next step up, which create quite difficult risks, not just for the utilities, but for their investors, and that is the moving target piece. Let me just focus on two things. First, it is perfectly clear that we have not yet come to grips with nitrogen oxides, and it is not at all clear that the scrubbers that are now being sold to us as the be-all and end-all and do-this-and-you-are-home-free are going to take care of that added problem.

Beyond that there are the problems that doubtless will come out of research, that there are some carcinogens that are here and they must be taken care of with new add-ons.

And, finally, there is the technology forcing us back to the Clean Air Act, which says, if there is a better technology five years from now, scrap your old investment and put on the new. I recognize that this is perhaps a way of internalizing some costs or some desires to be technically as good as we can be, but it certainly creates a problem that is extraordinarily difficult for the investor.

From the utility's point of view we have two sources to recover revenue. One is the customer. My question there is, is there full acceptance by the customers of the Ohio River valley utilities of the added costs of burning high-sulfur coal of the local variety? Second, if this is not accepted by the customers, how is the American investor going to react if he is told that this is a clear and present danger and it is going to be put off on him? The whole problem of burning high-sulfur fuel oil or burning any of the fuels that we know about is that we have a moving target in controls. But what we know today and can do today is not going to be enough for tomorrow. We talk about internalizing environmental costs, and let us assume that these are true and legitimate costs. Somebody has to pay them-- the consumer or the investor in the electric utility.

My first question is, from the experience in the Ohio River Valley, are the consumers happy about paying these internalized costs, and are they ready to step right up and do it? If not, and you put costs on the investor, what is our experience to date on the willingness of investors to keep plowing money into the Ohio River Valley utilities in order to internalize costs which the consumers reject?

HEIBERG: I am going to call on Mr. Cohn to help me with the moving target, but let me make a couple of comments. In the first place, with respect to the scrubber, which you mentioned--that was an area where the panel had very clear disagreement. We did not agree whether the scrubber is here and doing its job at the present time. Mr. Cohn might want to elaborate on that.

Before Mr. Cohn talks about the moving target aspect, I have to admit quickly that I share some of this concern. The Corps of Engineers, planning for the Inland Navigation Waterway System, is very deeply connected to the projections and the needs and the uses of the utilities, the barge companies, and so forth, for coal and other products on the Ohio River. I am having one heck of a time in making my projections, and that is just a reflection of the fact that the industry is having one heck of a time with the moving target, again with respect to how much coal comes out of the West, for example. That means trying to project where it is going to go through the locks and which lock is going to be busy next year or next decade.

I will just make one comment on the acceptance of the consumer in the Ohio Valley. We had some fun times the last couple of months, including paying our utility bills, and I can speak specifically for

Cincinnati. I do know that Cincinnati Gas and Electric was reported in the papers about a week ago as being ready to cut off some 1,200 local services because they hadn't been able to pay their February bill. So, apparently, the acceptance you are talking about is in question. It certainly is in Cincinnati.

Mr. Cohn, will you help me on this business of the moving target?

COHN: Let me say first that I would go back to a question that was raised by a preceding member of the Panel for Inquiry and that was related to educating the public. This is an area about which the public could use a great deal of education, and there is a very simple lesson to be learned, I think, and it is that the utility doesn't pay for anything. The customer has to pay for all of the costs that are loaded on to the utility. Now, where the utility is resisting costs, it is in fact resisting them on behalf of the customer. This is a lesson that I think ought to be gotten across more and more.

Coming back to Mr. Allen's direct question--first, I think he asked whether the customers were happy or were willing to pay the costs or whatever, and I think the simple answer to that one is we have never met a customer who was happy about paying an increased power bill. I don't know about your fellows in New England, Don, but not so in our system.

The mayor of Louisville did say, I think, that the people in Louisville were reconciled to paying the additional costs associated with the use of the scrubbers. I just have no way to test that. I suspect that most of them don't realize quite what is happening.

You ask whether the cost should be paid by the consumer or the investor, and I would say in no uncertain terms that you are not going to tag the investor more than once, because if you tag him once he is never coming back, nor is any other investor. If you comply with the Securities Act, you make full disclosure.

We ask the investor to invest his money in our company, and we tell him we will do our best to see to it that he gets a fair return on that investment, not that his investment will be confiscated to pay for operating costs that are part of the cost of producing the electric power that we are delivering to the customer. So I would say that it has got to be the customer who pays for any of the costs that are imposed.

This brings me to a word or two about the scrubber question, which I guess created the most lively controversy in our session. I was the target of the questions because the company that I am associated with has said that it does not believe the scrubber makes very much sense. I was asked to explain why that was so.

What I said was, number one, our engineering people, who have looked at every scrubber that has been developed in this country and abroad believe that the technology has not yet been developed. Two, the scrubber is wasteful. It uses a lot of energy. Three, if the scrubber breaks down more often than the power plant or at different times, the power plant is out of action. That has some very important economic effects, and it has important effects beyond that in terms of the reliability of power supply. Finally, the fact is that the technology of the scrubber, the technology of doing something more about the impurities in coal, particularly the SO_2, is in the process of development. The scrubber is a very, very expensive piece of equipment. The EPA used to quote $40 per kilowatt, but somebody told me that the latest figure is $147 per kilowatt, forgetting about some incidental costs. It is very expensive, $50 million a unit, a power plant unit. Once that money is sunk, it becomes extremely difficult to go in any other direction.

Under the applicable law today we are given the alternatives of burning low-sulfur coal, on the one hand, or using a scrubber to comply with all of the standards. We have chosen the former because we believe, one, that it is less costly today and therefore in the best interest of our consumer and, two, that it gives us the choice of taking advantage of the new developments as they come along. And this too we think is in the very best interest of the consumer.

WEINBERG: Are there any scrubber enthusiasts in the audience who would like equal time?

SWIGART: I can only speak from our experience in Louisville with the Louisville Gas and Electric scrubber and some reading that I have done on the subject, but in my knowledge, utilities have not been penalized when that scrubbing system has been down. It has been down. It took a few months to get it operating efficiently, but it is my understanding that they are getting around 92 percent efficiency out of it.

I think a similar situation is the recent spill with our Metropolitan Sewer District. They had to shut down and dump raw sewage into the river. It is not my understanding that EPA is going to move ahead and penalize these kinds of situations. I think they have a very understanding attitude about it.

I will not argue with Mr. Cohn on American Electric Power's attitude about the use of scrubbers. Fortunately, it is not universally accepted.

WEINBERG: I think the scrubber argument can probably occupy all of our time, and so perhaps we can turn back to the Panel for Inquiry and ask if the other members of the panel have some questions.

ROSE: I only heard the last part of the Ohio River discussion. The main thing that struck me was that the argument seemed to revolve around a multitude of very little issues, and very little was said about long-term effects of burning coal, widespread effects, rather more severe but not immediately evident environmental, ecological, epidemiological effects. I was wondering if those were discussed earlier or not at all.

HEIBERG: We did discuss the long-term effects of various policies. In fact, we focused on the projections that ORBES is using with respect to future needs. We talked about what is pushing those projections, what will either hold them to the low projection or push them to the high projection, and what that inferred for the future. We didn't come to any conclusions, nor did we try to.

I think I would respond to you by saying that no, we did not focus on the long-term questions because by our makeup and by our way of looking at things, we are sort of down where the water is flowing past. We are wrapped up with the decisions that are being made in the five- and ten-year time frame, both from the standpoint of industry and from the standpoint of political leaders. We are concerned about trying to wrestle with those kinds of decisions, and we are asking for the policy that is needed to give guidance to the Ohio Valley or to the nation from Washington. But we are also asking for clearer guidance to help us make those decisions on a timely basis. I think that is probably because of the makeup of our committee, and I think it was probably deliberately pushed that way. We are looking at the short-term decisions and the midterm decisions hardest, because that is where we are coming from.

WEINBERG: I would like to remind you that tomorrow will be devoted a good deal to the long-range questions. Professor Houthakker, do you have some comments?

HOUTHAKKER: I did not attend the Ohio River panel. I was in another one. My questions may be completely off, especially with reference to the response to the last question, but I have two main interests regarding the subject, and I don't know whether they were discussed.

One is what are the bottlenecks to increased coal production in the Ohio Valley region, which presently produces the bulk of our coal. The second one has to do with transportation, which is more immediately related to the subject of the Ohio River, since the Ohio River does serve a very important transportation function in coal.

In the first place, is the capacity of the Ohio River adequate to sustain a considerably enlarged volume of coal output, and, in the second place, what would happen if techniques such as cryogenic transmission became generally adopted? Would that change the transportation picture drastically in that area?

HEIBERG: The first question I will handle from the standpoint of the navigation system, and then I will ask for help on the other. We really did not get into the bottlenecks of coal production other than from the standpoint of trying to come up with the right short-term or midterm projection and the right kinds of decisions with respect to the environmental constraints or safeguards. That is what we focused on from our standpoint.

With respect to bottleneck on the Ohio transportation system, there very clearly are several bottlenecks. As I mentioned briefly this morning, the Ohio River itself, and in fact this is true of most of the inland system, essentially without damage to the water or without requiring large amounts of construction, can take a good deal more traffic. It has done so very easily with the quadrupling over the last twenty-five years, except for the specific bottlenecks, which are the locks on the river. And you move from bottleneck to bottleneck. Right now we are looking at several, specifically Gallipolis, the large main stem dam and lock system above Huntington. That is our current bottleneck. As we sit here today, there are about twenty to thirty tows, and those are towboats with about maybe nine barges each, waiting there now, because we have had some problems with Gallipolis. It is at the margin, with about all the traffic, all it can take. We can't push anything more through there. As I suggested this morning, I really need to be told now or be allowed now to move ahead, to have the new locks, as I see it needed for today, eight years from now, by 1985 or so.

We have other problems. Winfield lock on the Kanawha River is a bottleneck not quite as severe as Gallipolis. We have an old system on the Monongahela that is essential to the steel/coal industrial complex upstream of Pittsburgh. But most of those locks and dams, many of them, are seventy years old. They were built for fifty years. They are doing well, considering they are seventy years old, but they are challenged today. We could have the collapse of those essentially upon us, which would mean terrible results for the industry of the Pittsburgh/Monongahela area.

I need answers with respect to our projections for the future, which bounce us against the industry, both the coal industry and the transportation industry. I need answers with respect to what we need on the river. Tell me what we need in 1985 or 1995, and I will get my people to do the best they can to figure out what size locks are needed.

But we know we are challenged now. Yes, we have bottlenecks now. If we have a lot more movement of coal beyond the movement that we have on the Ohio now, we are going to have a great deal of difficulty putting it on the water, and we will have to find other ways to move it.

WEINBERG: I wonder if Mr. Cohn would like to say anything about the cryogenic transportation of electricity.

COHN: I question whether any increases in the efficiency of transmission would make a great deal of difference in the use of coal. Increases might help in a few cases in connection with mine mouth power plants, with using even longer-distance transmission. But we do have very high-voltage transmission that will take the power a long way.

WEINBERG: How long is a long way?

COHN: Oh, 150, 200 miles. Something of that order. And then by displacement a great deal further--interconnection of one company with another--by displacement a long distance.

I want to answer your question about coal. The two bottlenecks I think are very clear. There are large numbers of subsidiary bottlenecks, having to do with all kinds of local problems--instability in the United Mine Workers--and this is internal instability. But the two greatest bottlenecks are very obvious. They are self-imposed bottlenecks imposed by the government itself, and they have to do with the uncertainties and the rigidities of the rules that relate to the mining of coal, number one, and, number two, the uncertainties and the rigidities that relate to the using of coal.

WEINBERG: I think we will have to come to the end of this very interesting discussion. I must make a personal observation. As many of you know, my own experience has been in the nuclear business. We had a discussion with a number of utility executives, including Donald Allen, and the main conclusion that we came to as a result of an evening's spirited discussion was that nuclear energy was in great trouble because of large numbers of uncertainties. I had expected to come here to hear, okay, you have a few problems with nuclear fuel, but with coal everything is going to be simple. I guess that is just too simple-minded a view.

KAIPAROWITS

ORIENTATION

Calvin L. Rampton, *Chairman*
Attorney; Former Governor of the State of Utah

Kaiparowits has become widely known throughout the country, particularly in the state of Utah. It is an Indian word that, roughly translated, means "home of the mountain people." As the chairman of this case study panel I am sure that I am supposed to be objective in my presentation, and I am going to attempt to do so. But if I slip once in a while, please keep in mind that I was a proponent of the project, and it was a very important and sometimes emotion-filled subject for the entire twelve years during which I was governor of the state of Utah.

One of the very first major decisions that I had to make after being elected as governor in 1965 was whether or not to allocate part of Utah's entitlement to water from the Colorado River for the Kaiparowits project. That was in the first year. It was in the last year of a twelve-year term that the Kaiparowits project was finally abandoned. So it was, in a way, one of the dominating factors all of the years I was in office.

Even though the first major step, so far as the state of Utah was concerned, in Kaiparowits did not take place until 1965, the project had already been in the planning stage for a number of years, and a consortium consisting of Southern California Edison, San Diego Gas and Electric, and the Arizona Public Service Company had proceeded with drilling on the Kaiparowits plateau and other preparatory planning.

The selection of the Kaiparowits plateau as a prospective site for a

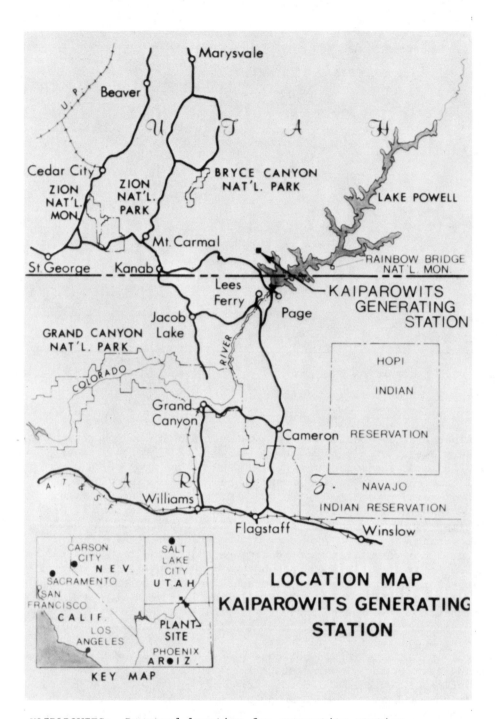

KAIPAROWITS: Proposed location for generating station.

major power plant of course was dictated by the fact that it has a very large deposit of quite good coal. The Kaiparowits coal has two advantages. It is fairly high in Btu's. It is comparatively very low in sulfur content. It has one disadvantage from an economic standpoint. The strata of coal is very deep, so you have to get the coal out by shafts and tunneling and drifting rather than by any open-pit methods, so increasing the cost of the coal.

But, nonetheless, the companies felt that they had a viable project. The proposal was to build a 5,000-megawatt coal-fired power plant. At that time, in 1965, the estimated cost of the plant was $500 million. By the time the plant was finally abandoned in April of 1976, it had been cut back to 3,000 megawatts, and the estimated cost had escalated to $3.5 billion dollars. In other words, the cost of this plant had gone up per unit some nine times in that period.

In 1965--you have to keep in mind the time frame--there was not a great deal of opposition to the Kaiparowits project on an ecological basis. Keep in mind that while there were many in the country, without doubt, who were concerned about ecological matters at that time, the problem of environmental protection did not have the broad public concern that it now has. In 1964, the first year in which I was elected governor of the state of Utah, neither major party, either on a national level or in the state of Utah, had in their platforms a plank on ecology. It was just an ignored matter. By 1968, both major parties, on national and state levels, devoted a substantial part of their platforms to ecological matters. By 1972, ecological matters dominated both platforms.

There was, however, some concern in my state regarding air pollution even at that date, because we had had an unfortunate experience in the state of Utah in the twenties and thirties with the coal-burning locomotives that operated the railroads through Utah. Further, the Salt Lake Valley is in a bowl, and we have an inversion tendency there, so we were concerned about air pollution, because we are aware that a plant that had been constructed at Fruitland in New Mexico was a very dirty plant indeed and emitting a great deal of black smoke. So, I and others in the state carefully questioned the sponsors of the project, and we were assured that the state of the art was such that this would not happen again and that they could build a clean plant.

On the basis of that, we did allocate to the companies for the operation of Kaiparowits 102,000 acre-feet per year of Utah's entitlement from the Colorado River. However, even though this water belonged to the state of Utah, and it was our entitlement to make, because they were going to take it from Lake Powell, which is a federally constructed reclamation project, it was necessary that the companies deal with the Department of the Interior in arriving at a contract for the price to be paid by them, which would go toward the amortization of the project. However, then Secretary of the Interior Stewart Udall seemed to want to do more than that. He wanted to have the federal government pass upon all aspects of the taking of the water, and that resulted in somewhat of a confrontation between Mr. Udall and myself.

For a period of some three years, the companies negotiated with the

Department of the Interior. I was continually telling Mr. Udall, that is none of your business. All you are supposed to do is fix the price. They have already got the water.

I called him on Christmas Eve in 1968 and said, look, Stewart, you are going to be out of there in less than thirty days from now. Let us get this done. He assured me that it would be done before he left office the next January twentieth. I called him quite a number of times in January and for some reason could never get through to him. But at any rate, the administration changed, and the project wasn't approved.

His successor, Walter Hickel, had been a good friend of mine, a former governor, so I immediately started getting on Mr. Hickel. He finally did approve the contract for the water in the fall of 1969, as I recall, probably quite late in the fall.

By this time, of course, there were other problems being raised with the project, ecological problems, financial problems, and so forth. The companies did begin the preparation of an environmental impact statement, which was more than two years in the making. The environmental impact statement was finally presented to the Department of the Interior, if my memory serves me right, in the early spring of 1973, and I assumed that it would be approved. However, after looking over the impact statement then Secretary of the Interior Rogers Morton disapproved the project. Now, just why it was disapproved I never did understand. I had a theory because just a week or two before, the building of the Alaskan pipeline was approved, and I think the secretary was dealing them off, one for the environmentalists, one for the developers, and we just happened to come in on that deal of the cards.

But, at any rate, while turning down the project, Mr. Morton said, if you move the prospective site of the plant some sixteen miles from a point called Nipple Bench, the proposed location, to a place called Four Mile Bench, we might take another look at it. I think a number of things were happening when Mr. Morton said that. I think the energy crisis was coming on us and the climate was changing, and I always felt that if we had moved it sixteen feet instead of sixteen miles, he would have still approved it on the second go-round.

They began another series of environmental impact studies, and these were substantially completed early in 1976. By now, Thomas Kleppe was secretary of the interior. He had promised us a decision on the Kaiparowits project by the first of May. On the fourteenth of April I was in my office in Salt Lake City, and I received a telephone call from Bill Gould, who was in charge of the project. He was in Washington here, and he said, I want to tell you so you don't have to read it in the newspapers that we are going to withdraw from the project. Now, why they withdrew then I have never been able to understand. Maybe Mr. Drake will enlighten us as to the timing of the withdrawal.

The Kaiparowits project as such is effectively dead. The companies still have their water allocation, but it will be up for renewal, and I would be very much surprised if it were renewed. That is not now my decision, however.

There are other projects of a similar magnitude that are being proposed fairly near. The Intermountain Power Project proposal will also be for 3,000 megawatts if it is approved and the financing is arranged. It will be some thirty to forty miles north of the Kaiparowits site.

I would say that while Kaiparowits as a project probably will not be revived, the issue of coal-burning plants in the southern part of the state of Utah is still very much alive and will continue to be a matter of controversy for some years.

Brant Calkin

President, Sierra Club

Kaiparowits is a good example of the wrong industry in the wrong place for the wrong reason. It has been surrounded by, or it was surrounded by, controversy and uncertainty, and it was likely to continue in that mode for a long time. While it was generally opposed by environmental groups, including the Sierra Club in which I am active, it was never completely supported by industry.

It is not my purpose today in these opening remarks to try to condemn the project or to replay the controversy about it, except in response to questions in part of the discussion. I would really prefer to examine some of the deeper dynamics of the whole matter, the issue of the disposition and utilization of and effect on public lands of national value. I would like to talk about energy needs, and I would like to talk about energy futures. The common concepts of economic development were not accepted by many environmental groups, and, therefore, the trade-offs were not acceptable. And that is a very profound change in our view of what economic development should be.

Social goals were all wrapped up in that controversy. And so was the question of alternatives. I think all those are better expanded through discussion and comment than through pronouncements or anything of that sort. Let me just list those as ideas that we might talk about, since I think they are much more basic than the numbers game of how many tons per day of this or that were emitted.

I would like to suggest that Kaiparowits disproves the idea that we ought to get as much as we can, where we can, as fast as we can. And I do not even have to take issue with the idea that we ought to do more. But I would suggest that if we try to do it the same way, we are going to suffer the same fate.

We do not need another Kaiparowits, in my estimation, but that philosophy will get us one just as sure as God made little green apples. We may be in the process of making another one, as Governor Rampton pointed out earlier, with the Intermountain Power Project. And I would very much like not to have to direct the resources of the environmental groups with which I am associated to going through another useless replay of that controversy if it can all be avoided. We have other things we would like to work on.

My organization is particularly interested in expanding its awareness and activities in more urban concerns. There are lots of other things for us to do as well, and we would like to avoid having that kind of thing come up again.

James H. Drake
_____

Vice-President of Engineering and Construction, Southern California Edison Company

I am not here to debate on the part of the participants. I am not here as a proponent of the power project, which is dead. Its participants, Arizona Public Service Company, San Diego Gas and Electric Company, and the Southern California Edison Company, expended a substantial sum of money, roughly $22 million, and then were forced to give up the site as a location for a power project.

We do perhaps hope to get a reduced amount of water, maybe something of the order of 30,000 acre-feet or thereabouts, for perhaps a coal gasification project, perhaps mining this coal and exporting it by slurry pipeline or some other method to some other location where we might be able to site a power project, or perhaps for coal liquefaction.

Many factors appeared to favor the project: (1) additional generating capacity could not easily be added to the south coastal basin in southern California because of air quality problems, (2) the economies of the project, with a coal-burning plant fired by fuel from an on-site coal reserve appeared to be excellent. While construction costs would be higher than for oil- and gas-fired plants, fuel costs would be low.

The first action of the participants was to secure Colorado River water rights from the state of Utah. This was accomplished (up to 102,000 acre-feet per year) in an elapsed time of twenty-one months from the original filing date, which turned out to be something of a negotiation record for the project.

Plans for the project were announced in 1965 by the Kaiparowits Coal Project participants. The plan called for a 5,000-megawatt power plant to be constructed over a period of fifteen years at an expected cost of $500 million. While Utah had authorized use of its water, it was impounded in a federal area that was to become Lake Powell, under control of the Department of the Interior's Bureau of Reclamation. Application was made to that agency in September 1965. At that point, the plant was to start construction in 1970 and to produce the first power in 1975.

The application to the Department of the Interior had a difficult time, renewing the fight between the federal government and Utah over Colorado River water rights. On the other hand, opposition from environmentalists was practically unheard of in the midsixties. Negotiations went on three years, and, finally, in late 1969, the secretary authorized the use of Colorado River water. In the meantime, inflation had driven the cost of the project up from $500 million for a 5,000-megawatt station to $750 million for 3,000 megawatts.

The project still looked economically feasible in 1969, particularly because the cost of other fuels was rising. In fact, some of the Kaiparowits project participants joined with other electric utilities to start construction of the coal-fired plants at Four Corners, New Mexico, and the Navaho plant near Page, Arizona, in that same year.

A series of events took place that caused further delay. This was to happen at a time when each month's delay cost the participants over $6 million. The first of these was that the secretary of the interior announced that decisions on further power plant construction would be delayed until a task force had reviewed all the factors involved in energy development in the Southwest. This represented, in effect, a one-year moratorium on planning for the project.

The next development was that in 1970 the Sierra Club filed a lawsuit designed to stop further power plant construction in the Southwest by seeking to bar federal cooperation with power plant developers until the Department of the Interior agreed to comply with provisions of the Environmental Policy Act of 1969. The act required that projects of the scale of Kaiparowits file environmental impact statements.

Finally, in 1971, the U.S. Senate opened public hearings on power generation projects in the Southwest. These led to a two-year study that concluded that there was no alternative to building some coal-fired plants in the Southwest to meet growing energy demands, and that the ultimate decision on construction should be with the secretary of the interior. The participants felt they could live with this. Work on an environmental impact study already had begun, and preliminary engineering studies were underway.

The Sierra Club suit was dismissed by a U.S. District Court judge in early 1972, a setback for the environmentalists. The outcome of this decision was to be a greater demand by environmentalists for the government to adhere strictly to all environmental safeguards.

Then came a spate of delays over environmental considerations. Only a month before the participants' environmental impact report was to be submitted, the secretary denied the applications for the project for environmental reasons. The next day, June 14, 1973, preliminary engineering work was stopped. Eventually, it was decided the Four Mile Bench site would receive greater emphasis, the environmental impact report would be submitted, and the secretary would reconsider his opinion.

In December 1973, the Department of the Interior announced that it would accept the renewed application and that environmental reviews would begin based on consideration of the Four Mile site. By this time the estimated cost of the project had risen to $1.7 billion, an increase of over three times the original estimate ten years earlier, and the planned capacity of the plant had been reduced from 5,000 to 3,000 megawatts. The oil embargo (the "energy crisis" of late 1973) caused the project to remain economically feasible even in the face of escalating costs. The price of low sulfur oil rose to nearly eight times its pre-1970 price, energy independence was the watchword, and Kaiparowits looked better than ever.

What happened then to eventually lead to the project's cancellation? Engineering plans were designed to minimize the environmental impact of the project, particularly upon air quality. Emission control systems would remove 99.9 percent of the particulates and 90 percent of the sulfur dioxide. The coal mining would be done with a minimum of disruption to the terrain. Deep underground mines, not strip mines, were to be located close to the plant site. The coal conveyor system would be covered and the dust suppressed.

During the period following acceptance of the renewed application by Interior, plans for the new town of about 14,000 people were developed and publicized. The siting and layout of the new town was developed as a cooperative effort between local, state, and federal officials, as well as the participating companies. Simultaneously, plans were moving ahead for a mining school, offering the promise of full-time jobs for many of southern Utah's unemployed.

Generally, the plans for the project were well received in Utah. The Bureau of Land Management's (BLM) draft environmental impact statement was expected to be ready for public hearings in March 1975, and it was believed construction would begin in late 1975.

In mid-1975 the Salt River Project (which had joined in the project participation agreement in mid-1973), after reviewing its energy options for generating resources, announced it would withdraw from the project. In retrospect, this may be viewed as an indication that perhaps the project was becoming too costly, even in the midst of the "energy crisis." Salt River's withdrawal from the project left 18.6 percent of the project unsold, their 10 percent plus the original 8.6 percent the participants had left open for other utilities. And the proposed costs were continuing to rise. By September 1975, the project budget was $3.5 billion; $3.0 billion more than the original estimate (in current dollars), and 40 percent less capacity than originally proposed.

Hearings on the draft environmental impact statement began in September 1975, a month behind schedule. At the hearings (at which project participants were not encouraged to participate), objections were raised concerning increased salinity of the lower Colorado River basin and the impact of a new town of 14,000 in an area where a major city, Kanab, had a population in 1970 of only 1,400. A forty-five-day extension of the public comment period was announced at the end of the public hearings in response to requests from environmental organizations that they had not had ample opportunity to review the entire EIS. The date for the final submission of the EIS was moved forward to March 1976.

A rash of further delays ensued. For example, between mid-October and mid-November 1975, there were a total of nine legal motions, petitions, and official protests filed against the project. While quick resolution could be expected for some, many could be expected to keep the project tied up in litigation for an extended period of time. A few examples are as follows:

1. The Sierra Club petitioned the California Public Utilities Commission to compel the California-based participants to obtain a Certificate of Public Convenience and Necessity prior to starting construction, not for the transmission system which would be constructed partly in California, but for the generating plant in Utah.
2. The Escalante Wilderness Committee, backed by the Sierra Club and the Environmental Defense Fund, persuaded the Utah state engineer to continue hearings on the extension of water rights, making it appear that this permit would have to rejustified.
3. The National Park Service, in December 1975, belatedly announced it would conduct a three-month study of the impacts of Kaiparowits upon air quality at three neighboring parks: Bryce Canyon, Capital Reef, and the Glen Canyon Recreation Area.

On December 30, 1975, the participants announced that delays in regulatory approvals caused by the objections of environmental groups, and lengthy approval processes, forced deferral of the project for one year, in effect forcing a one-year halt of further financial outlays.

In March 1976, the National Park Service's three-month study was released. It maintained that pollution from the project would be seen for 100 miles at all three parks, contrary to the results of the participants' studies. Once again, there were protests. Calls were made for a regional survey of power in the Southwest before making a final decision on the project. The EPA claimed that the power group had not documented the need for the power and would advise the secretary of the interior not to approve the project. In Congress, the advisability of revising the standards of the 1970 Clean Air Act was being considered, standards that would have placed such severe restrictions on the Kaiparowits project that it is possible it could never have been built.

On April 5, 1976, an organization called Save Needed Environmental Levels League (SNELL) filed a complaint seeking an injunction to stop the Kaiparowits project until an adequate environmental impact statement complying with the National Environmental Policy Act (NEPA) of 1969 had been prepared and filed.

The participants filed an answer to the complaint on April 7, 1976, seeking dismissal of the complaint. Also sought was a declaratory judgment that the environmental impact statement on the Kaiparowits project fully complied in all respects with NEPA and all other legal requirements applicable to the project and that the participants might proceed with construction. Relief was also sought enjoining all environmental groups and individuals who were parties to the suit from initiating and pursuing any litigation that would delay the project.

Confronted with the fact that the financial viability of the project was seriously in question, given the cost of past delays and inflationary pressures, and with no real prospect that either project delays or rising costs would moderate in the future, the participants decided that, as prudent managers, the sum of $3.5 billion could not be committed to a project attended by so many uncertainties. The project was abandoned on April 14, 1976.

On August 31, 1976, the court granted the motions to dismiss the SNELL lawsuit.

Daniel Dreyfus

Deputy Staff Director, Senate Committee on Interior and Insular Affairs

I come to this discussion with a peculiar stance because I am an engineer. I spent many years in water resource development, which I know is--always has been and today especially is--anathema to most environmentalists. On the other hand, I did a lot of the basic staff work on the National Environmental Policy Act, which is anathema to almost everybody. I have a tendency to get a little bit upset with both sides of the argument and never know exactly where I am.

I was involved in Kaiparowits as an interested observer. The Senate Interior Committee, which is now the Energy and National Resources Committee, held hearings on the whole complex of power plants in the Pacific Southwest, sometime back during the time when Kaiparowits was hanging on the ropes and its fate was a question of federal attitude.

One of the things that we should examine when we talk about Kaiparowits is how it fits into the setting of the greater development of that region. Now, the first question is, why did anybody ever think about Kaiparowits?

Between 1940 and 1970, there was extreme regional growth in the Southwest. Population growth in that region was increasing at about 3 or 4 percent annually, as compared to 1.4 percent nationally. Energy demand was over 8 percent annually during that time and was considerably higher than what it was nationally, which itself was a picture of very great growth.

Essentially, you start out with electrical plants; you start out trying to site them near the load center, because line losses are a very significant question. If you look at the Southwest, the load center is the southern California metropolitan area, which utilizes about 60 percent of everything generated in that area.

In the sixties, we began to place constraints on siting of power plants. And some of the early environmental constraints were the air quality constraints applied locally, through the federal act, of course, but by local jurisdictions in the southern California area. And, essentially, they foreclosed coal and oil development in southern California.

Nuclear generation in that area has some serious problems also. The coast is seismic and the California desert has not got the water for cooling of power plants.

The coal in the Colorado Basin began to get to be an attractive development. There is a great deal of coal throughout the basin. The power companies began to move into the desert region. They began to look for opportunities to develop coal-fired power plants to serve the metropolitan districts in Utah, in southern California, in Phoenix, and in Albuquerque.

Now, one note that has something to do with power plant siting is that a thousand-megawatt power plant uses about 3.5 million tons per year of coal. That is not too relevant, but remember 3.5 million tons. The water to cool that same power plant is about 20,000 acre-feet of water, and that water weighs 27.0 million tons. So you see that almost ten times as much weight is involved in moving the water to the power plant as in moving the coal.

Given this kind of a situation, and given the peculiarities of western water law, very often power plants were sited on the basis of finding a water supply. Coal is fairly prevalent. You can find the coal or move the coal or do whatever you must do with the coal, but you site power plants in the western deserts on the basis of available water supply.

And so they did with the Four Corners plant in New Mexico and the Mojave plant in southern Nevada and the Navajo plant, which was in Arizona. These plants predated Kaiparowits. A picture began to form of several major power plants in this otherwise untouched desert area. The Four Corners plant was a particular horror story because the Four Corners plant started in about 1957. They leased coal from the Navajos. The plant went on line in 1963. In 1971 it was 2,000 megawatts, and it was producing about two times the particulates of the New York City metropolitan area.

It got to be a cause. *Life* magazine came out in 1971 with a landmark article entitled "Hello Energy and Goodbye Big Sky." People began to realize where these sites were on the map and the interrelation among them. Politicians got into the game, and environmental groups, nationally, rather than locally, got into the game. We began to look at the situation, both in the executive branch and in the Congress.

When you look at that region, you find that 80 percent of the total acreage is in federal ownership--predominantly federal lands for recreational, grazing, and timber purposes, Indian trusteeship lands, six national parks, twenty-eight national monuments, three national recreational areas, and a water supply that is provided by federally sponsored management projects.

There was vast federal concern. The National Environmental Policy Act was enacted in 1970 and provided that the federal government and the Department of the Interior in particular were supposed to look at the cumulative impact of what they did.

For the first time, instead of looking at one water contract, one Indian coal lease, one national park decision, the secretary of the interior had to look at the cumulative impact of what he did.

At that time, the other power plants were pretty well committed, and Kaiparowits was the next one coming up. It was still only in the formative stage. No ground breaking had occurred. It was also one of the biggest. It was up to 5 million kilowatts. It was in a hitherto untouched area.

In my judgment, the Kaiparowits site is a bad place for a power plant both from the regional point of view and from the national point of view. If you go to the Colorado Basin to site 5 million kilowatts, the last place you should go is the Kaiparowits plateau. The reasons why these people were involved in Kaiparowits were, primarily, because they had some coal leases and they had the water. The state of Utah did have a remaining water allotment and it was willing to provide for industrial growth. The people in the Kaiparowits plateau welcomed this kind of development, while in southern Colorado, which was an industrialized mining area, they were exceedingly hostile to this kind of growth.

A general attitude prevailed in Utah that the state's business was the export of natural resources, whether it be mining resources, agricultural resources, or power resources. There was a reception in Utah that was missing in some other places, and Kaiparowits was where the pin went in the map.

The impacts in Kaiparowits are greater than they might be elsewhere. We are talking about power lines where there were no power lines, roads where there were no roads. A power plant is a little industrialized area, but power lines go for hundreds of miles across the desert and they are a problem for hundreds of miles of right of way.

There has to be an industrialized area. There would be a metropolitan area of 14,000 people, something like three times the total population of the counties involved. The infrastructure that went with this created an industrial area where there was none before.

The arguments began to rage. The lesson at Kaiparowits is that the site was ill-chosen by virtue of the fact that the way we do business forces us into ill-chosen sites. Therefore, when the argument started, the site was indefensible. I think that is why it collapsed. The *coup de grace*, I believe, was the Air Quality Law—the uncertainties revolving around the new amendments and the court cases on air quality, the nondegradation. Now, on that note, I want to change my hat and take one last parting shot at the Air Quality Law.

It would have been better to put that power plant, for example, in some of the old mining regions in Colorado, where there already are industrial impacts. But I do not think it would have been any easier to put it there in the kind of environment we live in today.

What the Air Quality Law—if it comes along the way it is developing—is going to say is, if you can find a site, you can build about a thousand megawatts. Five thousand megawatts is going to be out of the question anywhere. We are going to have to find a lot of sites; we are going to have a lot of power plants, each with its own infrastructure, metropolitan development, power lines, and coal transportation facilities. While we may do something about air quality, we are going to mess up our remaining undeveloped areas.

I think reconsideration is necessary on both sides. It was a poor site because there were other sites in the region that could have more easily been developed with less net impact on the region. We did not choose those other sites because of our governmental structure and the way we go about choosing sites.

Steven E. Plotkin

Office of Energy, Minerals, and Industry, U.S. Environmental Protection Agency

Although some of the speakers discussed the potential for developing new technologies for burning coal in a clean way, and although my agency is spending a hundred million dollars a year or so to do just that, I do not really believe that we are, in the short-term foreseeable future, going to place a major power facility in any area without substantial environmental impacts.

In arguing about a specific site like Kaiparowits, you cannot argue in a vacuum. The major question about Kaiparowits that has to be answered before you can decide whether it is acceptable or not is, what constitutes an unacceptable impact?

If you know that you cannot place a power plant anywhere without an impact, what do you mean by "unacceptable," given a variety of sites? How far should a developer have to go to satisfy environmental concerns, knowing that he cannot go far enough to satisfy everybody. If the people who live where the power plant will be located are environmentalists, they are going to see substantial impacts and be forced by their sense of environmental concern to fight that power facility.

These questions cannot be answered in the context of one site or one technology. The Kaiparowits argument is the archetype of industrial-environmental conflict all across the country. Industry sees the needs for power and the benefits of that power both to the nation and to the locale. The environmentalists see damages, and they will always be seeing damages at any site you can choose.

Environmental-industrial arguments create an automatic, insoluble adversary position, as long as environmental planning is done in the way it is done today. Alternatives never can get looked at properly.

What is a reasonable context for looking at the impacts of any power facility? We need a clear picture of the benefits of the energy--and this is certainly a difficult point because of the uncertainties surrounding demand projections--and we also have to understand the full range of impacts when siting a facility, given the potential for using different technologies, both as the base power generators and as the environmental controls, and looking at a variety of locations.

There is an enormous variation of costs, an enormous variation of impacts in all sectors--in air quality, in water quality, in socio-economic impact--depending on where that power plant is located or what the technology is.

For instance, every major power facility, whether it be a synthetic fuel plant or an electric power plant, uses a tremendous amount of water. But you can show that there is a very substantial range of water use depending on the technology.

You can show that electric power plants probably use the most water. You might be able to show that there are several kinds of gasification plants that, for the value of Btu's that you get out, use the least water. Even given one technology, you will have a different water impact depending on what area of the country you select.

For instance, there may be a few sites in an area or in a region where dry cooling could be used, although there are few sites like that. On the other hand, there are a great many sites where a combination of wet and dry cooling could be used and the water impact would be substantially less than if you select a site where the only feasible alternative is wet cooling.

Of course, the water availability makes a tremendous difference. If you go into parts of North Dakota, there is plenty of water. If you go into the upper Missouri Basin, there is a moderate amount of water. If you go into Colorado, there is far less. So, depending on where we locate these sites, we can expect to see substantially different water impacts--and with long-range transmission of electricity, you actually have the alternative of siting in a variety of locations.

If you are shipping power east, instead of west, you have a substantial range of sites from which to choose.

The key here is that if you can look at a variety of sites and a variety of different technologies and if you can foresee the impacts, you can begin to get a handle on what is acceptable.

A reasonable person, even one who is an ardent conservationist, accepts the fact that major impacts will occur. What he wants to know is whether an industry has made a good faith effort to look at the alternatives and whether it is selecting an alternative that will come close to minimizing impacts.

Industry, on the other hand, has got to be in a position to know if it has hit the mark. It has to know, if it examines a site, whether the impacts it comes up with indicate that it can do better elsewhere.

Where do we get this information? There has been a push on the part of the Sierra Club and other environmental organizations to get that information into environmental impact statements, to examine alternatives within those impact statements--alternatives in environmental control technology, in locations, and in technologies.

I feel that it is impossible to do that, at least in terms of the broader scale of alternatives. I believe that, in an environmental impact statement, a company should have a look at a wide range of locations within a given area, or different ways to get the water to a plant, or different environmental control technologies.

Given the expense of looking at a great variety of different sites, I question whether a single power company has the resources, even when it has the desire, to look at the full range of impacts. The Sierra Club sued the Department of the Interior to force it to do a generic environmental impact statement for the northern Great Plains. The argument was that since the nation was moving in the direction of developing that whole section of the country, there was a program to develop that part of the country, and the government had to satisfy the Environmental Policy Act by filing an environmental impact statement. They lost.

I think that is regrettable because the government is the only body with the resources to do the kind of study necessary to properly evaluate energy development. Until we do it, and it is going to be expensive, we are going to end up with one Kaiparowits after another. We are going to have one site after another where we cannot judge what is acceptable and what is unacceptable.

My agency has a program designed to do regional assessment. We tend to spend about two million dollars per project and the projects cover a broad area. We have a project in the Ohio River basin, ORBES, that is being discussed in one of the other sessions. We have a study of the West that is looking both at the Four Corners area and at the northern Great Plains.

We have a study that will begin soon in Appalachia. The reason we spend about $2 million per project is not because we think we can do the job with that amount of money, but basically, that is what our resource base is. The EPA does not have the research funds to carry out $10 and $20 and $30 million projects, but I believe that is what it is going to take to examine a range of sites, to be able to educate people about the impacts of development.

Until that is done, these studies help. Otherwise, I would not be involved with them. But they are insufficient; it is the government's responsibility to pick up that ball and it has not done it.

DISCUSSION

RAMPTON: I was disturbed a little bit by Mr. Calkin's statement regarding the Intermountain Power Project. I would have expected that statement from the Friends of the Earth, who are a little more militant than you are, but I was surprised that it came from the Sierra Club, which I have always regarded as highly responsible.

We have not done an EIS on the IPP Project yet. Are you not doing like the Friends of the Earth seem to do, arriving at a position intuitively and then looking for a basis to sustain the position already arrived at?

CALKIN: I do not think so. We are not totally without information on the Intermountain Power Project. We have accumulated a fair amount of data and the utilities and the state of Utah have their own data. Our position is that it has the same dynamics, the same general area, many of the same players.

RAMPTON: But in an entirely different air shed.

CALKIN: That changed very quickly, too, you see? It is not an entirely different air shed, and it has many of the same public land values that go with it. The prospect is not good. One of the reasons that I am here is to look at what happened at Kaiparowits and what the parallels are.

I would suspect, for example, Governor, that many people in here have heard about the Four Corners power plant but not so many have heard about the San Juan generating station. The San Juan generating station is only twelve miles from the Four Corners plant, and it is going ahead, up to 1,500 megawatts. We have learned from Four Corners, we and the utility companies and the government, and we are not making those same mistakes again. I am not sure we are learning that lesson in this Intermountain Power Project.

DRAKE: You have a different set of players. Neither Southern California Edison, San Diego Gas and Electric, nor Arizona Public Service are involved in the Intermountain Power Project.

CALKIN: We still have the federal agencies that we have to deal with. Mr. Dreyfus is talking about the transmission right of ways. We have still got the federal officials in BLM and Utah and so on. There is quite a bit of parallel.

SPEAKER: What fuel is used for that power plant?

CALKIN: The San Juan?

SPEAKER: Is that the one that the Sierra Club is approving?

CALKIN: Yes, that is the same coal that is used at Four Corners. It comes out twelve miles away.

DREYFUS: It comes out of a Utah construction company mine at Four Corners, and they haul it a little ways. It is a New Mexico Public Service power plant.

SPEAKER: How big is that plant?

CALKIN: Right now it is at 700. It is going to 1,500 megawatts.

SPEAKER: What are the differences between San Juan and Four Corners?

CALKIN: There are several. First of all, and perhaps key, is you have a different managing agent and company. The Four Corners power plant is run by the Arizona Public Service Company (APS) which was a participant in Kaiparowits. We do not consider them progressive by any stretch of the imagination.

But I will say here, in front of God and everybody, that we do think the Public Service Company of New Mexico (PNM) is a good, responsible company, with whom we have occasional or even intermittent disagreements.

RAMPTON: How do you get along with Utah Power and Light?

CALKIN: I wrote to Utah Power and Light for information on Kaiparowits. I called them. I made several inquiries, and I got absolutely no response whatsoever.

RAMPTON: Well, they were not one of the parties to Kaiparowits. They did not have the information.

CALKIN: They had information on Kaiparowits because, at that time, they were contemplating participation. But at any rate, my inquiries at Utah Power and Light--and I have the correspondence in my files going back to 1970--got zero response.

One of the key differences between the San Juan generating station and the Four Corners power plant is the credibility of the operator.

SPEAKER: Do you have a staff of engineers that have operating experience in assessing the credibility of an operator?

CALKIN: One of the best indications of the credibility of an operator is the history of the operation. We do have people who are reasonably knowledgeable in power plant operations. We are not strong in that. We do have people who are very strong in pollution control, so we have some independent judgment there.

The second thing is how PNM phased in their project. They started out with a 345-megawatt unit, with a relatively quick move up to the next unit. They indicated that they were not totally committed to the full expansion program early on, and we were able to grow with them. We learned as we went along. That is a lot different than someone saying, "We are going to put a much larger facility in, and we are committed to the whole thing at once."

Another thing is that PNM learned from APS and from the Department of the Interior. The original contract for the Four Corners power plant with the Department of the Interior was not very good. But when they started dealing with PNM, there was much more in the way of teeth in terms of water use restrictions being applied back up through the system.

I do not think I could credit PNM with being brilliant. They just came along a little later, and the arguments had been somewhat refined, and they had more help. But there are significant differences. Phased-in commitment, compared to Kaiparowits, more credible operators--that type of thing.

DREYFUS: I wanted to point out that the San Juan plant is a dramatic example in the sense that it did get developed during the time that this controversy was raging over Kaiparowits. It went along without very much notice and without too much difficulty.

Some of the physical circumstances, however, are as follows: Farmington, New Mexico, is an industrial area. The Four Corners plant was already there. They are expanding an existing surface mine. The power grid is largely in place. The new logistics and infrastructure that went with the San Juan power plant are minimal for the additional capacity. It is not in a new place. I think that has a lot to do with the impact and a lot to do with the acceptability.

CALKIN: Let me just correct Dan, if I may. There are two mines. The Western Coal Company mine is twelve miles away and is completely separate from Four Corners, but it does have the same operator, Utah International.

SPEAKER: Assuming that the economics favor a Kaiparowits-type power unit, what would you consider necessary--or what constraints would you like to see applied to it to make it acceptable to you?

CALKIN: The constraints can reflect either technological feasibility, economic feasibility, or impact, and it is not clear that the first two always add up to an impact feasibility. Constraints are site specific and they change with other factors. In the case of Kaiparowits, I think that was almost a suicidal site selection and compounded all the other problems. I am not sure that with 99.5 on particulates, even assuming some later control of submicron fines, and 90 percent on sulfur, that the impact, in addition to the existing Navajo plant, could have been acceptable.

But as a practical matter, the plant did not fold for that reason. The plant folded, in my estimation, for economic reasons, which were partly, but not totally, compounded by environmental considerations. No matter what the technological feasibility is, in that area it just simply may not have fit.

RAMPTON: Take the San Juan site, for example. I gather it is acceptable in terms of the proper environmental controls. A large part of the arguments of the Sierra Club and others was based on what I would call a no-growth policy for the areas to be served. I will call your attention to the statement filed by the San Diego chapter of the Sierra Club, which says, "The EIS does not specifically identify the growth policies of the San Diego area." I would call your attention to the statement that, at that time, represented the official position of the state of California. It was made by an environmental writer for the *Los Angeles Times*. Moreover, testimony was made in San Bernadino that the Kaiparowits project would cause air pollution in southern California.

This sounds rather ridiculous, since winds in the troposphere move from west to east. But the argument was--and I think part of the argument of the Sierra Club was--that making an abundant source of energy available to the southern California ari basin, or to San Diego for that matter, would encourage more people to move there, more industry to move to an overburdened area.

In your opinion, is this a legitimate issue to inject into siting a power plant, or should land use be controlled by local and state statute?

CALKIN: The question of what is an acceptable growth rate is one of the basic dynamics. If we are going to trade off the public lands, parks, and monuments, what are we trading for? There is a healthy growing debate in the public and in the Congress today as to what we want to be ten years from now. Do we want to be a neon strip from San Diego to San Francisco? Perhaps. Do we want to make that decision, or do we want to back into it by not making decisions along the way?

That we are going to trade off those lands and say they are necessary for some purpose may be the decision honestly arrived at. But we cannot do it by saying everybody is going to grow. If you add up growth figures around the country, they add up to about $2\frac{1}{2}$ times what the national growth is.

Where are we going to have growth? Where is it possible? What is it good for? Then we can figure out whether we have to trade off park lands to have it. We can do it, but answering these questions is one of the basic dynamics.

DRAKE: I do not disagree with the issue you pose. The question I am asking is, is a San Juan, a Pacific Intermountain, or a Kaiparowits project a legitimate area in which to debate that issue and hold up a project?

CALKIN: Absolutely.

DRAKE: You think so?

CALKIN: If you do not, there is no debate.

DRAKE: Why not debate through the California state legislature? Why not through such acts as the Coastal Zone Act, which is restricting growth within a thousand yards of the coastline?

This is a very undemocratic and obtuse way of trying to control growth, specifically, to try to cut off the water and power to a region. I see this argument coming from the Sierra Club, over and over again: stop a power project or stop a water project as a way of controlling growth.

A more democratic way to go at it is through land use, through local statutes, through planning commissions, through comprehensive plans approved by the California state legislature, and perhaps through the federal government. I think that a power project is an improper area in which to debate that issue.

RAMPTON: Isn't it also true that you can control growth only to a limited degree because, in a free society, people are going to continue to have babies and they are going to continue to move where they want to live? If you make an assumption, for planning, that you are going to have no growth when, as a matter of fact, there is going to be growth, your basic planning is faulty and you are going to get growth anyway and not be ready for it.

PLOTKIN: Governor, we just recently did a study at EPA that examined the impact of extending sewer facilities--and we discovered that when you put in infrastructure, growth follows. When you expand sewer facilities into an area, then that is where growth occurs because you make it economical for developers to move into that area.

We have seen the same thing with highways, and we should see the same thing with power and water facilities. I do not think it is true that people move where they want to move. They move where they can move, and the only place they can move is where there are facilities.

For instance, there are huge projects in Arizona, where roads are laid out and sites are plotted. People have bought land, but nobody lives there. And the reason is, no water or electricity.

SPEAKER: You are talking about the reallocation, and you are talking about growth--absolute growth. It is not the same thing.

PLOTKIN: I think we are talking about patterns.

RAMPTON: I think they are related.

SPEAKER: They are related, but are you not talking about absolute growth when you are speaking about zero growth or low economic growth nationally? And you are talking about people shifting from one place to another place--the same number of people.

PLOTKIN: I interpreted Brant Calkin's remarks completely differently than you. I interpreted them as concerning the pattern of growth. For instance, this business about a single neon strip from San Diego along the coast is not just a matter of absolute growth. It is a matter of where that growth goes and what the pattern is. And that is what I am responding to, and I suspect that that was what Mr. Calkin was talking about.

SPEAKER: One sentence of explanation. I moved from Udall through Morton, first as an economist and then as an environmental officer. We made a contract with Denver on the subject of managing growth, in which we looked at various examples. And it is my feeling that we have to do socio-economics and land use analyses and that they are includable in EIS and they are justified.

RAMPTON: Do you want to respond?

CALKIN: I think the gentleman raised a good point there, but it is not clear that patterns and absolute growth are mutually exclusive. If, for example, we have a policy of insulation in the San Diego area, then the effect on an absolute growth would be less, or it might be less. And then on top of that, we would have the pattern that Steve Plotkin brought up.

RAMPTON: As far as absolute growth goes and as far as patterns go, I am of the opinion that you can just do so much and no more. Two years ago this summer, I had a chance to visit mainland China. That is probably the most regimented society that exists in the world, and yet, the leaders there continually complained that while they were trying to hold growth, they were getting growth.

In a place where most people are assigned to a job, they were still getting rural to urban movement. If it cannot be controlled or influenced in a regimented society like China, how much less are we going to be able to do it in a free society?

DREYFUS: When we start choosing tools by which we are going to control growth, we do not want to be simplistic about it. You talk about sewer hookups. If you start out with a zoning practice that says you cannot build a house until you have a sewer hookup, and you do not provide the sewer hookup, nobody builds a house.

If you start out, however, in a primitive society where people build houses and you say, "We are not going to provide the sewer hookups," then you have a public health problem of great consequence but the houses get built anyway.

We are talking about electricity and water. If you want to say nobody can hook up at the downstream end, fine. But that is a local zoning decision, just as the gentleman was talking about. If you say, "We are not going to let you build a power plant at the upstream end," eventually things will get so bad people will move out of Los Angeles. That is the kind of a governmental decision that is unacceptable.

So, we have to examine the system. The point is that if you want to keep people from hooking up and taking electricity out of the Southern California Edison system, that has got to be done in southern California. If you do not have control over those hookups, then to act to frustrate the company's ability to build a generating station is a highly irresponsible way to get at the problem.

DRAKE: That is exactly the issue I was raising.

DREYFUS: I think that is one of the things that troubles me about the methods we now have. We cannot make anybody do the right thing, we can just frustrate them when they try to do the wrong thing. It is a matter of control by pushing on the end of a long rope. It never seems to get us the right solution.

DAVID MYHRA: I want to comment on the point that our friend from EPA raised about power plants and growth. Perhaps he is aware of a study that I am not aware of, but based on my research and site studies and site visits, if a power plant area lacks industrial location features, such as markets, transportation, labor force, and other factors that attract growth, the mere fact that you build a power plant is not going to make it a growth area.

It will still lack transportation. It will still lack the markets. It will still lack the proper labor force for industry or commerce. You can look at the hydro facilities at the Chief Joseph Dam in Washington State, the Green Peter Dam in Oregon. You can look at a number of sites in North Dakota and Montana where power plants have been built. Even Four Corners, for example, has not grown that much as a result of the Four Corners budget.

So, the mere fact that you build a power project does not mean that you will have heavy, intense urbanization.

RAMPTON: We are talking about two different things. You are talking about the plant site, and they are talking about secondary growth at the point where the electricity is delivered.

MYHRA: The point was also raised that some of the sites would attract large quantities of urbanization growth.

RAMPTON: As far as the Kaiparowits site is concerned, we did as good a job of planning as can be done. We had our studies made. We knew what schools were going to cost, roads, sewers, and water. We had means available for financing them. We were not going to have unacceptable growth there as far as the economy of that area, or the society of the area, was concerned.

But what the 3,000 megawatts of electricity dumped into Los Angeles is going to do to future growth, I do not know because I did not regard that as my business. But I did regard what was going to happen in Kane County as my business.

PLOTKIN: I certainly do not disagree with your point. In fact, I guess it is the one I was trying to make earlier when I said that the range of impacts in every area shows a tremendous difference, depending on what the site location or technology is.

I have read your paper and I agree with it. Basically, the socioeconomic impact is very dependent on location. Rock Springs does not have to be duplicated everywhere, and it is not just a question of what the power company does. It is a question of what infrastructure is there and whether the people are willing to work with the company and the power company is willing to work with the people, whether there are roads, and whether there are people who are unemployed, who have the necessary skills to take over some of the jobs.

I think it is up to the federal government to identify what the impacts will be at different sites so that companies can make reasonable choices and environmentalists will know when they are being dealt with in good faith. Your point about variation and impact is correct. You do not have to have a boomtown when you have a power plant.

STUART CARLSON, Federal Energy Administration: From 1968 through about 1973, I was employed by the Bureau of Land Management in Kanab, Utah. I was in charge of their lands and minerals program, which encompasses this whole activity.

The first point I would like to make is that we had some rather unique pressures applied to us as a land management agency in that, although this project was started in the late sixties, it was not until about 1973 that we got Rogers Morton to identify the lead agency for writing an EIS. We were continually challenged on the grounds that, ostensibly, we lacked a national energy policy. We heard this continuously. I would like to say that in this check-and-balance society, we also, in my judgment, lack an environmental policy as well. When we were critical of environmental groups at times, we would ask, "If you are not in favor of Kaiparowits, what are you in favor of?"

The answer was, "We cannot reply to this. We are only in a review capacity." And I have heard you use the same defense in response to questions here. If you would like to remain in a review capacity, that is your own role. But this places a terrible burden upon anybody who is in a land management agency. When industry comes in and by law has a *bona fide* application and wants to do something, we are shotgunned continually, with no solution in sight.

You used the word alternative. I would like to know what the alternative was to Kaiparowits. I am not sure there was one; it is the principle I am talking about.

DREYFUS: As it turned out, there did not need to be an alternative, did there? They walked out of the plant, and I am not aware that they built anything else.

CARLSON: I am just reflecting on the difficulties we would have in a Bureau of Land Management District proceeding.

One other point I would like to make--and the Governor, out of charity, did not get into this--but we were told that we had a national park to defend down there. This has been one of the leading arguments.

Just prior to the submission of the final plans for the Kaiparowits project, the Glen Canyon National Recreation Bill was passed. It addresses Lake Powell and about a million and a half acres surrounding it and it was, like everything in southern Utah, fraught with emotion and dialogue. The bill was sponsored primarily by Senator Frank Church. The wording called for mineral and industrial development compatible with the spirit of the act. The moment it was passed, suddenly, the focus seemed to shift, and now we are defending, to the detriment of the mineral and industrial development. This really bothered us because the Bureau of Land Management happened to have mineral management responsibility within the Glen Canyon national recreation area. I personally doubt if we will ever see any development.

DRAKE: Dan Dreyfus says he is not aware that we built anything else. I cannot speak for the other participants. Subsequent to the Kaiparowits project, we went into the Palo Verde nuclear project, which is being constructed by Arizona entities. We only got in because Tucson Gas and Electric got out.

That is replacing, in part, the power from the Kaiparowits project. We have a project called Lucerne, which is a combined cycle project for a 1984-85 time frame. This is a project that will use distillate, and this project, if the Jackson bill goes through, will be precluded.

If we cannot build that, we will have nothing. One of the principal issues in my mind concerning the Kaiparowits project is that for the next ten to fifteen years, I do not see anything that will meet the energy needs of this country except coal and nuclear power. The Kaiparowits project would have displaced more than 30 million barrels of fuel oil in the Pacific Southwest.

Even at reduced rates of growth, we need something. We can argue about what the rate of growth is. One of the biggest lessons to be learned here from the Kaiparowits project concerns the legal issue. I do not know what the President's energy message is going to contain. Let us assume that it is favorable to coal with reasonable environmental constraints. Let us assume that we will have an all-out national policy to go to coal. Let us say that we get over 300 permits. Let us assume that we interpret the Code of Federal Regulations and other regulations with alacrity.

The fact remains that anybody with a twelve-dollar filing fee can go down and hang up any project for months or years in spite of the best intentions of the executive branch of this government and the good intentions of people in federal, state, and local agencies.

Let us say that the Sierra Club is in favor of the San Juan project. If you are backed up by a team of law students, you can go down and hang up the San Juan project. Maybe it will not be the Sierra Club, but somebody else can do it.

If we want to get free of the OPEC countries, this has got to be changed. This is perhaps the biggest lesson we can learn from the demise of the Kaiparowits project.

RICHARD GREEN, UCLA: I would like to go back to the question, which I think is a fascinating philosophical one, of whether or not there should be local control of land use or whether or not remote sources of power should be used as a device.

We did a five-volume study of energy and growth in the Los Angeles metropolitan area several years ago. At the present, the zone density of population in the city of Los Angeles is about 2½ times anything we could foresee.

The same kind of misjudgments were made in all the surrounding, relatively undeveloped, communities. What does this mean? Planning is fully responsive. All you have is physical planning for details-- who gets the high rise and who gets another kind of development.

You have no control with this kind of planning. When you say local control, it is a farce. I have been an unpaid adviser to air quality maintenance planning activity. Controls as to where growth occurs and how it occurs--the pattern of development, what kind of growth it is--is essential if we are ever to solve basic environmental problems in this large, important metropolitan area.

How do you get control? Through provision of necessary utilities and services. For example, air quality maintenance planning has had one beneficial effect. For the first time, people who do water planning have started to talk to air quality people. I am pleased to hear the EPA spokesman talking about integrated assessment. I believe this is essential in two areas: in the area where you are planning the power facility and in the area where the power is received and utilized. Without this dialogue we are heading for disaster.

RAMPTON: As far as Kaiparowits is concerned we think we planned the impact of the plant site well. The impact on Los Angeles was something I could not assess and would not know anything about. I had to assume, or we in Utah had to assume, that there was going to be a market for the power. If there were no market the companies, presumably, would not build the plant.

You are talking about two different things so far as impact at the plant site is concerned; that has got to be a local decision. That has got to be done by local decisionmakers. That is why I resent the effort of people, who are not competent to make a judgment, entering so largely into the decision-making process there.

CALKIN: Governor, could I ask if you were not giving people in that area the option of making a decision over national interest lands?

RAMPTON: To some extent, yes. I certainly do not quarrel at all with the national involvement in the environmental impact statement. I think the federal government should make it. I am not certain it should be making an economic impact study for the local area. I think that should be more a local decision and determination.

While I respect you as a planner, I do not think you know anything about Kanab, Utah. In fact, I have to give you safe passage every time you go through there.

CALKIN: Governor, that brings up a basic question that we ought to talk a little bit about. All of us have an interest in the national land, but those of us who specialize in a study of it would be willing to concede, if you will, some degradation in that land for really good substantial benefit. We are in the trade-off business like everyone else.

That means we have to be able to see what the trade-off is. To suggest that we should examine environmental concerns and not figure out what the real economic benefits are, means we are trading off in the dark and we must not do that. We are going to back into economics no matter how you look at it. And we are in it to stay.

RAMPTON: I know. But there was a part of that study that infuriated me. It was devoted to the question of whether or not a small, provincial Mormon town could stand the impact of the culture that would be brought in.

I think the guy that has got to make a decision on that is the state president of the Mormon Church. He knows what his people can stand and what they cannot. I doubt that the person who prepared this part of the report had any basis upon which to make his conclusion. I think the Mormon towns can absorb more than some of you people think. I recall that when I first became governor, there was a proposal to put a Job Corps camp in a little town called Milford. The local members of the Church opposed it.

I went to see the president of the Church then, David O. McKay who was ninety years old, to find out whether the Church was opposed to the Job Corps camp. If they were against it, I was planning to tell Sargent Shriver that this was not a healthy place to put the camp because I did not want to bring the Job Corps youths into a hostile community.

But anyway, President McKay said, "No, I think it is a pretty good idea." He said, "Why are the local people opposed? Why are the local Church authorities opposed?" I said, "Well, they are fearful of bringing young men of uncertain background in among the Mormon girls." He thought for a moment. Keep in mind that he is ninety years old. He said, "Oh, I do not think I would worry about that. Unless young men have changed since I was a boy, if there is a local girl inclined to get in trouble, there will probably be a local boy to take care of the situation."

SPEAKER: This discussion has become more a discussion of national policy related to growth and what the impacts of growth may be than of Kaiparowits.

If we can discuss Kaiparowits again for a moment, I wonder whether it is the position of those who opposed the power plant at Kaiparowits that there is no way in which resources in that region can be exploited satisfactorily and compatibly with little harm to the national parks?

Or is it a specific thing? Is it that power plants cannot coexist with parkland? Is this a no-growth region because there are parks there--no growth of any kind, no power plants, no gasification plants, no mines, no what?

CALKIN: I do not think it means no growth. I think it does mean no impacts of certain kinds. It also means that unfavorable impacts are going to have to be balanced by an objective view of what we are getting in return.

It is conceivable that we would simply write off a national park. That is heresy from a Sierra Club member. It is conceivable we would do that if there was a genuine and a real benefit of overwhelming importance. It is possible that in southeast Utah, coal would have a value for some purpose greater than the parks. That might be demonstrated to us.

But we are not going to trade them away cheaply. We are not going to trade them away for uncertain benefits, and we would prefer to look for other uses of the coal or other resources that are more compatible with those values. It is not an absolute ceiling, a lid, a no-growth deal. It means you have to pick and choose carefully, you have to show what we are getting for what we are giving up.

SPEAKER: Can I follow up on that for just a moment? If I understand the answer, the environmentalists who view their role as protectors of the parks or whatever other citizen interests exist would like some quantified benefit to be described for whatever development may go on, and that benefit has to be related in some way to the benefit of the park.

CALKIN: Not necessarily. I did not mean that. The benefit could be elsewhere if it serves a genuine need elsewhere.

SPEAKER: Let us say somebody built a coal mine and shipped the coal out. There is *prima facie* evidence that there is substantial benefit because somebody is buying the coal at the other end. And in the marketplace, which determines whether things are sold--commodities are bought and sold--there is a fairly basic economic law that decides whether that is beneficial compared to a mine somewhere else. The guy at the other end buys coal, and tries to buy the best coal as cheaply as he can buy it. If he can buy it more cheaply and the character of the coal is better for him than somebody else's coal, he will.

What I perceive in this discussion is an attempt to create a new requirement for systems analysis, which goes beyond what anybody has ever claimed for systems analysis, a global analysis that says that one is able to define some measures of gain that society can compare in an abstract way with equally valid measures of penalty. Then gain and penalty should be balanced off so that you can decide whether the gain is worth the penalty.

This is an interesting discussion exercise but not one likely to lead to a reasonable outcome. Perhaps that is why we go twelve years and 300 permits and, at the end of the several hundred permits, decide we have not really learned anything.

DREYFUS: You are touching upon my own basic interest in this whole business, which is a political science interest. The problem that you have got in a lot of these cases of frustration of proposals is this: There is no such thing as a free market that makes people do the right thing in a global sense, even economically, to say nothing of the unmeasurable factors.

The proponents of Kaiparowits did not go in there because it was the world's best place to build a power plant. They went in there because it was the best place they knew of to build a power plant under the circumstances. First of all, the decision was dictated by the air pollution people in southern California. It was influenced by the water rights on the river, which meant, for example, that the Mojave plant in southern Nevada, which is an infinitely better place to have a power plant, was foreclosed. They were out of water at the Mojave site, and there was getting to be a little hostility about additional capacity at that place.

Colorado is hostile to development. They have coal in Colorado and they have water in Colorado. But they were hostile at that time to development. In Utah, development was well thought of. They want this plant in southern Utah. The people that live there want the development. They are comfortable with the idea of the development, and if you give them the full knowledge of what to expect and they still want it, you cannot quarrel with that decision.

The proponents went there for a variety of reasons that did not measure all of the alternatives and that probably did not measure the national concerns with regard to those lands or the national involvement in the region in terms of that laundry list of things I mentioned--the parks and the monuments and the natural resource lands.

The opponents come to this thing with a different viewpoint. They say there are better places to build this power plant, if indeed it must be built, and maybe it does not have to be built. That is the classic argument on all the power plants and nuclear plants in the East and in the West.

First of all, "We do not think you need it." That tends to be one of those nebulous kinds of things. You are dealing with the future and nobody knows what the future is going to bring. "But if you need it, there are better opportunities that you did not look at." There usually are, simply because the decision maker has a limited viewpoint.

Our problem is that our governmental institutions get mixed up in this game, and they take sides. They have an infinite number of ways of screwing things up but almost no way to influence the initial decision on either side.

They cannot tell the Sierra Club when to lay off, and they really cannot tell the power company where to build a power plant. They can just frustrate everybody's attempts in the hopes that sooner or later an accommodation will be met. And that is a very inefficient way from everybody's viewpoint to get at the heart of the matter. Now, what do you do about it? Land use planning, of course, is one thing we have often talked about.

SPEAKER: I would like to make a comment and then ask a question. If you disapprove the Kaiparowits project, but you think that there are other locations where you would allow a plant, once that kilowatt-hour gets into the lines, you cannot prevent growth in southern California because that does not identify one kilowatt-hour from another.

So, either we stop building all plants and limit the growth and movement of people in California, or, if you are allowing construction of a plant anywhere for delivery of power to that area, you lose your ability to control something that has to be controlled locally, the question of where growth in California takes place?

PLOTKIN: I think you are taking my remarks and carrying them much farther than I really intended. I was trying to make the point that, in fact, provision of utilities can control growth. I was not trying to make the point that that is the way we ought to do it.

I think that for some examples, for instance, with sewer lines, perhaps that is the way we ought to do it but not with electricity.

SPEAKER: What I was saying was that the decision of Kaiparowits should not include land management in California. Because if you are going to build a plant in Colorado, then, for California, you will have lost that point.

PLOTKIN: I happen personally to agree with you.

SPEAKER: Then I have a question for Mr. Drake. Have you looked at the Kanab, where there is a great deal of strippable coal, I understand, about fifty to sixty miles west off the Kaiparowits--excellent strippable coal, about half a billion tons. What is the water situation there?

DRAKE: The water consideration is not good. I will let Mr. Calkin talk about that. That could impact the Escalante wilderness area. I am sure, from the Kaiparowits hearings, that the Sierra Club would be opposed. But from an environmental standpoint, I think it is out. I will let Mr. Calkin discuss that. But we in southern California have not tried to make any effort to acquire water in that area.

CALKIN: I suspect that if the plant was desirable in that location for other reasons, we could probably haul the water in there, probably pipe it in. Piping it in is not as optimal from a water point of view as building the plant right next to the puddle. But I suspect that if it were good for every other reason, we would be able to handle that water problem and pipe it in.

Kanab may or may not be a good place. The region is sensitive for a number of reasons, and it is conceivable that you could build a plant. I would say, for example, it is conceivable that you could build a gasification facility there--and I am just speculating-- much more reasonably than you could build a conventional fossil fuel power plant.

DRAKE: Of course, we were using or proposing to use part of the allocation of Utah's share of the Upper Basin--they obtained through the Upper Basin compact, the Colorado River water.

I think Dan Dreyfus, from the national point of view--the Sierra Club versus Morton, or the regional impact study--has addressed the question of where all these plants should be put. I think there is one resource we have not talked about, one that we cannot be unmindful of in this country: the wealth of this country is finite.

We were already 600 miles away from the power markets of southern California by transmission lines. Now, when we talk about piping water around the countryside, when we talk about locating plants in Colorado and shipping power by transmission lines to southern California, we come to the question, how is this going to be financed, by the federal government or others?

There are no free lunches in this world. The rate at which this country is generating wealth has gone down drastically in the last ten years. And we have got to remember that power costs get added on to the cost of doing business in this country, get added on to the householder's bill, as everybody knows by looking at their electric bills.

The cost of oil and energy in our system is half our operating expenses today, and the cost of oil has gone up about 800 percent in the last five or six years.

We have got to remember that this country is running out of wealth. So, when we talk about generating energy by wind or solar energy or locating plants here and there, let us not forget that the economic resources of this country are limited.

SPEAKER: You have given some alternate suggestions for the Kaiparowits facility and you said that each one was better than Kaiparowits except that there was local hostility.

DREYFUS: Not just local hostility. In some cases, legal inabilities. The compact, for example. One can only use that 102,000 acre-feet of water in Utah.

SPEAKER: If a regional study comes up with a site possibility where the benefits are diffuse and far away and the adverse impacts are very local and concentrated, what sort of a suggestion is that to a utility to adopt that site?

DREYFUS: In the world we live in, none at all. Essentially, we only have political responsibility at two levels in this country--at the state level and at the federal level. There is no regional government that can make you do something you do not want to do because there is no electorate at the regional level.

When the state or its local entities say that something shall not happen within the powers that they have, which are infinitely greater really than the federal powers when it comes to land use, that is it. The federal override is getting stronger every time they amend the Clean Air Act. But the federal override is still relatively arms length.

When you get into, as you say, "the national interest," it would be better, for example, to add additional capacity at some existing power plant than to build in this otherwise untouched area. But there is no federal ability to get that done. And a utility, of course, has no ability to get that done. The multiplicity of local entities say, "You will not build a fossil plant in southern California. You will not build a nuclear plant in a great many places. You will not do this. You will not do that." And a siting comes down where it happens to be acceptable.

DRAKE: I think we have gotten to the situation, from the standpoint of balance of trade and national policy, in which there is going to have to be some federal override. So, I suppose in that question, you are quite right. Control of land use, at the present time, is reserved by the states.

The way we are doing it, by acts of Congress, is obtuse, the amendments to the Clean Air Act being an example. But should we get at this more directly by federal preemption of states' rights on land use? I think that is one of the key questions we are dealing with here. What is your opinion?

DREYFUS: My opinion is that we are doing it by indirection. My opinion is that when the secretary of the interior decided not to sign a water contract, he in fact zoned that county of Utah. He simply has not admitted he zoned it.

DRAKE: It was an indirect and inefficient way to do this. Or should there be direct federal preemption?

DREYFUS: That is what this case is all about. It is clearly not efficient. And there is the philosophical question, do the people in southern Utah have a right to choose what happens to those in southern Utah, or is the national interest, as the Sierra Club sees it, involving the federal lands more profound?

That is the problem. We talk about coal and we talk about power plant siting, but the fact is, we have got a political problem here that is bigger than any of this.

SPEAKER: It is not only indirect. It is negative.

DREYFUS: It is always negative. In order for the federal government to have a positive influence over siting, it has got to assert a federal right, which it does not constitutionally have. Otherwise, it can only be negative. It has the ability to prevent, through exercising these very specific powers. It has not got the ability to dictate by exercising broad powers that are reserved from the federal government.

CALKIN: I do not think that is necessarily quite right because the federal government can distribute assets--lands, rights of way, water, or even cash, and--

DREYFUS: It can also bribe, that is true.

SPEAKER: I have two observations. One concerns the environment. I worked on the alternatives in Interior, and we do them on Kaiparowits, and the thing that I remember most is that you were going--I do not know whether it was the Sierra Club--I do not remember who it was-- but, in any case, they were going to let people into a pristine area with their trucks and so forth. And that was quite substantial.

The other observation has to do with utilities. We have come quite a ways. I did participate in the Southwest energy study, which related to several locations--I believe Kaiparowits also--and a North Central survey. You were talking about land use. The utilities headed this North Central survey, and we in Interior and others were going to use all of Gillette's coal and absorb all of the water, and the ranchers and others got upset.

And one of the reasons, as I see it, looking inside of Interior, that the northern Great Plains survey was undertaken was to assure the ranchers. Some of my associates said that some of these ranchers were out with their guns and that the water, the land use, had been upsetting.

I make the observation that sometimes we miss the little guy.

SPEAKER: I would like to follow up on the earlier question that I asked about other uses for the Kaiparowits resource. Now, as I understand the answer that I received, we would have to judge whether the development of that resource was better than equivalent development somewhere else of an equivalent amount of coal.

But I do not know of any mechanism in the United States for anybody to do that other than the market. Now--and I think, really, this is a statement as much as a question to you, Mr. Dreyfus--I find something very ominous in the fact that there is no court of adjudication here that anybody can identify.

When the man who is an advocate gives a list and says, "I have 300 permits that I have to fill out, and when I get all done with that, I have not accomplished anything because somebody else can go and block me," really what I think he is saying is that we have a balance in this society of the gadfly, if you want, or the person who is opposed to something, and the advocate, but we do not have the court that normally makes that decision.

And if there is anything that we really seem to need in order to be able to meet our national objectives, it is a place where the people who are protagonists can go and argue out their case and get an answer, without having to wait twelve years and then decide the game is not worth the candle.

DRAKE: The EIS of course, and the hearings on the EIS, you might say, in theory, preside in that forum. And let us assume that the secretary made a favorable decision after that hearing. The point I am making is we would still face litigation. And that is the problem you are raising.

Even after we have an EIS hearing, we have the government study it, environmentalists study it. Then Interior weighs and makes a decision, favorable or unfavorable. However it may be, you still face the courts and the interpretation on the basis for the decision.

SPEAKER: I think the point just made is rather critical to everything we have said today. It is a fact, I guess, that there is no satisfactory mechanism in existence today for the resolution of these critical questions. And they are all around. It is not just energy. It is recombinant DNA. It is many areas of abstracts--civil rights and so forth.

I hope nobody will take exception to the comment, but the way in which those who have spoken have attacked the Sierra Club and those who speak for the environmental movement--and I do not pretend to speak for anybody--does not make it conducive to development of a mechanism.

To suggest that because the Sierra Club goes to court and exercises its rights under the National Environmental Policy Act, and otherwise, is undemocratic, or that the statements that they make in a litigation are self-serving--that is what we all do in litigation, make self-serving statements, and we make them here-- is not likely to have the Sierra Club do other than Winston Churchill did and say, "We will fight them in the cities and we will fight them in the water and we will fight them in the farms and so forth."

We need to develop a way in which we can speak to each other, in which we can hear Mr. Calkin saying, "I could not get a response from Utah Power and Light." I think that is shocking. I think that is just calculated to make him certain that he is going into court because, as he said, he has lost all confidence in their credibility.

And none of us in this room, I believe, is in bad faith or is lying or is distorting the facts. We are all trying to do something in our own ways. We all have different values that will lead us in the way we go.

If he cannot get an answer from a utility--and he represents a large movement, far larger than any of us in this room, as far as the public is concerned and the newspapers are concerned--then it seems to me something is wrong with the way in which those in industry look at this problem and are treating this.

Now, last night, Mr. Decker, from the Dow Chemical Company, was describing a national coal policy project. I know a little about this, and I understand a great many in industry, including many utilities and possibly Southern California Edison, are involved, and the Sierra Club and many others in the environmental movement are involved. And they are trying to come together in an *ad hoc* kind of way and with no power of any kind to see if there are ways in which they can identify the areas of difference and the areas of compromise and develop methods by which we can, together as a public, come to solutions of these critical problems.

What happens in Kanab is not isolated. It may have an effect on New York City and it may have an effect on Saudi Arabia for all I know. So, we have got to look at these things in a broad way.
And I think what we ought to be doing here, and, I guess, everywhere that we deal with these kinds of questions, is to see what mechanisms we can develop among ourselves--among the intellectual leadership of the country--to make it possible for these widely disparate and divergent views to come together, to at least identify the areas of compromise and the areas of difference, and somehow then present that to the courts, the public, the Congress, and whatever other mechanisms we have in the country for finally making decisions where we can agree with each other.

CALKIN: Before this gentleman spoke, there was another comment to the effect that we do not have any place to go where we can resolve the issues. We may have a very basic disagreement on whether we do or not. It is possible that, some day, we will have a national court of scientific truth, but it is unlikely. But you are suggesting that we are seeking a national court to make political decisions. Resource decisions are not exclusively economic decisions or environmental decisions. They are inherently a part of the democratic process as political decisions.

We accept that. That is just part of the system. That means that sometimes we are adversaries, sometimes we are collaborators. The court is, if you will, the Congress, and the Congress takes a long time to work. In the case of the Kaiparowits thing, as Mr. Drake points out, in a sense, it took twelve years, and they did not like the decision that came down through the executive branch of government twelve years after they began considering a project.

But I have a much easier time, quite frankly, grappling with the resource allocation decisions, the cutting and fitting of where is the best place to do this and that, than I do with the perhaps more profound implications of who is going to make them.

Now, you mentioned that we needed a place where we could figure out what the national objectives were. I think the place where the national objectives are determined, the place where the objectives that are mirrored in the Kaiparowits controversy were made, is quite frankly, the Congress.

Just as an example, in 1971, at a Senate Interior meeting in Albuquerque, I said--and it was not an original idea at that time-- I said in 1971, on a hearing on Southwestern power plants, we need some kind of a national energy policy, not a perfect policy, but something.

And I am embarrassed that six years later, I have not been successful in convincing people we need it done. We really do not lack the mechanism. The mechanism has been built in as part of it-- disagreement, agreement, confusion, and so on. The Big Brother concept frightens me in many ways and I, quite frankly, got a little chilly when I heard you talking about a national court where we determine what the national objectives are. That scares the hell out of me.

DREYFUS: I certainly would agree with Brant Calkin's remarks. We do need a national energy policy and maybe we will get one on April twentieth. Maybe we will not. But whether it is Kaiparowits, or San Juan, or coal gasification, we have got to get behind something in this country and get behind it pretty quickly. We cannot tolerate the debate, the delays, and the litigation that this project went through.

One thing we ought to keep in mind in all this is that this is a transitional problem. Think back and put yourself in the frame of mind that the country was in in 1968, as compared to where it was in 1973 with regard to the whole environmental question. Two very important things happened in between these times.

First of all, we developed one or two national policies. One of them was a policy of public involvement in decision making, both governmental decision making and industrial decision making that did not exist before. We have new things like the environmental impact statement. We have information now going out to the public that did not get out before for several reasons.

Some of this information was not compiled before because it is very expensive and nobody thought that it needed to be compiled. Second, some was not published because, quite frankly, nobody was concerned and the media did not present the controversies over these kinds of developments in the early sixties.

So, here is a project where the early planning and the original decision making took place in one situation and the final decisions took place in another situation. It is calculated to be one of the most difficult ones. It is probable that a similar decision process starting today would be far more sophisticated at the outset. By the time you got to the hard fight, it may well be that people would have thought about it.

Maybe this is a naive judgment, but I really believe that the proponents of Kaiparowits found themselves trying to address questions of alternatives, questions of growth, and questions of impact at the high point of this controversy that had never entered their minds when they first picked the site.

As a result, to a certain extent, they were rationalizing. They had to be rationalizing. They could simply never have foreseen what they were going to go through. I think that for the next thousand-megawatt plant that they enter into, they will foresee this kind of thing and at least will be able to defend the proposal better. They may possibly choose the site better, but certainly, at least be able to point out that they tried to find alternatives and could not.

DRAKE: I do not disagree with you. I would like your suggestions on where that next thousand-megawatt plant is going, outside of Colorado.

SPEAKER: But you have no certainty at all that it will not be banned up to another ten or twelve years because, by then, we will find some trace elements in the coal that need to be banned, or sulfates will be bad once they get a good program going, or something else is going to happen that is going to prohibit the next plant anyway. So, you never have that assurance.

I think some comment was made that you tried to contact the Utah Power utility and there was no comment. It really is disturbing. I have some questions--a two-part question. Number one, have you known whether your organization or any other organization contacted the utility to make a constructive criticism?

DRAKE: Can I speak to that, just a moment, on Utah power? I would like to make a point--Utah Power and Light was never a participant in this project. They never put the 20 million bucks into this project. They were not responsible for preparing the EIS. They were contemplating, as Brant Calkin says, possibly getting into the project.

So if we are talking about Brant Calkin talking to Utah Power and Light about the environmental concerns, the commitments that were being made, Utah Power and Light had no business answering them even if they knew, because they were not a participant in this project and they did not put one dime into it.

SPEAKER: Then they really should have said that.

DRAKE: Maybe so. I think Utah Power and Light was playing, frankly, a waiting game as to whether they would get into this project, or go on using coal resources in Huntington Canyon, or use some coal leases they had farther north. But they never were in the project.

CALKIN: I think the example of that correspondence and those inquiries is unnecessarily harsh. I did not bring that up on my own. It was in response to a question. And I would like to tell you quite frankly, if I called Jerry Geist of the Public Service Company of New Mexico, I could ask him anything and get a straight answer. It may not be the one I want, it may not be the one that tells me what I want to know, but it will be Jerry's best information. And I quite frankly do not have any doubt that if I call Mr. Drake, I will get it. We have both grown a little bit since that time.

We still have occasions where we do not communicate very well, but I think we are in a much better position since Kaiparowits on exchanging very frank views on that kind of thing than we were before. And I would prefer, quite frankly, to leave that as an anomaly, a very irritating one. There is a little phrase that runs around in my outfit: "Nothing sustains us like the arrogance of our enemies."

But be that as it may, I do not think that is symptomatic of the whole problem and it is just a minor thing in my estimation.

SPEAKER: The second part of my question is, have you ever known of a situation where you made a constructive comment?

CALKIN: Yes. And again, let me go back to the Public Service Company of New Mexico. We sat down with their engineers and said, "Listen"-- and this was partly on their initiative too--"if you want to expand San Juan, you have got to do better than they are doing across the way and because they are already there, you have got to do better than you are already doing with your first 345-megawatt unit. So, how about this, that, and this?" We worked out sort of a negotiated settlement with what we hoped were our good ideas and theirs. We went before the EPA and requested a variance, or supported their variance, and so, given the right occasion, we can do it.

Now, there was another comment that came from Mr. Weinberg that I would like to talk about, and that is working together on things. Again, using my local example, New Mexico has a strip-mining law and it is a pretty good one, although, on its face, by environmental standards, it is a turkey. There are no standards in the bill.

But we got together with the coal industry, before the legislature, and hammered out what we thought was a fair bill, and we had faith in the local company officials that they would stand by it. We offered the bill jointly to the New Mexico legislature as the product of the New Mexico Mining Association and the Sierra Club--an un-holier alliance, there never was.

It went through every committee of both houses to the governor's desk without a single dissenting vote. And today, I think, that is a relatively effective bill. It resulted, quite frankly, from some growth on both sides.

Now what I am concerned about is, having had Kaiparowits, what growth are we going to have?

DRAKE: Where, not what.

CALKIN: I mean, what growth between us--between the utility companies and environmentalists and local interests. Is it going to remain a classic example of a confrontation that somebody won or lost? It may. I hope for something better.

SPEAKER: You mentioned, Mr. Dreyfus, that this was a transition project. It started in one era and wound up in another era. Something bothers me. How do we know that the projects, the ideas that get started now, will not wind up in a different era?

DREYFUS: We never know that. We never know about tomorrow, let alone about eras. But in retrospect, it is very clear that a lot of the complications on this whole operation were transitional. As I said before, there were several power plants built and a complex of power plants led to a lot of interest in the region. And this one plant got caught.

You take the Navajo plant at Page. Very many of the same things that are said about Kaiparowits could have been said about Page. But by the time the federal government began to reevaluate whether or not it had a comprehensive role in this regional development of a power plant complex, the plant at Page was a commitment.

Furthermore, there was a federal involvement in Page that kind of locked them in. They were one of the power companies, for example. The Department of the Interior was one of the participants. And so, Kaiparowits is the one that took the rap for the whole developmental scheme.

It is a little unlikely that you will get this same kind of situation again.

Something similar occurred in the nuclear business with Calvert Cliffs power plant. Calvert Cliffs got more grief than most plants have since because it happened to be in progress at the threshold, when the nuclear people were brought into the environmental policy arena. These kinds of things are exceedingly painful. I am not saying that if you started the plant at Kaiparowits, today, it would go through. What I think I am saying is you would not plan Kaiparowits today. I do not know whether Mr. Drake would agree with me. But I do not think you would plan Kaiparowits today at all. And, yet, when it was planned, nobody saw the problem. It looked pretty good.

SPEAKER: I do not think many people realize that the Bureau of Reclamation in Salt Lake City did a regional assessment in San Juan, El

Paso, Wesco, and APS. Somewhat after the fact, but at least at their expense, they did a regional assessment. They say the same words in all four of them.

What Brant Calkin said earlier about APS--it had to be done because they had 70 percent ambient and were going to hold it until that was done. This brings in the point of integrated assessment. This was a very primitive, early beginning of that kind of thing. It shows up in the EISs that were developed.

SPEAKER: We hear frequently about environmental impact statements. Does anyone make an economic impact statement on the cost of delays, the cost of all these environmental impact statements, the cost of moving a plant fifty miles in one direction and then paying all that money to pump water for the next twenty years? Is this up to the Congress?

DREYFUS: There is no general requirement to do it, except, as you may know, there have been requirements for economic impact statements written into a few special programs. Essentially, it is not a national requirement as it is with the EIS.

On the other hand, I do not ever recall that economic information was not made public. Look at the philosophy of the EIS when we first got into it. The proponent of a project, after all, is using a federal service or a federal resource. If the federal government is out of it, there is no EIS.

The National Environmental Policy Act only captures you when you are doing business with the federal government. So, you presumably come to the federal government with a proposal. Now the assumption is, you know why you have got the proposal. The assumption is, you are an advocate and you are going to make statements justifying that proposal.

The problem was, nobody was making statements about the adverse environmental impacts. The proponents were not apt to go out and do that on their own, and the federal agencies were not required or responsible to do it in any but the most parochial cases.

Of course, if you wanted to put something in a national park, the Park Service had a responsibility to protect the park, but the BLM did not have a similar responsibility to describe the broad range of environmental impacts upon public lands. The act simply said, if you are going to have a proposal and if the federal government is going to act, whether it be a federal proposal or somebody else's proposal that involves a federal action, the federal agency has got a responsibility to describe to the public what the environmental impact is. That is all NEPA does.

From there on out, there is the question of litigation, which generally turns upon whether the EIS has been adequately done or not. Regarding a good many of the delays, you know, reasonable people would agree, the EIS has not been adequately done. In the early days especially they were not.

Some of them are adequately done. Then you get into other aspects of litigation. You get into clean air litigation, a lot of peripheral litigation. It gets to be a war of attrition.

All I can say about that war of attrition is that these are, after all, vital public services. When an individual comes forth and says, "I want to build a power plant," you say, "I question whether you need it," and you argue that. And then you question whether his site is the best place to put it and you run him out.

If in fact he does need it, sooner or later, that will become evident to the public in a very painful way. In our political system, at that point, the war of attrition goes away.

SPEAKER: Yes, but then you have a ten-year period in which you are in a bad situation.

DREYFUS: Exactly. There is the question of the lead time. There is the question of the unnecessary crisis. It is a cybernetic system, and it swings back and forth. It would be nice if we could dampen the swings a little bit, but that requires reason on both sides. We are now in the worst swing. I hope it will not go back all the way the other way and that reason will prevail.

I think that we will leave that up to Brant Calkin in his closing remarks. Perhaps he would like to express how the environmental community sees this, whether they think this is the final swing and whether they see the necessity for responsibility and leadership in terms of the real developmental needs.

DRAKE: I would just like to comment on the alternative costs in the case, assuming the power project was going on the Kaiparowits Plateau. I mentioned two sites. One was Nipple Bench. The other was Four Mile Bench, approximately fifteen miles further from the lake.

We certainly brought to the attention of Interior the difference in cost between these two sites. It was about 120 million dollars first cost, about 240 million over the life of the project if you present worth the cost of pumping power over the life of the project --pumping power for water.

The principal difference in the two sites cost-wise was the additional transmission line distance and the fact that we had to convey some forty-four thousand acre-feet of water another fifteen miles and pump it.

SPEAKER: Where do we go from here? The growth in electricity demand has certainly slackened from the sixties to the seventies, but I do not think that, first of all, it is really completely clear. There is a lot of uncertainty. It could certainly increase back into the 7- or 8-percent range.

Are the mechanisms going to be available so that these discussions on engineering lead times can be collapsed and so that there will be the flexibility to respond to the growth, whatever that growth is going to be? I think the engineering lead times are long enough to make it a critical question.

DREYFUS: That is a good summary question--for everybody to say where we do go from here.

CALKIN: If we assume that it is a pendulum motion, I think we have an assumption of the worst arrangement. I would look at a more continuing process.

In environmental concerns, the ones that were exemplified at Kaiparowits are not going to go away, but they can be ameliorated. The environmental impact statement process can be improved, and I have worked with the Bureau of Land Management to help shorten it.

But the thing that is installed in government and in the public mind is that we have to make decisions like this better and more clearly. It is too much to expect that we are going to be perfect at it or that the system is going to clear that crystal ball totally. We are just going to keep working on it and try to avoid things like Kaiparowits.

As I said, we do not need another Kaiparowits experience. And I do not think we are going to be in a pendulum situation, where now we will go back and forth and hit those peaks again. We will damp them as we go down.

I think the basic question of energy growth, national goals, and some of the other things that were very basic to the discussion today remain less clear than the site specific controversies that we had at Kaiparowits. This leads me to think that there will be others like it although, I hope, not so extreme.

That is not a good way to make the change productive in my estimation. So, we are still going to pursue some more generic view of what is appropriate. I hope we will pursue an energy policy. Kaiparowits some day may fit in that. But right now, we are all running in the dark, and it is not surprising that we collided on some relatively remote plateau in southeast Utah.

The next time, the stakes could be a lot higher, and I hope that we do not have to resolve it the same way.

DRAKE: I think I have largely answered my single biggest concern. Personally, I do not feel that we are going to experience 7- to 8-percent growth in our territory. We are going to experience some growth, even if not another customer moves on our lines, even if there is an embargo on new electrical connections. We are going to experience some growth, or the industry in our area is going to get throttled.

The question of how much growth certainly is a forward question at the present time. I do not speak for my company or the other Kaiparowits participants, but we should address ourselves to these legal issues, we should address ourselves to the fact that maybe the San Juan project is a beauty, but--I still say--anybody with a twelve-dollar filing fee can hold that one up. I think, frankly, we are getting into a position here where maybe some of the states' rights are going to have to be, I will say, eroded, and possibly we are entering a situation where power plant siting--and I say this regretfully, coming from a private segment of the industry--is going to have to be handled by an act of Congress to forestall all these legal issues. Either that, or we are going to face interminable delays, which I do not think are in the national interest of this country.

PLOTKIN: I guess when a utility today goes about making its siting choice for a new electric power station or synthetic fuel plant or whatever, although it would like to incorporate all of the variables that affect that plant and that ought to, ideally, affect that siting decision, I think most utilities are forced deliberately to put blinders on their siting choice.

I think it becomes almost just a matter of the resources available to them. And given factors such as the availability of water, the length of transmission lines, the cost of coal, those are the crucial factors to the economics of their generation of power, and those variables always get factored into their siting decisions.

The other variables, the variables that control the externalities, the environmental impacts, and the socio-economic impacts and, to a very real extent, also control the feasibility of that plant because they control the amount of political opposition that may arise to that plant--very often, the utilities are in a very poor position to factor those variables into their siting decisions.

Certainly, many of the power companies now are beginning to do that. The environmental impact statement process is forcing them to do that. But I still think there is a tremendous information gap, and I do not think the individual companies are in a position to fill that information gap.

I summarize by just saying that the only way that information gap can get filled, that sites can be identified appropriately, is for a larger agency--and I would say that it has got to be the federal government--to assist in that process. And that assistance has got to come in the form of what I call integrated assessments.

Those assessments not only assist the power companies, they also assist the environmentalists because the environmentalists are in the same very, very difficult position. The nation is asking the environmental movement to make relative, not absolute, choices. Given that every site has an impact, every power facility has an impact, we are asking the environmentalists to be reasonable about accepting impacts in exchange for power, in exchange for energy.

But the environmentalists are in a very poor position to make those relative choices because they do not know what the alternatives are. I mean, they are working in a vacuum, and unless we can fill that information vacuum, we are in a lot of trouble.

Although I see the Energy Research and Development Administration embarking on integrated assessments of a sort, and EPA embarking on integrated assessments of a sort, I really think that the magnitude of those assessments falls so short of what is actually needed that we are going to fail unless we pump more resources into the effort.

DREYFUS: To face up to some of the questions of what we should do and what might come next, I probably ought to trot out the land use bill. My boss, Senator Jackson, is responsible for some of the problems, I think, in terms of the fact that he was a driving force on the National Environmental Policy Act and several other colorful acts involved in this issue.

He thought he had the answer. Recognizing that what you really have here are these dialogues that take place after the fact with fragmentary evidence, he believed that the way to do this was to force the dialogue to take place ahead of the curve--look at and anticipate the growth pattern and then, having decided what you want to do, use that comprehensive decision as a basis for approving something like a specific power plant.

He had a National Land Use Policy Bill, which was supposed to provoke and coerce and entice the states--and also finance them-- into doing rather broad-scope land use planning. The planning process presumably would provoke these public arguments as to how big Los Angeles should be, after which that would no longer be a factor in the question of whether or not they need electricity. If it is going to be bigger, it needs electricity. Then, you are down to options.

The land use bill failed in the Congress several times. There apparently is not a body of public opinion that favors the federal government having a broader hand in the business of planning land use. Land use is viewed in this country as being a very intimate matter of private property rights. Nobody wants it dealt with much further away than the county courthouse. Why, I am not sure, but it is clearly a pervasive American feeling.

And it is pretty easy to kill the land use planning concept by appealing to the public's distrust of decisions being made in the abstract many thousands of miles away, decisions that will affect them.

Essentially, what we intended to do in this country was to manage environmental impacts, not prohibit them. Now, to a great extent, we have come around through many environmental laws to the question of prohibition. The pending Clean Air Act amendment is the latest one, and therefore is the best example, and it very clearly intends to prohibit, not manage, environmental impacts.

There are concessions made in the near term, but essentially, it prohibits, not manages. This is the kind of concept which may not be physically possible of achievement, and there may have to be a pendulum swing to some acceptance of environmental degradation because, technologically, I think, we cannot prohibit further environmental impacts.

One of the things that might mitigate this decision-making problem is experience. The huge economic cost of this frustration is largely brought about by the fact that the financial commitments are being made while the dialogue is going on. I think that, given foresight, and once we get into the mode, it is possible for the business and financial communities to involve themselves in the decision-making dialogue before they start investing in concrete.

This does not necessarily contemplate a shorter process. Maybe it contemplates a ten-year process in which five years are spent in discussion and in making the choice and then the investment is made. So, you are not paying interest for the whole ten years, and you have the flexibility to change your plans for the first five years. You do not have these benefits when you are committed early to the proposal.

I think it is undeniable that we are going to do less, and talk about it a lot more in the future. We are already doing that. Whatever growth takes place in this country in the future is going to be a lot more expensive than it has been in the past.

In the past, we have in fact, invested resources in rapid growth and postponed the environmental bill. And now, the bill has come due in several respects, and it has proven to be a fairly heavy burden. It has placed, I think, a good many constraints on the aspirations of the average American already. There are a lot of people not going to college today who thought they were going to go. There are a lot of people not buying single-family homes who thought they were going to. These people in effect are now paying the bill for those who spent the resources earlier and deferred part of the total cost.

This is a philosophical question. Can you go back to deferring the charges for somebody else, or can you not? Do you want to? How much of a burden can you pick up all at once?

I hope this Forum will answer those questions because I will not be here when they write up the final results.

SUMMARY REPORT TO THE PLENARY SESSION

Calvin L. Rampton

Mr. Chairman, I am sorry to have to report that although we tried very hard, we never did reach the place of perfect agreement and understanding. There were certain areas, however, in which there was considerable unanimity. We didn't attempt a full-scale post mortem of Kaiparowits beyond trying to look at what went wrong and what could we do in the future to avoid a situation in which after a long period of planning the decision would finally be made not to go ahead.

We discussed the question of what is the government's role in planning for future coal-fired plants, both the role of the local government and that of the national government. The principal problem, or one of the principal problems in regard to the decision-making procedure and in regard to Kaiparowits, as I think we almost all saw it, was that the thing was being decided in a vacuum, it was being decided on Kaiparowits alone. We had no figures that could be agreed upon as to marketing needs and need for electricity. We had no other project or projects being considered alternative to Kaiparowits. We merely had Kaiparowits there, with the companies themselves, in view of changing conditions, having to scale down their estimates as to growth and demand over a period of years.

Mr. Calkin, who is president of the Sierra Club, indicated that his organization was willing to support some coal-fired projects and, in fact, was supporting a particular coal-fired project at the present time.

We talked about the question of the assumptions that we have got to make in order to determine whether or not projects of considerable magnitude can go ahead. We have got to make a determination as to what our policy is going to be if, indeed, the government should make a policy as to growth. Here we broke down growth into two segments: one, the growth side of the plant itself, that is, in terms of the new towns, the boomtowns, and so forth, and, second, an independent problem of growth, where the electricity is going to be used. Should we in our national planning or even in our local planning attempt to use the utilities, the availability of utilities, whether it be electrical power, water, sewers, gas, or what, as a means of controlling growth? Or have we got to assume that the growth is going to be there and attempt to meet it? On this we got very little agreement among the members of the panel.

I had to leave unexpectedly during the panel, so I am going to ask Dan Dreyfus to report on what happened the last hour after I left.

DREYFUS: I think the Kaiparowits discussion was probably different than the Ohio River discussion in the sense that if I summarize it correctly, we were discussing a land use problem at Kaiparowits rather than a regional energy development problem. The Kaiparowits site was a specific proposal. The proposers came in with a set of criteria upon which they had selected the site. The critics, who came somewhat later, and as a matter of fact after a watershed in environmental policy, approached the site with an entirely different set of criteria, and I think the fundamental discussion toward the end of the session was that there really is no forum today in which these two viewpoints could be brought together and one specific authority could make a ruling. The local entities helped to some extent to form the opinion to go into Kaiparowits by barring development elsewhere and inviting development at Kaiparowits. The federal government ultimately blocked the power plant for an entirely different set of reasons. Yet, there really is no mechanism by which all of these authorities or by which any supreme authority can listen to all the arguments, balance and weigh all of the considerations, and come to a conclusion.

So, essentially, I think we concluded that we have a decision-making crisis here which is somewhat larger than the question of using coal.

RAMPTON: Brant, are you here in the audience? Would Brant Calkin, president of the Sierra Club, step to the microphone and supplement what Dan Dreyfus and I have had to say?

CALKIN: I would just like to point out that I thought the discussion was very fruitful, and there was a considerable amount of agreement on the deficiencies of the present decision-making process. The site-specific characteristics of Kaiparowits led us into a much broader discussion of how we can handle these sorts of issues in the future, and I hope that is the benefit to be derived.

It was unfortunate that Governor Rampton wasn't there for the whole time, but quite frankly and fairly, Governor, we did try to consider what the benefits to Utah are, the complications of public land issues when they are mixed with energy issues, and the issue of growth. All in all, I think it was a relatively cerebral approach to a site-specific issue, and I appreciated very much being in it.

RAMPTON: I would like to ask Mr. Drake, who is a vice-president of one of the companies sponsoring Kaiparowits, to give his impression of the panel.

DRAKE: There is one thing that we did discuss that hasn't been mentioned either by you, Governor, or Brant Calkin of the Sierra Club. Even had the Sierra Club or any other powerful environmental organization or a combination of them, like the Audubon Society and the Sierra Club, been in favor of this project at this location, or a coal project at a different location, we still faced the fact that through the courts and through the interpretation of environmental law at the federal, state, and local level, a project like this could be hung up for two or three years. As a matter of fact there was, at the closing phase of the project, a suit brought by an outfit called SNELL, Save Needed Environmental Levels League, some obscure outfit in Arizona, which, as far as I know, had nothing to do with the Sierra Club, had no liaison at all with the Sierra Club, and seemed to be disgruntled with the Arizona Public Service Company. They attacked the Kaiparowits project, the environmental impact statement, and all the federal agencies involved with input into that statement on the grounds that the statement, in spite of being 2,600 pages long and occupying six volumes, was deficient under the statute promulgated by the National Environmental Protection Act of 1969.

So my conclusion, which I think is an important one, is that even if the executive branch of the government cooperates, even if all the bureaucracy that the General has referred to and the thousands of people that get involved in these projects try to put a project through, even if you have the major environmental interests behind a particular project, and even if you have adequate dialogue, the facts of the matter are that under the law anybody with a twelve-dollar filing fee can hang a project up for months, if not years. If he hasn't got the resources he can get them out of some law students, out of some college someplace like Provo, and do it.

RAMPTON: I agree with Mr. Drake that that has been happening, but it doesn't have to happen because the Federal Rules of Procedure now and for a long time past have provided that the court, the court of original jurisdiction, can require a substantial bond to get a temporary restraining order and injunction. Just why they haven't been doing it I don't know. I agree with you that the courts have let those who have challenged projects go ahead, even with frivolous lawsuits. Mr. Calkin, I am not calling your lawsuit against Central Utah project, the Kaiparowits project, frivolous, but there are frivolous lawsuits brought, and the courts let them go ahead.

Now, last year as governor, I was faced with a bill that our legislature passed that said any time an environmental challenge was brought against a project, the court had to demand a restraining order. I had to veto the bill not because I didn't agree with it, but because they singled out environmental causes, which I regard as an unconstitutional classification.

But I think certainly the courts, and particularly our federal district courts where most of these things occur, have got to look a little more critically at the nature of a lawsuit, the substance of a lawsuit. If in fact it appears that there is not great substance the courts should require not just a nominal bond but a substantial bond that can really compensate the company for the delay if in fact it ultimately is determined that it is a frivolous lawsuit.

PLOTKIN: I think the thing that struck me most about our meeting was how ambivalent all of us on the panel were--and perhaps the people in the audience were also--with regard to the nature of the whole decision-making process that took place at Kaiparowits. That is, everybody seems to be making decisions for other people. I think all of us have a hard time living with putting that into perspective. For instance, the developer came in and talked to the local people and generally came to an agreement that affected their lives most of all, and then the outsiders and the Sierra Club basically came in as an outside organization, fighting the power plant.

Mr. Drake was obviously very distressed at the fact--and I believe you were also, Governor--that the local people did not have the opportunity to make the decisions that affected the way their own land was used.

RAMPTON: I would say Mr. Drake had a right to be distressed. The price of that project was going up $6 million a month, for every month that that was delayed.

PLOTKIN: I am making no value judgments. I think that looking at it from a number of different perspectives, the Sierra Club came in and was concerned with the impact that the decision made by local people and by the developer would have on national lands, which belong to the American people. It was a problem of just who has the right to make these decisions that impact people at every level of government and of geographical area, in terms of the local people, the state, the region, the nation. There was some thought that perhaps it ought to be made by Congress, yet Congress was obviously not willing to grapple with it. It took them twelve years, really, to reach any decisions, and then really the decision was made on an economic basis only because of that twelve-year period.

I think we really are in trouble in our energy policy, and I think you could reach that conclusion from our panel, just with regard to who is going to make these decisions.

WEINBERG: I would say from my observations of discussions of this sort, and I have been involved with them more or less for many years now, that that somehow seems to me to be one of the most central questions, a question I guess that almost goes to the very heart of how our democratic system is evolving. There are some who shake their heads and say that our democratic system may actually be being subverted in the sense that the question of how power is legitimated is being questioned in a very, very basic and subtle way. My own feeling about it is that according to the way in which our democratic system was conceived, we had a way of legitimating power --the Constitution, the ballot box, and so on. Now we see trends that as far as I am concerned tend to raise real questions as to whether the methods that our founding fathers had established for legitimating power are in fact appropriate, effective, and sufficient methods for dealing with the issues that face us now.

Now, if the panel will allow me to continue for a bit to speculate, my own feeling about it is that it goes back to a problem of technology assessment that I don't think our Office of Technology Assessment or in fact any of our technology assessors have really gone into fully. It has to do with the question of the side effects of the tremendous technological developments in the means of communication.

My own feeling about the matter--and this is a personal feeling, but one which I have often come back to--is that when we invented the transistor and when we invented the electron gun, when we invent laser communciation--we probably ought to try to do some kind of technology assessment that would get very much to the heart of the question of how you legitimate power in a modern democracy, with its tremendously powerful means of communication.

RAMPTON: Mr. Chairman, I don't care how good your means of communication, your techniques, are. If everybody is talking and nobody is listening, it doesn't matter what kind of wire you have between you. Nobody is going to learn anything.

That is our problem up to now, not means of communication but an assignment of a decision-making role, letting the decision makers in the various areas make decisions with the proper input, not having everybody make a speech and nobody resolve the problem. That is what happened at Kaiparowits.

WEINBERG: I guess my only point is--and again this is a personal intrusion on the part of the chairman--that I think the new technologies of communication make exactly that possible. Everybody can talk now, and nobody has the time to listen.

INQUIRIES FROM THE PANEL

Having gone beyond my prerogative as a chairman, I will now turn to the Panel of Inquiry to ask whether they would like to put questions to Governor Rampton.

BOYD: I wanted to ask two questions that I think are pertinent, but let me make one statement first, that the science of technology assessment is just as infantile as the science of communications may seem to you at the moment, Mr. Chairman.

But there are two questions I would like to ask. When you are dealing with this kind of a question, as you did at Kaiparowits, do you have difficulty in the coordination between the federal departments? Is there any mechanism by which you get the federal departments coordinated, or do they operate entirely separately, and do you have any suggestions?

RAMPTON: They operate not only separately from the state government but separately from each other.

BOYD: The latter point is what I am talking about. Do you have any thought about how coordination can be done, and can the proposals for the reorganization of the government in the energy field accomplish that as you see them in the public field today?

RAMPTON: Yes, it can, I believe. I suppose that it has to be the President that will do it, or maybe it can be done at cabinet level. Coordination can be accomplished if a lead agency will be very early indicated that is supposed to carry the thing for all federal government agencies.

In regard to Kaiparowits, up to that time, not only every department, but every division within the Department of the Interior, was going its own way on the Kaiparowits and the EIS. Of course, what we need is some sort of a national energy policy. Those of us in state and local government, if we know that there is a national policy upon which we can depend, and know that it won't be changed on us from month to month or week to week, can make the local decisions consistent with that national policy. But we have had no national policy, so it is awfully hard to make those decisions that can and should be made on the state or local governmental basis.

ALLEN: At the risk of putting the chairman back in the spotlight, I happen to know that our chairman is very much intrigued by the idea of concentrating power into power parks. So my question to the panel is this: Kaiparowits was big; it was interstate. Are there any lessons here that would say, perhaps, that the inability to decide things is much greater and much more of a hurdle with a large interstate nuclear power park project than it would be, let us say, if the Californians had come and taken a small amount of coal home to just a little old neighborhood park near Los Angeles and built a power plant near Los Angeles.

RAMPTON: In the first place I don't know whether we would have permitted that in the state of Utah. We want the industrial base as well as to supply the raw materials, so there would have been considerable resistance to that. But the bigger the project, the wider spread it is. The bigger the problem, the more difficult the decision. There is no question about that.

ROSE: My impression of the western lands is that they tend to be very fragile. I am reminded of the writings of John Wesley Powell, the first man to travel down the Colorado River, after whom they named Lake Powell. If he were alive today, I think he would turn over in his grave at the thought of it. But what do we learn from Kaiparowits and other plants about the limits to the use of western lands? We hear about development and industrialization. Are there some limits, and can we put it in a broader framework, or does each of these cases have to be argued just by itself on some kind of marginal cost-benefit and by the usual theory of the tragedy of the commons? We will end up then in more trouble than before.

RAMPTON: Of course, what you say is true--some of the western lands are fragile, and it is the result of the fact that it is dry out there, and we have a great deal of natural background so far as pollution goes. There are some areas, and Kaiparowits was one of them, that wouldn't have met the EPA standards on the day Christ was born. So you can't believe that it is going to meet them now. The question is, how much can you load on the natural background there?

As far as the fragile ecology, yes, there are some areas there, but you would have to try awfully hard to hurt the top of that plateau very much. The good Lord did as bad by that as could be done by man, so I just wouldn't worry about that.

HOUTHAKKER: I would like to raise one of those questions that is not going to make me any more popular around here, but I suppose it is one of the functions of a panel of inquiry, and this has to do with the effects of the environmental movement on our present energy situation. The Kaiparowits outcome, if you call it that, was just one of a long string of cases where as a result of court actions of various kinds, certain energy projects were either abandoned or were long postponed.

Are the various groups that exercise their unquestioned right to undertake court actions happy with the outcome of these actions? The fact that the price of oil went up by so much has led to further desires to increase domestic energy production, which I believe to be contrary to the underlying wishes of the environmental groups, with whom I have a lot of sympathy. I am also an environmentalist in my own way, but I believe that by adopting this piecemeal approach, in which one project is knocked off after another or at least postponed by several years, the net effect of all these actions is to make us more and more dependent on precisely the kind of production that I believe most of the proponents of these court actions want to avoid.

RAMPTON: I am going to ask Brant Calkin if he will answer that. First, I would like to say, however, that although I have sounded critical of some environmental movements, I certainly feel that the contribution that has been made to this country, particularly in the last decade, by environmental groups, and even the most militant environmental groups, has been of great value. I know that sometimes some of the things they have done have been for shock purposes.

I think Mr. Calkin said as we neared the end that he doesn't want to see a pendulum motion. He doesn't want to see it swing up and back. Just let go of it, Brant, so it can get down to the middle.

CALKIN: We are not happy with the process of opposing some plants and not others. It puts us in a position of being nay sayers, which none of us like to be and which is dangerous politically, if you will. It is the result of a lack of adequate considerations elsewhere in the process. We would like never to go to court.

One thing that Governor Rampton and I, for example, agree on is that we both personally requested a national energy policy years before the Kaiparowits decision. And what happens is you find that, generally speaking, it is all right to preach, but nothing happens until you meddle. We don't like it. We have been unable to get acceptance for many things that are now agreed upon by all to have been necessary because we went out and meddled. We don't like it.

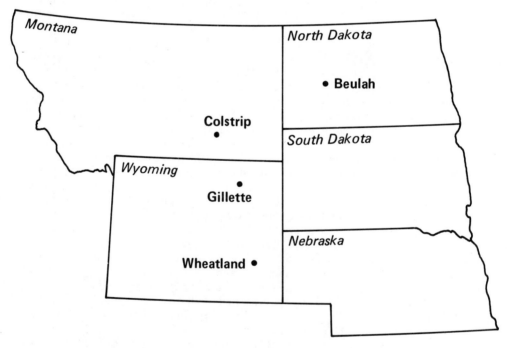

NORTHERN GREAT PLAINS: Location of impacted communities discussed by the Forum.

NORTHERN GREAT PLAINS

ORIENTATION

L. John Hoover, *Chairman*

Assistant Director, Energy and Environmental Systems Division, Argonne National Laboratory

I am not going to give you an overview of the northern Great Plains and its role in solving our coal problems. I think many of you are probably quite familiar with this region's large reserve of low-sulfur coal and of the potential role that the northern Great Plains can play in meeting our energy needs.

I would like to briefly tell you something about who our discussants are and what our focus will be. We have brought together a group of discussants who have pragmatic experience in dealing with coal conflict in the northern Great Plains. These are representatives from state and local government and industry, an individual who is a consultant and educator and who has had experience with the Indian tribes, and a representative from a regional agency.

Our focus is on social, economic, governmental, and political issues. While issues such as water resource shortages and reclamation are indeed important, we felt that we had to have a narrower focus because of limited time. The issues that you are going to hear discussed include capital shortages, underprovision of services, and the role of the private versus the public sector. You are going to hear questions asked about the federal government. What has it done, and what hasn't it done? You are going to hear about the effect of the energy development on the regional economy.

Finally, our participants will discuss solutions, including front-end financing and ways to get capital and the role of preplanning, an integrated assessment process, in which the community and industry jointly work together. You are going to hear some words said about limits to growth and, finally, public awareness as a possible tool to either deter or encourage growth.

In addition to discussions on energy development at Colstrip and Gillette and the solutions that are being proposed for these areas, we are going to discuss Wheatland, Wyoming, which is in the southeastern part of Wyoming, and Mercer County, North Dakota. We are also going to discuss the implications of coal development for the regional economy of the northern Great Plains. In these case analyses we hope to provide not just presentations by our discussants but also dialogue between the participants and the audience.

Jeannette Studer

Economist, Old West Regional Commission

My remarks relate to the effects of coal development on the economy of the areas in the states where coal is being developed. I am an economist for the Old West Regional Commission, which includes the states of Montana, Nebraska, North Dakota, South Dakota, and Wyoming.

I have chosen to address some broad economic questions, such as when does coal development cease to be a stimulus to the economy of the area and become the economy of the area, and what are some of the implications of this?

The northern Great Plains is easily characterized as an area of low-population density with few major cities. The economic base is predominantly agriculture and has been throughout history. There is little manufacturing, some tourism, and in certain areas extractive industries, which have been mainly petroleum and gas production in the past.

The labor force is highly educated, and while unemployment has not been a major problem, the area reflects very low per capita incomes. Thus, although the labor force has been employed, wages are generally low. Energy development, besides providing more jobs, would provide much higher-paying jobs.

Table 1 shows the differential between wages in some counties in the Powder River basin of Wyoming, which is the northeastern section of Wyoming covering the Gillette area. Gillette is in Campbell County. As you can see, there is quite a differential between the wages paid by mining and by the other major employment sectors. These were 1970 data, before major coal production had begun. Today's earnings are even greater,

TABLE 1 1970 Average Earning for Laborers in Selected Industries, Powder River Basin Area

Industry	Campbell	Converse	Johnson	Sheridan
Farm[a]	$ 4,475	$ 4,375	$ 4,437	$ 4,451
Mining	10,022	10,451	10,198	9,670
Manufacturing	6,370	4,789	6,877	7,492
Other Private Nonfarm	7,037	8,396	6,741	5,986
Total Private Nonfarm	7,918	8,601	7,040	7,120

Industry	Crook	Natrona	Niobrara	Weston
Farm	$ 4,451	$ 4,455	$ 4,476	$ 4,288
Mining	8,458	10,199	10,536	8,886
Manufacturing	6,608	10,825	5,242	10,118
Other Private Nonfarm	7,343	7,675	6,458	6,993
Total Private Nonfarm	7,460	8,477	6,835	7,782

SOURCE: Bureau of Economic Analysis, Regional Economics Division.
[a]Average earnings for laborers only; does not include proprietors.

and a sector such as mining is much more likely to have kept pace with the rate of inflation than has a sector such as agriculture.

As I mentioned before, the economy of the area is generally based on agriculture, but agriculture is a low-wage, declining employment sector. The sectors that have produced the greatest growth throughout the region over the past twenty years have been mining, services, and, generally, government services.

The effect of coal development on the economy of the Great Plains is difficult to separate from the effect of other types of energy development, since the region is rich in many energy resources, not just coal. Oil and gas have been major economic sectors in Montana, Wyoming, and North Dakota, and increased exploration activity is taking place in many of the same areas in which coal is being developed. Uranium is also being mined and milled in the region, especially in Wyoming. Eleven uranium mining and milling operations were announced in Wyoming in 1976 alone, and that is in addition to six new coal mines being started and two mines expanding to increase coal production by 30 percent in that one year. So the general point is that many of the problems that have developed cannot be strictly related to coal development. When you are talking about Rock Springs, you are talking about a lot more than just coal development.

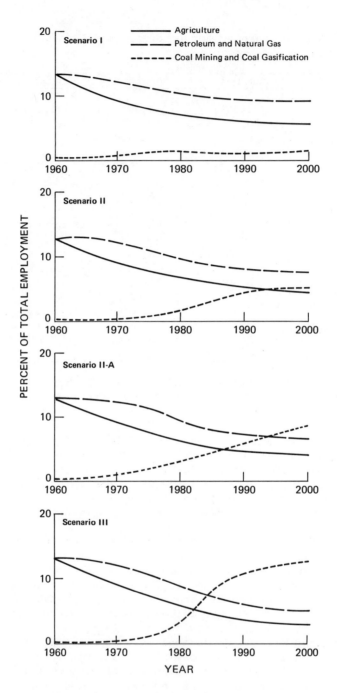

FIGURE 1 Selected industries as a percent of total employment over time, Powder River basin, Wyoming.

Figure 1 was taken from some work that was done for the northern Great Plains program on the effects of coal development on the Powder River basin of Wyoming. The graph represents employment by sector for only the Powder River basin of Wyoming. As you will notice, this area is already a large employer in petroleum and natural gas sectors.

The four different graphs represent different levels of energy development. The first graph is a very limited growth scenario, which has already been surpassed. The bottom, or fourth graph, represents the highest level of development, which is probably unrealistic, since we have seen very little development of gasification. The middle scenarios are the most likely. That labeled II-A was considered the most likely by the researchers.

The point of this graph, and the reason I show it to you, is the emerging dominance of coal as an employment and an economic sector. This only includes direct employment in coal-related mining and processing. It does not include railroad and other transportation sectors, and it does not include the increased employment that would come in the service sectors due to coal.

As you can easily see, coal becomes a very dominant part of the economy of this area in a very short number of years. Also, if you consider the mining and the petroleum and natural gas employment sectors together, agriculture becomes dwarfed to the extractive industry sectors.

What does this mean? Is coal mining and processing as a dominant economic sector necessarily bad? No, I would not say it necessarily is so. However, what makes the dominance of coal different from the dominance of agriculture is the fact that the coal industry is mainly dominated by a few industrial giants, multinational corporations, while agriculture is the closest thing we have left to a purely competitive industry.

Coal development in the West generally means absentee ownership and foreign--by which I mean extraregional--control of key resources. This means that jobs, earnings, and the economy of the area are controlled by people who do not live in the area and do not have any particular ties to the area other than vested interests through their industrial sector.

Mason Gafney spoke at a symposium in Bozeman, Montana, last fall on the colonialization of Montana. And in defining colonialization, the critical factors are absentee ownership and control of key resources, including capital, with local affairs dominated by a class of people with a vested interest in the existing state of affairs. He finds Montana and other western resource states prime examples of colonialization with coal development only increasing the dependence of the region on foreign capital.

The region has long been seeking to expand its economic base, but the question is, at what point will coal no longer be an expansion of the economic base and become the dominant sector, influencing not only the employment and earnings but also the social and political structure of the area? For economic control and political control are very closely related.

A secondary question, then, is, is it desirable to have a dominant economic sector that is controlled by multinational corporations headquartered outside the region and subject to the whims of the world market for coal as well as the world financial market?

In addition, the fear exists that the coal boom may be short-lived. It may be or it may not be. I don't think anybody can really predict that at this point. But it certainly is a valid concern, because this region has had a very long history of boom and bust cycles for extractive industries. Gold, copper, oil, and coal have all thrived and declined in the area in the past, and there are many communities and regions within this area that are still trying to adjust to the bust cycle of an extractive industry.

A possible approach for communities that are going through the rapid development of coal resources is to work for a broader expansion of their economic bases. It is understandably very difficult for a community that is having a hard time coping with the massive population influx from coal to consider going out and seeking new industries. However, this is the prime time for them to expand their economic growth. Industry is always attracted to growing areas, and areas that are in a state of decline have a very difficult time attracting any new industries.

As far as solutions are concerned, there aren't any real solutions relating to my topic. This is an issue that needs to be raised to the point of public awareness, and the affected public will have to be the ones to judge the far-reaching consequences of coal development at different levels and at different phases of their lives.

HOUTHAKKER: I wonder whether the speaker could give a clearer picture of the effect that coal development and mineral development generally would have on the income level in the states she is talking about, especially, I gather, the state of Wyoming. As she has indicated in the excellent charts that were projected, the present income level of employees in agriculture is very much lower, indeed only half or less than the income level in mining and manufacturing. Now, I gather there is very little manufacturing anyway so that doesn't matter. But there is already some mining and likely to be a lot more of that.

Now, this means that the income level in this region might indeed rise very considerably. Apart from any fear of the multinationals--of course, that is a horrible term, something you can scare children to bed with--there is another side to this picture, which I believe she has not fully explained. I wonder whether it could be quantified to a greater extent. In other words, you have made these projections of the growth of various industries. Have you also made projections of the income level and, especially, the share of the poorer people in Wyoming as it would be affected by greater mineral development?

STUDER: I don't have any exact figures with me, but I would say a couple of things. One, it has a great affect on the particular area where coal mining is taking place. In other words, right around Gillette, right around Colstrip, very high income levels develop. However, when you take a whole state, that obviously dissipates quite quickly, and you don't see a large per capita growth rate for the entire state. But, certainly, it shows in certain pockets.

It also affects a much broader area than those directly employed in mining. It affects the service sector, it affects the agricultural sector, because they must raise their wages in order to compete for employment with the mining sector. So there is a general rise in wages.

SINGER: I noticed several interesting things about your presentation. The statistics, first of all, seem to show an alarming decline in agricultural employment, until I realized that this alarming decline came about only because there was a rapid increase in some other kind of employment. In other words, these were percentages that you plotted, indicating that agriculture should not be appreciably affected by whatever else goes on in that region.

The second thing I noticed was that you used this buzz word, a rather emotional word called "multinational corporations," a word borrowed, I think, from the Third World and the less-developed countries. I don't think that is very appropriate in a discussion within the United States, particularly when we recall the fact that the largest owner of coal in the Great Plains region is Burlington Railroad, part of a multinational corporation.

The implication also was that these corporations who would mine coal would not be competitive, because you kept talking about the competitiveness of agriculture. I agree that agriculture is competitive, but I think we would all agree that coal is competitive too. At least, I have seen no evidence that would indicate otherwise.

Next, I noticed that you referred to the boom and bust economy of coal. I think that is quite misplaced, particularly if we compare it to the boom and bust economy of oil and gas, to which no one seems to object at all. Now, anyone who invests in coal in the Great Plains, let us say who puts up a power plant, a gasification plant, or a liquefaction plant for coal, is not going to walk away after a few years. And the coal potential of the region is such that they would be exploiting it for decades, maybe centuries, in comparison to the exploitation of oil and gas, to which no one seems to raise any objection. The reason objections are not raised to oil and gas, of course, is that large royalties go to the landowners, the farmers, the ranchers, and severance taxes go to the state. My impression is that these royalties and severance taxes would exist also for coal and would give considerable income to the governments of the states and to the people of the states who are the fee owners of the land.

STUDER: First of all, I think you missed my point on several questions. I am not saying that. The point I was not trying to make is that the coal industry companies are not competitive. The point is that it is a different type of economic sector than agriculture, which is controlled by much larger conglomerates of companies.

Second, you mentioned Burlington Northern. It may be that "multinational corporation" is not the correct term. Burlington Northern is controlled by the Chase Manhattan Bank of New York, which I don't really consider as being totally regionally controlled. Even though they actually sit there, their actual control is not from the region either.

As far as boom and bust cycles, the point I am trying to make there is that this is a fear that people have. Whether it is real or not real, I don't know. As I say, I don't think anybody can predict. I agree with you; I think there may be new technologies for coal to keep it as a major sector for many years. However, people have gone through the boom and bust cycles of not only coal but also oil and gas. Gillette is a prime example. It went through an extreme growth due to oil and gas exploration. People don't really care where the boom and bust comes from. It is just as bad whether you are talking about copper or whether you are talking about coal. But this is certainly a fear that exists among the people.

SINGER: But you don't believe it?

STUDER: I am not going to say. I mean, I don't know whether there is going to be a bust at the end. I think we need to prepare ourselves for the alternative in case it happens. That is why I would encourage them to try to develop as broad an economic base in their communities as they possibly can.

Second, on the severance tax--it is a little different, in that most of the coal, of course, is controlled by the federal government, whereas the oil and gas reserves were not.

EDWARD MALECKI, University of Oklahoma: I would like to make a comment primarily to Dr. Houthakker with regard to income growth or income gains to individuals.

You have to remember that in a small economy like the northern Great Plains, probably the majority of the high-income jobs go to outsiders, people who are brought in because of their skills, which are not available locally, particularly during the boom part of the cycle, that is, construction, and very frequently later on during operation. This isn't always the case, but a very significant and sometimes primary number of people involved in the higher-income jobs are in fact people who are not native to the region.

The local people, who might increase their incomes, increase them less, primarily because they can get higher incomes in service jobs, but not, generally, in the energy industries themselves. There are some cases where locals, especially in Indian tribes, are able to require through lease terms that a certain percentage of local people can gain through energy employment and income growth there. But I still would say, and I think Jeanette Studer would agree with me, that the vast majority of the people who gain most in the West are not westerners.

HOUTHAKKER: Let me briefly respond to this point. It does get at what is apparently a basic concern, namely, fear of outsiders. Now, I can understand that the Indian tribes were unhappy to see their traditional way of life disturbed by the coming of the white man, and I can sympathize with that. But the white men who are there now and who are now the large majority, of course, haven't been there long enough to have a valid claim to keep out outsiders, which seems to be the main thrust of the comment that was just made.

I happen to look at things from a national or even from a global point of view, and if some people can improve their situation by moving to Wyoming to drive a truck in the coal mine, then I regard it as an advance, and not something to which the inhabitants of Wyoming can legitimately object on the ground that these people didn't live there before. If you adopted that criterion, then only the Indians would be left.

STUDER: I would just like to make one point. I think we are getting around to some of the central issues and problems that people are talking about in the western states. They have this conflict. They would like to see more jobs, higher-paying jobs. They would like to particularly see their children have a good education. They would like to see them be able to stay up there and have good jobs. And they would like to see some growth.

On the other hand, the reason they are living there is because they like few people, beautiful scenery, et cetera. And they know that massive coal development means an end to that way of life. They have chosen that way of life. As you say, we do not have fences. They could move to New York if they wanted a high-paying job and were willing to sacrifice beautiful scenery and take on two-hour commutes.

So this is a trade-off, and the majority, I think, of the people in this area are going through this mental debate of what is the trade-off between the higher incomes and jobs and the fact that I like this area of low population and good environment.

THEODORE CLACK, Office of Budget and Program Planning, Montana: There is one more issue that Jeanette Studer alluded to and didn't address directly. But perhaps Ray Gold will get into it later on today. The research Ray and other sociologists in the area have done indicates that newcomers don't like it that much either. They don't like the great, open spaces. They are unhappy. There is nothing for them to do. They simply do not enjoy the things that we enjoy.

This is, of course, a parochial viewpoint, I am sure, but we feel we are as able to determine what we like, need, and value as an outsider is. We see no need to bring people in who are just going to be miserable, and who, ultimately, if enough come in, will make us miserable, permanently.

Michael B. Enzi

Mayor, City of Gillette, Wyoming

Gillette has been called a lot of things. We prefer to call it the Frontier City, the Pioneer City, the cattle center of northeast Wyoming, and the energy center.

Gillette is in Campbell County and is the only incorporated town there. Under that land is the world's largest single coal deposit. That coal deposit could supply one third of the nation's total coal production. In other words, if Campbell County were the only place in the United States with coal, the United States could still be the fourth-largest coal producer in the world. Campbell County has more coal than any country except the USSR, China, West Germany, and, of course, the United States, of which we are still a part.

Gillette is a community experiencing the exhilaration of growth. We are probably welcoming a higher percentage of newcomers than any other community in the nation. In 1967 Gillette was a ranching community of some 3,500 people, and oil was struck. By 1970, 7,000 people were recorded to live in Gillette. We now have 11,000 people living there, and by 1980 we expect 20,000 people, and by 1985, 35,000 people as a minimum. Gillette is Wyoming's youngest community, with an average age of 25.

According to the Federal Energy Administration projections that are done monthly and are based on presently existing federal coal leases and the diligent development requirement, Wyoming by 1985 should be producing a little bit over 200 million tons of coal per year. Now, to put that in perspective with some of the other states, the second-largest, western, coal-producing state will be Montana, with a little under 90 million tons of coal a year. Campbell County, by itself, will be producing a little over 130 million tons of coal by 1985.

How fast is Gillette growing? Gillette has a growth rate of about 20 percent. The average community grows at approximately 3 to 5 percent a year. To give you a better idea of how much growth that is, if the same amount of impact were happening to Chicago, they would have an additional 675,000 each year. If you are curious as to what that would do to your community, divide the population of your own community by five, and ask yourself if your community could handle that many new people with all the facilities that are required each and every year for the next ten years.

So what is our biggest problem? In a growth community, everything is the biggest problem. Everything we need is a basic facility. The average community put up its infrastructure over the last fifty years, and for the last fifteen years has had to concentrate on replacement and maintenance. In Gillette we are building a new town every five years.

The city council has set some priorities and is trying to handle one major area of basic facilities each year, plus beat out the brush fires in the other areas. This is the year of water concentration in Gillette.

In July of last year we increased our water supply by 20 percent. In August we let bids that will triple the water storage capacity of the community. This winter we completed wells that have increased the water capacity another 25 percent, and we are looking for a new source of water so that we will be able to handle at least 35,000 people, and we would like to have a little bit of water beyond that.

The cost? Seventeen million to thirty million dollars, depending on the distance of the new source.

What is the federal government doing to help Gillette? Right now a two-year human services project is being funded by the Economic Development Agency. A financial management study in the amount of $50,000 is being funded by the Federal Energy Administration. The city is participating in an area-wide water quality planning program funded by the EPA, and we have also received those federal revenues given to the state and divided up by population to all Wyoming communities, such as 701 money and revenue sharing.

The city of Gillette does not qualify for most federal programs because we don't have high unemployment. We don't have any large concentrations of minorities. We are not listed as overcrowded, because everything relies on the 1970 census, or worse yet, the Bureau of the Census estimates. We worked to have population estimates recognized until we found out they were projecting Gillette with a population decrease.

Our school population is increasing in excess of 300 children per year, so we obviously chase all parents away when their kids are old enough to go to school.

The city of Gillette has also just received word that it has been turned down for the third year in a row for a HUD Community Development Block Grant. In growth communities, construction workers normally live in recreational vehicles, campers, or small trailer houses. In the case of many workers living in recreational vehicles, their families have accompanied them. Our only facility for accommodating this type of living was at a place we call the "fishing lake," which is a playground and fishing area. We were certain that the community could be improved by a facility that would truly accommodate recreational vehicles, complete with restrooms, showers, and laundries.

We had to question to what extent our application was even read when we were turned down because Gillette supposedly does not need another playground. Incidentally, that was turned down at the regional level, so that gives you an idea of how extensively our problems are recognized.

The federal bureaucracy has not yet understood the meaning of "growth community." We are denied grants on the basis that the construction of these additional facilities will encourage the growth and that that will cause adverse conditions. Instead, we are going to have adverse conditions regardless of whether we have the facilities, and without the facilities the adverse effects are considerably greater.

The biggest problem, though, with the federal government is probably the waste of time caused to employees and city officials by talking to federal employees who are in town studying Gillette. Virtually every phase of Gillette's social life has been studied. Our best guess is that more than a million dollars a year is spent studying Gillette.

The usual procedure is to send a team to Gillette who visits with the city officials and employees and listens to what problems there are, and a few weeks later, or a few months later, or in some cases a few years later, we get back a pamphlet that of course has to be considered accurate because it lists what we said were the problems. That doesn't mean that we have all the information that we need. We lack information on the basic facilities area, and we lack designs to complete the facilities.

Few studies of our area or our town get beyond the academic approach to research. We have been generally researched enough. We need some specific information, and we are not able to get that information included in the research being conducted because the research designs are done without any local input.

Environmental impact statement requirements should be revised so that those areas that receive the most immediate and largest environmental effect have the most concentration and therefore provide the most information. That area is municipalities affected by impact. It is hard to get the people out there worried about the life span of the antelope when they are on water rationing.

If half of the environmental impact statement money that is spent studying animals and grass were spent studying human problems, western municipalities would be in a much better position to handle the kind of growth that they are facing.

So I have complained about the federal government. Maybe you would like to know what the people of Gillette are doing to help themselves.

Gillette has increased the facilities investment fees and the user rates of all city services astronomically. Gillette has some of the highest-priced water in the state of Wyoming, and Gillette has the highest fees in the state of Wyoming for hooking on to sewer and water.

We raised our sewer investment fees from $150, which is average for Wyoming, to $1,081 so that we could assure sewage treatment. The water investment fee was raised from $250 to $950 plus costs. Councils don't take that kind of action and the accompanying flak they get in an election year unless it is absolutely necessary.

It should be mentioned that bonds are out of the question, since the bond market is based on the community's past history of paying off bonds. A community of 3,500 can't show a potential for paying off the bonds of a community of 25,000. Gillette residents have passed an additional 1-percent sales tax referendum to help pay for capital improvements.

So where is Gillette going to get its money? So far, we have eliminated the federal government, and Gillette has already increased the things over which the city has control. Property taxes are at the maximum level allowed by the state constitution. A little bit later Dan Sullivan will talk about a state source that seems to be a light in the tunnel for us.

What are Gillette people's attitudes toward growth? Quite frankly, the people in Gillette haven't spent a lot of time worrying about whether to have growth or not but rather have accepted the inevitability of growth and are working to solve the problems that could accompany growth.

We realize we are sitting on the world's largest coal deposit. We would rather have the growth spread out over as long a period of time as possible, but we realize that should the ever-increasing shortage of natural gas cause the lights to go out in Chicago, or New York, or New Jersey, or any of the heavily populated areas in the United States, the fever would be so great to develop coal that Gillette would be inundated and destroyed in the rush.

Can newcomers participate in city plans? Gillette's population is young, with an average age of 25. Two out of three people have moved there in the last five years. Since most of the people have moved to Gillette recently, there is little developed power structure. Gillette is virtually in the process of building a new town, and there is room for everyone's participation. In Gillette the only newcomer is the spectator who is reluctant to help out, and the pioneer seems to be anyone who is willing to pitch in.

We do have more problems than a lot of places, but we have more people individually, actively pursuing solutions than anywhere else. People who live in Gillette now and those who come in the next few months will be the pioneers of a new community, the homesteaders of the future.

I hope I haven't made it sound too simple. In the preceding questions, all that has been covered is the need for basic facilities. Nothing has been said of the problems of crowding, having a large portion of the community under construction at any given time, mental problems, isolation problems, family problems, youth problems, and all the other range of human problems that are the result of rapid growth.

SINGER: Do you tax on an average basis or on an incremental basis? In other words, when a newcomer wants a service such as sewage or water, does he pay the full cost, or is his part of this cost borne by the people who are already there?

ENZI: I am glad you asked that question. That is a big controversy in Gillette. Since most of the people are newcomers, they are rather opposed to having a fee that discriminates against newcomers, such as our tap fee. We have no choice but to discriminate against the newcomers and place virtually all of the burden on the newcomers. If you have a community that is growing at 20 percent per year, in order to place the cost on the entire system through user fees, you have to triple the rates every two years. And we just don't think that those newcomers, as they become residents, would put up with tripling of the rates every two years.

HOOVER: Do we have a researcher or a federal government employee that has a question?

JOHN N. DICKIE, Department of Housing and Urban Development: I didn't have anything to do with the turning down of the bloc grant application for Gillette, but I would like to ask you, what cooperation are you getting from the state, and were you disappointed in the way that the legislation came about in terms of redistributing the federal mineral-leasing revenue? I understand that the legislature went back to a distribution based on the 1970 population rather than some other formula that would have better reflected the intention of the program. I am talking about the increase from 37.5 percent to 50 percent. This differential was supposed to go back to impacted areas. But it seems like in Wyoming most of it is going back to Cheyenne and Casper.

ENZI: In regard to the federal mineral royalties, of course, there are several formulas for the distribution of that additional multimillion-dollar return to the state of Wyoming that we would have liked to have had. Unfortunately, growth communities or any communities have to live with the political structure and the amount of political clout they have got. In Wyoming, while all communities are claiming to be impacted, there are only two or three of us who would fight to have the money directly allocated, in total, to impacted projects. So the legislators have come up with a formula--which of course we have to agree with in order to get any money at all--that distributes the money more equally around the state. We would like for the state to be concentrating more on the immediate needs too, and I have a speech that I give for the state organizations as well. As long as I was in Washington, I thought I would concentrate on the federal government, but we do seem to do a little runaround as the bucks pass between companies, state government, and federal government. And all of them feel that the others should be carrying the bulk of the load.

HOUTHAKKER: I would like to ask the Mayor, how much of a city is Gillette actually becoming as a result of all of this? Is it still primarily a community of people who work in the coal mines, or is there now also a substantial base of retail trades, services, et cetera? Have you any figures that you can give us on the distribution of occupations? I can imagine that in a rapidly growing town you have no time to collect data on things like that, but you could go by personal impression, though.

ENZI: It changes so rapidly that you are right, we don't try to get specific data on it. The coal mining is just beginning in Gillette. Somebody mentioned earlier the oil production. At the moment that has plateaued. Some people would say that it has busted, but we have just as many people there who are employed in oil as we had at the peak of the boom. So the population has kind of plateaued in oil production.
    I think that it has become citified. In fact, every newcomer who comes, or a good many of them who come, have ideas of things that they have seen in the community where they lived before that they would

like to see in Gillette. And those who are willing to work on them
find that while in a normal community they would be going through
months of committee meetings to achieve that same thing, they can
have the facility in place in a matter of weeks through private dona-
tions and private work. We are getting a lot of facilities that we
never would have dreamed of having seven or eight years ago.

We are taking on a little bit more of a city aspect. Although we
do have a great many newcomers, I think you will find a very positive
growth attitude in Gillette, because even the old-timers in Gillette
say that it is better than it has ever been before, even though to
the outside world it isn't going to look as good as what they think
a community should.

STANLEY SCHWEINFURTH, Office of Technology Assessment: I would like to
ask the Mayor, what resources does the county have for collecting
revenues from mineral development, and what, if any, problems you
have in getting capital improvements constructed? Is there any conflict or any competition between you and the mineral industry in getting workers, hiring workers, and getting jobs completed?

ENZI: I will address the problem of the city versus the county resources.
All of the development is taking place in Campbell County. All of the
big capital facilities are outside of the city limits. So we don't
derive any tax base from the mines themselves, whereas the county tax
base has risen from some $57 million to $322 million in the last four
years and, by the turn of the decade, should be at about $1.8 billion.
We are at about $13 million in our tax base.

The county, as a result, levies half of the mill levy that it is
allowed to by state constitution. However, there get to be a number
of conflicts between the city and the county in regard to getting the
county dollars into the city. The ranchers are still, populationwise,
out in the county, the dominant factor. All of our county commission-
ers are ranchers, and as a result they participate to the degree that
they politically can. But the ranchers don't like their taxes raised
any more than you do. Yet, the big pitch, of course, is that industry
pays 80 percent of the tax tab in the county, so let us build everything we can while we can in the city. It just isn't politically
feasible.

In regard to the worker conflict, the wages in Campbell County have
risen considerably in the last several years. Again, we don't spend
the time to keep track of it, but we do have tremendous competition
from industry hiring away those competent people that we are able to
hire. Even within the industries there is competition in hiring away.
The oldest coal company is Wyadac Coal Resources. They have been in
operation in Campbell County for almost fifty years. When they send
someone to a technical school, they say they send three--two for other
companies and one for themselves.

We run into some of those same problems as do, particularly, the
retailers in the community. The small businessmen have a tremendous
time competing for skills and talent with industry.

Gary Payne

Director of Platte County Branch, Southeast Wyoming Mental Health Center

I am an applied psychologist working in the area of rural community mental health. For the past five years I have been living in and studying Wheatland, a small community in southeastern Wyoming. Three years ago this community was selected as the site for a large-scale, coal-burning power plant by Missouri Basin Power Project, and today what I share with you is based on that community's response to rapid growth.

The Wheatland experience is a positive one. It involves to the best of my knowledge an unprecedented level of planning for rapid growth and cooperation between a responsible industry and a far-sighted community. In concrete terms, this has meant anticipating a three-fold increase in population and providing adequate housing, public utilities, a comprehensive city plan, schools constructed in anticipation of students, social service agencies with adequate staff and facilities, and programs for integrating newcomers, minimizing the effect of inflation on senior citizens, and treatment for the higher incidence of alcohol and drug disorders associated with a rapidly growing community.

In general terms, the development of a community plan to cope with rapid growth involves three steps: first, how to anticipate the effect of growth; second, how to involve a community in the process of planning; and third, how to find funding and financial resources to implement a community's plan for coping with rapid growth.

The anticipation of growth effects involves several underlying issues. These issues and my recommendations are as follows.

Issues
- Lack of conception of community.
- No standard parameters for assessing community.
- No accepted definition to determine when growth becomes rapid growth.
- Uniqueness of community versus commonality. Objective outsider versus community participation.

Recommendations
- Include social indicators in impact assessment.
- Translate social data into cost figures.
- Develop a definition of rapid growth based on rate of change.
- Involve community from outset in assessment process.
- Apply predictions individually to community.

Neither the conception of community nor the technology for assessing aspects of it are well developed in the social sciences. For example, efforts to assess the so-called quality of life in a community are underway but generally are accompanied by uneasiness even on the part of the investigators doing the work. That is not to suggest that quality of life is not an important community variable, or that community assessment is unattainable, but rather that it is in the initial stages of development.

A second issue is that of choosing the parameters used for measurement. A strictly economic assessment of rapid growth causes less concern for a community than assessment that involves social indicators such as alcoholism, agency case loads, public service expansion, child abuse and neglect, and other less palatable measures. In addition, there is no accepted definition to determine when a community leaves a normal growth situation and enters a crucial stage of rapid growth. Funds designated for impact areas tend to be dispersed to many communities undergoing moderate growth and not necessarily focused on the few communities where the need is most crucial.

An additional issue addresses itself to the relevance of studies of one community when applied to another. The number of variables that affect the level of impact is almost legion. Proximity to population centers, size of original community, assessed valuation, level of indigenous construction skills, and the nature of the incoming industry are some of the parameters that can markedly alter the rate and quality of growth. To some degree there is a need for a unique projection of growth effects for each community.

Researchers flock to impacted areas like vultures sensing a fresh kill. Community members tend to be threatened by the process and unhappy with the published results that reduce family, friends, and community to so many problem situations.

In order to put these fears to rest, a community needs to recognize its own requirement to study itself and become an active participant in the design of the research. Such studies commissioned by and done for a community are far more likely to be useful than those done in the search for a general formula for the course of growth.

On the basis of these issues, my recommendations for a community perched on the brink of rapid growth and attempting to anticipate the effects of that growth would be to use, in addition to the economic, housing, and utility needs, measures of social problem areas and translate those social problem areas into a dollars-and-cents figure. Also, a clear definition of impact or rapid growth is needed based on the rate of growth, if for no other reason than to help determine where critical growth is occurring within a state.

Anticipation of the effects of rapid growth is best done when it is initiated by the community who wants to know what the future will bring. In a major way the community is most aware of its blind spots when it anticipates the growth process, and research of this kind has strong community support. Also, communities are sufficiently individualized, that it is perhaps more important to help the community develop its own ability to anticipate the future rather than to look for a general formula for predicting the effects of rapid growth.

The second major problem is how to involve the community in planning for rapid growth. The issues and my recommendations are outlined below.

Issues
- Should the planning be done by professionals, local government, industry, or new community organization?
- Is development imposed upon a community or sought by it?
- Planning as a measure of community's capacity to cope with rapid growth.
- Making technical experts responsible to the community.
- How much indeterminancy can a plan tolerate?

Recommendations
- Develop a special community organization for planning.
- Place responsibilities upon the community to prepare a plan for coping with rapid growth.
- Allow ample time for the planning process to occur.
- Develop a technical team to assist the process.
- The community plan should include population projections, agency case load preventative programs, housing, public utilities, revenue projections, and commitments from local, state, and federal governments and industry.

First is the issue of who should do the planning. My biases lean strongly toward the creation of a new community entity that has between twenty and sixty members and represents the breadth of concerns within a community. Organized into subcommittees and assisted by technical consultants, such an organization of community gatekeepers and policymakers has the potential to do the necessary planning.

I also propose that impact alleviation plans be a community responsibility. In the Wheatland case it was primarily an industrial responsibility, and that was kind of nice, to let them do the planning for us, and yet in actuality the responsibility is clearly the community's responsibility to prepare impact alleviation plans. As such the plans become a measure of a community's interest in growth and its ability to cope with the ongoing effects of rapid growth. No doubt there are some communities who will not be able to achieve such a task, and most likely they are the very communities destined for boomtown disintegration.

A community plan for coping with rapid growth requires time, probably at least a year to complete. In my opinion such a plan would be a prerequisite to the actual onset of development. Fortunately, in this area of coal development there seems to be ample time for planning. Obviously, a community requires some technical assistance in the development of such a plan, and funds and/or qualified personnel need to be available to communities anticipating rapid growth. Very little of either is currently available.

Similarly, the content of the plan needs to follow a structured format. For example, population projections, induced labor, social service agency case load projections, housing, roads, public utilities, revenue projections, and federal, state, local, and industrial responsibilities are some of the items that need to be included within a community plan. The real crux of the issue, however, is how to develop these financial commitments in a meaningful way. The problem, the issues, and some of my recommendations follow.

Issues
- What are the relative financial responsibilities of local, state and federal government and industry?
- Income to a community is distributed so that the greatest income is at the time of least need.

Recommendations
- Guidelines are needed for areas of responsibility. For example, the federal role may be technical assistance and facility expansion. The state may be responsible for staff, and industry for front-end financing to be reimbursed.
- Joint powers boards should be used to distribute revenue over time.

Clearly, some guidelines are needed for assessing the relative roles of federal, state, and local government. One way or another, rapid growth becomes a costly business. Given a *laissez-faire* policy and the resultant boomtown, the cost is high job turnover, resistance by other communities to future development, and the creation of a new class of marginally adjusted people with their attendant expensive social problems.

A structure I tentatively propose would involve the federal government in developing programing or funds for technical assistance and establishing a priority for impact areas in terms of construction and development grants. State government would assume, under my tentative proposal, the major burden for staff expansion. Local government would be expected, as have most impacted communities in the area I come from, to tax themselves to some limit before an area would be designated as an impact area. Industry would be in the position of putting up front-end money to solve the problem, caused by rapid growth, of having the least financial resources at the beginning of growth when the need for those financial resources is the greatest. I expect that an appropriate tax reform or some kind of equitable agreement might encourage industries to put up front-end money and lessen the burden of financing of this kind of development.

In sum, my experience in Wheatland leads me to believe that the problem of rapid growth induced by energy development is solvable. It is being solved by some very active communities out in the western part of the United States. I do believe there is a need for a more active role on the part of both federal and state government, and I suppose that in some ways that is the purpose of my talking to you today.

GARY COLEMAN, Department of Health, Education, and Welfare: I have a question for Mayor Enzi and for Dr. Payne. What are the principal sources of revenue in a one-two-three way for financing your infrastructure of facilities for providing social, health, and educational services.

PAYNE: The standard way is through a state office that receives and channels federal funds through the state government with some degree of matching. In our own case, in the Wheatland experience, the only way that some services were able to expand was through subcontracting with the industry that was coming in. So, the industry, recognizing that state government was not responsive to providing higher levels of social services and realizing that the federal government was not responsive either, assumed the burden in our own particular experience. But this was in the absence of normal channels--filling the gap for an impacted community.

COLEMAN: In some time period, in essence, was the industry itself really the major and most important source of revenues?

PAYNE: In terms of the necessary expansion, yes.

ENZI: I think that we probably get a lot less cooperation from industry in Gillette compared to Wheatland because we are impacted by a number of companies, at the present time at least seven, whereas Wheatland is impacted by virtually one company. One town impacted by one company is a one-on-one situation, where you can take industry on.

Our biggest source of revenue is from sales tax, and our second biggest source of revenue is really not a source of revenue. It is loans from the state. So we have to come up with revenue to pay those back. But that is virtually what is paying for all of the improvements.

ROBERT L. VALEU, Basin Electric Power Cooperative: I would like to make a comment at this time. I happen to represent the company that is that industry in Platte County relative to Wheatland, Wyoming. In regard to the funding question, unfortunately, the federal government does not have a one-two-three system for delivering revenues or funding capability to communities faced with impact problems.

The primary source for the community, basically, is its own tax structure, through the *ad valorem* tax base or the per capita taxes that are generated on the population basis. The problem that develops is that, even though the impacting industry pays that tax through the normal taxing mechanism, the revenues do not flow back in a sufficient time frame to deal with impact. So, consequently, you have got to go through a series of gymnastics to try and establish a reasonable financing mechanism. The funding is there. The timing is the problem.

I might add that in the Wheatland, Wyoming, case there wasn't a single federal program in 1975 that was applicable to the impact problems of that community that were anticipated to occur in 1976 and 1977.

DAVID M. WHITE, Texas Energy Advisory Council: Were the payments you made to Wheatland in the form of prepayment of taxes? Are you within the corporate city limits of Wheatland?

VALEU: First of all, prepayment of taxes was not possible under the constitutional provisions of the state or through any legislative mechanism in the state. The plant site facility itself is located in the county. Therefore, the significant revenue generation from taxing is derived through the *ad valorem* tax base, which accrues to the county rather than the municipality, which in this case was Wheatland, five miles from the plant site facility, but which was going to incur the actual impact in terms of people.

WHITE: Was there any formal requirement that you provide any sort of compensation to Wheatland, or was this an obligation the company took on itself of its own free will?

VALEU: The initial obligation on the part of Basin Electric Power Cooperative was to work with the community and try and resolve and find reasonable public funding mechanisms that could deal with public problems. The caveat that developed was the Wyoming Industrial Siting Law, which is a siting permit law that placed significant contingencies and requirements on the applicant, in this case the company, to fund, either with or without recovery mechanisms, those front-end impacts if the community could not respond. Since there were no federal revenues, and because of the inadequacies of the taxing mechanisms of the state, we had to respond, to the tune of approximately $19 million in contingent liability.

MITCH WALDROP, *Chemical and Engineering News* Magazine: Dr. Payne, a key to your recommendation seems to be very careful advance planning. If there was one theme in all the discussion last night, it was the word "uncertainty," due to just about everything under the sun. Given this, how can communities plan to the degree which you suggest? Could you comment on this?

PAYNE: Yes. It is possible to do that kind of planning, recognizing the uncertainties, the vagaries of fate, construction schedules, and one thing and another. And, at least in our experience, we became comfortable with that kind of thing. It wasn't that we knew absolutely how many people were going to be there in some given month in 1978 and what that would mean to us, but it was that somehow we became oriented to the future. We realized we had to be responsive in terms of some general set of parameters. This is, to my way of thinking, a real shift in a community attitude.

The initial reaction to the proposed development is, my gosh, what does it mean? You hear rumors, you know--there are going to be 1,500; there are going to be 20,000; this is going to ruin us; this is going to have this effect, that effect. And a community that goes through the planning process begins to feel comfortable with the uncertainties. And I think also it begins to fill in some of those uncertainties. They may not be right, but at least we become a little more comfortable with it. And to me that is one of the important functions of preplanning.

Daniel B. Sullivan

Executive Director, Wyoming Community Development Authority

I would like to make a general comment. Reference has been made to the increased federal mineral royalties that are being returned to federal land states, including Wyoming. Of course, this is something that the state of Wyoming and other states, I am sure, are looking forward to very greatly. Wyoming is going to receive an increase of from 37.5 to 50 percent this year, or some $14 million, heavily concentrated in oil and gas, as opposed to coal. It is interesting that the original form of Senate File Seven on the federal strip mine legislation, which is presently under amendment consideration, has or had included a provision for an abandoned mine reclamation section that would have taken thirty-five cents a ton from the royalty prior to the fifty-fifty split. Now, assuming that Wyoming was to produce 107 tons of federal coal, just federal coal, in the year 1985, Wyoming could have had a net loss of some $18 million in its mineral royalty return because of this abandoned mine reclamation section.

The other interesting aspect of that particular provision in Senate File Seven is that that money is to be used principally and almost exclusively for mine reclamation, which was apparent and prevalent prior to the passage of the legislation. This certainly would preclude any money being returned to western states in any great amounts at all. It would all go to the Midwest, the East, and Appalachia.

I represent the Wyoming Community Development Authority. It is not a state agency; it is an independent political subdivision of the state. Wyoming in 1970 had a population of some 340,000 people. That is less than probably work in downtown Washington, D.C., on any given day. Wyoming's population is currently estimated to be 400,000 people, and it is going to approach 500,000 people, the estimates say, by 1990. That is significant growth in a state that actually lost population between 1960 and 1970.

Wyoming has probably done more in the last three years as a state, in terms of cooperation and implementation of programs to help its growth communities, than in previous years. I think, personally, Wyoming is very lucky in that, as Mayor Enzi alluded, the whole state feels it is impacted. To a degree, the whole state is impacted. Coal production is new and very prevalent; oil and gas is still there; we have uranium; we have copper; we have soda ash; it goes on and on. You can go all over the state, and there is impact; there is growth in virtually all communities.

It is estimated that in the next ten years there is going to be a need of from $250 million to $500 million worth of infrastructure needs in growth areas in the state of Wyoming. Those are estimates, and they are dependent upon a lot of things. They are dependent on the planners' estimates; they are dependent on the federal strip mine regulations, plant siting, and all the rest of it. Certainly, we know that there is going to be a tremendous need for infrastructure and socio-economic financing to meet the public facility needs in a state like Wyoming and the other western states.

Wyoming took the bull by the horns, beginning in 1973 and continuing through 1975 and decided that we had to first attempt to identify the key problems in financing that kind of need, that kind of high growth rate. They came up with two realizations. One was you have to find an ability to capture debt, to import debt. If you don't have that in-house capability, you have the age-old problem of debt limitations, of property tax evaluations, and their restrictions. In Wheatland, Wyoming, you have large-scale development in one area, a $2.5 billion power plant that is not going to add to the tax base of that community until it is completed, well after the primary construction impact has hit. Number two, if you find a way of financing this need in a conventional sense, through the sale of municipal revenue bonds at a local level, there is no way that you can repay or amortize that debt based on the rate of growth in that community or the present growth in that community. The figures would just simply be too unrealistic.

So, number one, the question was, how can Wyoming amass, import, or gather enough capital in the state to finance the need? And the second part of that question was, once we have amassed it, once we have found a way of financing these kinds of facilities, how can we repay it? How can the communities repay it?

The question was asked, how would you rank the abilities of the communities to pay for the problems? Mayor Enzi and others gave an answer.

The citizens of Wyoming want to exhaust all their resources and all their technology to solve their own problems first. They have made the conscious decision that the federal government and federal programs are not going to solve their problems. So the thinking is that we have to do everything we can to solve our own problems internally and certainly maximize any efforts through state and federal programs that we can to bring up the rear.

The Community Development Authority has the ability to go out and issue its own revenue bonds without a state guarantee. We have the ability to package as many projects as we would like into one specific bond issue. We feel the optimum size at this point is around $20 million for public facilities, $20 to $30 million for housing in any one issue. Now, the interesting part of that is that we are taking very high risk loans, or loans that don't have the ability to generate all of the amortization and debt service on their own behalf, and we are packaging these into public revenue bond sales. The state has come up with a couple of innovative ways to secure those bond issues so as to make them more marketable to the potential investors.

One thing they did in 1975 was to pass what is known as the Coal Impact Tax. This tax is an increasing, incremental tax on the production of coal, and it was amended this year to collect $160 million. That money will be split fifty-fifty in terms of leveraged grants back to impacted communities for water and sewer projects and for state highway, county road, and city street projects. Now, the whole idea of the Coal Impact Tax is to leverage the debt recapturing capability of the state and of its communities. Briefly, what is meant by that is simply that as communities are eligible for loans from the Community Development Authority, and they have then given to us a pledge of whatever revenues are justified and realistic, based on very conservative figures and projections for that particular system--a water or sewer system, as an example--we then have the ability to go to the state and get large, sizable grants to help amortize that debt, based on the projected tax collections, not on the real tax collections in a given year, but on the basis of the projected tax collections over the next ten to fifteen years. The state has already allocated some $33 million of that Coal Impact Tax for grants to do exactly what I have said. So we may have a $3 million project in the town of Glen Rock, Wyoming, including water, sewer, and street paving, a total infrastructure type system in a community of 2,500 people, and you may have as much as 50 or 60 percent of the total debt service and amortization of that debt coming from a grant of tax money to be collected in future years, which is then turned around and pledged to us to service our landholders' debt.

Secondarily, the Authority was given the ability to create secondary reserve funds to secure their bondholders' risk. Out of the state's mineral severance tax we were given one half of 1 percent, which today amounts to some $4.5 to $5 million a year to create secondary reserve funds for the protection of our bond holders. What this means is that we capitalize a primary reserve fund in the bond issue itself that equals one year's debt service, then create a second reserve fund that equals the first out of the state severance tax collections designed for that purpose.

Should the Authority default on a bond or not be able to meet its obligation in any given year, it would have the ability to go to the severance tax fund (the secondary reserve fund) first. The law further states that the state will replenish that fund, *pro rata,* with the next collected severance taxes available. So, in fact, there is a replenishing aspect to that bond guarantee. We also have the ability of doing the same things in housing.

So now Wyoming has created a mechanism not only to import capital to the state of Wyoming from the high-user areas where the taxes are being collected, and where the primary bond holders and the investors reside, but we have also found a way of importing portions of the debt service as well.

There are a lot of attendant risks of course. Some risks we are all aware of, such as those of overbuilding or missing projections. We have to ask ourselves the question, what happens if technology, or federal policy and regulations, or administration changes, or de-emphasis on western coal--what happens if five years down the road those changes accumulate to mean that the western coal and Wyoming energy reserves are de-emphasized, or at least the projections that everyone is making now and for which we are planning are not realized? All of a sudden there goes our great and grandiose scheme. We no longer have the importation of tax dollars to pay off that debt that we have created. We have the increased mineral royalty taxes, and 7.5 percent of that total tax is going to be administered in the same way, virtually, as the coal impact tax, with the exception that it is not going to be restricted to water, sewer, and highways, nor specifically to coal impact areas, but to all growth areas of the state. This 7.5 percent is going to amount initially to some $3.5 to $4 million per year, which is going to increase as the exploitation of those minerals increases. That 7.5 percent of the federal mineral royalty is going to be administered as leveraged grant funds, again, for the repayment of the captured debt. But, again, we have the same problem. That is, if we have missed our projections or if, in fact, everything does not occur as we expect it to, we have the opportunity of having created $50 million to $75 million to $100 million to $200 million to half a billion dollars worth of debt, which the citizens of the state of Wyoming are going to have to absorb on their own.

The scenario that I build hopes to broaden that risk by offering an investment to the country that we feel will be very marketable, very secure, and will have some charisma to it. We are hoping to get very good ratings. We hope to be able to pass along low-interest money or at least comparably priced money, and if all of these things happen, we will then be able to share that risk throughout the country.

One of the main ingredients that I think is lacking in a state like Wyoming is technical expertise at the local level, identification of problems. I don't mean that in a planning sense; I simply mean that if we build a water system for $5 million, how grandiose do we get? Do we build a $17 million water system? Wyoming is playing as much a catch-up game as they are a game of the future. We had $100 million worth of public facility problems before the coal boom began a few years ago simply because of stagnated and levelized population points. Rawlins, Wyoming, as an example, still has wooden water lines. Those wooden water lines should have been replaced ten or fifteen years ago. A lot of towns don't have paved streets. Those paved streets should have come along twenty years ago. So we are playing games of catch-up as much as we are planning for the future. Local expertise is a big problem and a big need.

We need to have more professionalism and more responsibility, both fiscally and financially, in terms of how the money is being imported to the state, how it is being used, so that we don't overbuild and have the kinds of problems that I have alluded to.

SPEAKER: Wyoming has taken some very progressive steps in terms of impact mitigation. The Wyoming Development Authority is one example, and the Joint Powers Authority is another. But the Development Authority has been tied up in legal limbo for several years now. I was wondering if you could briefly tell us what the problems are and what the prospects are for the Development Authority to really get going and to implement all of the programs that they have been chartered with?

SULLIVAN: Yes, I will try to answer that very broad question as briefly as possible. The Wyoming Community Development Authority was created in 1975. It actually got its start in September of 1975, so in fact we have been in our legal limbo or in business for about a year and a half. We had to undergo a constitutional test suit that started in Laramie County District Court in Cheyenne. I can recall the days when our attorneys said we would be less than one week in District Court and it would all be certified to the Supreme Court. It actually took us just under a year to get out of District Court, which happened five weeks ago.

The District Court decision was strongly in favor of the Authority, and there were some fifty-six constitutional questions that were eventually asked in the District Court proceedings. The opinion of the District Court was very strongly in favor of the Wyoming Community Development Authority concepts but has now been appealed to the State Supreme Court, and we are hopeful that they will put it on a very high priority. We hope to be out of there this summer and selling our first bond issues.

We presently have $50 million in bonds ready for sale, $30 million in housing, and $20 million in public facilities. One of the big problems that I visualized when I returned to Wyoming was the idea of going out to communities, and promising, and talking a good game, and not being able to come through. So we have found some monies in other coffers, and we have actually under construction at the present time $12 million of that $20 million in approved projects financed from other sources.

The other part of your question was to explain some of the problems of the constitutionality question. Let me briefly say that one of the big problems we have constitutionally is the ability to capture debt and distribute it to communities who don't have the ability to do it themselves. This is in lieu of elections; it is in lieu of their statutory debt limitations. We are trying to convince the courts that there is not an election requirement if there is not a general obligation taxing authority of a specific subdivision. If it is a revenue debt, we feel that there is not a specific need for an election, but a hearing process and the rest of it should suffice.

A major area of concern regards the joint powers mechanism. The Joint Powers Act specifically states that you can finance cooperative projects among political subdivisions without the requirement of an election and without regard to specific debt limits. That would seem to be one of the highly critical portions of our court case; it won't affect the Authority nearly as much as it will affect what is in place in the state right now, because they have, in fact, made over $30 million in loans under the Joint Powers Act on their own.

MARTIN A. WHITE, Western Energy Company: With this rapid development the state is going to need a great deal of water, both for its communities and for industry. Do you anticipate any problem with that in Wyoming?

SULLIVAN: With the development of water?

WHITE: Water sources.

SULLIVAN: The Wyoming legislature dealt with that in its most recent session and in fact increased the tax on coal, trona, and uranium by approximately 3 percent to finance water project.
    Water is always a most critical issue in the West, and specifically in Wyoming, and I'm sure that we will have future problems. Hopefully, we will have enough money to deal with them.

JOHN H. ASHWORTH, MIT Energy Lab: I just have one question--and this was raised earlier by Ms. Studer, I believe--which is the parallel to the less-developed countries, and the question is simply this: it appears that the financial structure that you are building is based in part upon certain levels of revenue, revenues to the severance tax. Now, this leads to the interesting question that somewhere down the line the state cannot cut back at that point on the level of production without the entire financial structure, which is based on certain levels of revenue, falling apart. Is this a question, is this a problem, for your particular agency?

SULLIVAN: Yes, it is a problem, but I think you have to realize too that the traditional capital facility debt structure would dictate that you finance projects for a ten- to fifteen-year period. Assuming that your oil and gas declines or at least stays stable, the rest of these resources are generating increased revenues too. We think the severance taxes are going to continue to increase. It is just a matter of who replaces who, and coal is going to replace oil and gas in the next fifteen years. We feel that by structuring the capital debt in the next five years to amortize out in less than twenty years, we shouldn't have that problem. Again, we are using very, very conservative figures in terms of revenues generated and population estimates.

CLACK: One thing that may be of interest is that Montana has taken two approaches with its taxes, both legislative in origin. First is our Resource Identity Trust Fund, which is based upon a 5-percent gross proceeds tax on any mineral development. That is put in a trust fund for the state. As I remember, the interest may not be used; nothing can be used until $10 million is in the fund. Only the interest can be used up to $100 million of fund accrual. Then over the fund balance of $100 million, both principal and interest can be used.
    Also, in the last election we passed a referendum wherein our gross proceeds tax on coal will be redistributed and 50 percent of that will be saved for future needs, in the event of a bust or in the event of nonreclamation or whatever.

Martin A. White

Project Manager for Western Energy Company's Colstrip Development, Colstrip, Montana

A few years ago my dad and I were fishing a remote stretch of the Pattingale River in eastern Montana, and we were quite startled when another fisherman rounded the bend below us. This was the first time that anyone had imposed on our fishing domain in the ten years that we had frequented that stretch of river. Dad looked at me, stating, "Too many people around here, son. We will have to find a new fishing spot."

The issues of coal development in Montana are similar to the issues Dad and I faced on the Pattingale that day. We enjoyed the Pattingale the way it had been. A change, no matter how small, was upsetting. We didn't want more people on our fishing stream.

The ranchers and farmers who oppose coal development see it as Dad and I saw that intruding fisherman, someone imposing on their domain. They have enjoyed their lifestyle for many years and do not want it to change. Some statistics will help explain the sparsely populated nature of the area surrounding Colstrip, which lies in Rosebud County in eastern Montana. This county is home to many of these farmers and ranchers.

Rosebud County is slightly larger than the state of Connecticut. If you were to divide the county equally between its residents, each resident would get 400 acres. If you were to divide the state of Connecticut, each resident would get 1 acre.

Other statistics tell a different story. In 1920 there were 58,000 farms and ranches in Montana. Today there are 24,500. Rosebud County has seen the number of farms and ranches dwindle from 940 in 1930 to the present 320. Evidently, Montana agriculture is providing employment for fewer families each year.

The Colstrip development has employed many of the farm and ranch families who were forced from agriculture. This is evidenced by the fact that approximately 270 of the 300 miners in the Colstrip mines are native Montanans and about 90 are native to Rosebud County.

A question was asked in the discussion about who receives the money. The average wage at our Colstrip mine last year was $20,000. So the local people do receive some of this money.

A question raised by these statistics is this: is the impact of coal development having a positive or negative effect on the people of Rosebud County in eastern Montana? The few large ranchers and farmers might feel the impact is negative. However, the continually growing number of families who were forced from agriculture find the coal development impact positive. Although many families may have moved only 50 or 100 miles to Colstrip, when they are concentrated in this one area, there are problems. A recently arrived family will not be financially prepared to dig their own well; install a septic tank; build a house, a road, or power lines to the house; build a school and a church; hire a schoolteacher and a minister; build a house for the teacher and the minister;

build recreation facilities and stores; and hire a policeman to protect their property. You would not expect one family or group of families to accomplish all of these projects in a short period of time.

However, when a family arrives at Colstrip, that family expects to find those facilities and services ready for use. Providing adequate facilities and services when they are needed is the main socio-economic problem related to energy development in eastern Montana. Who will provide these facilities and services? Certainly not the local farmers and ranchers who wish the development would all go away. The federal government has many assistance programs, such as HUD 701 Comprehensive Planning Program, EPA 208 Planning for Wastewater Systems, HUD Community Development Bloc Grants, FDA grants, loans for public works under development, Bureau of Outdoor Recreation grants, and many others. I can only speak with experience about a few of these grants.

Preparation of an application generally takes two to six months, depending on the complexity of the project. The EPA Wastewater Treatment Grant secured for Colstrip was applied for in June of 1971. Approval of the grant was received in early 1972. The plant construction was completed in 1972, two years before payment was received from the EPA. The interest on construction financing added 20 percent to the cost of the project. This is 20 percent that cannot be recovered.

Colstrip received a Bureau of Outdoor Recreation grant. The BOR people at the state and federal level were exceptionally cooperative and capable. However, by the time the project was conceived and designed, the budget set, the funds applied for, the application approved, and the various elements of the project put out for bids, inflation had wreaked havoc with the budget. Once the amount of project funding is set, it is difficult to secure more money. Therefore, the scope of the project had to be cut.

Montana levies a substantial severance tax on produced coal. A portion of this tax money is earmarked for handling energy development impact problems. This has been somewhat successful, but applying for and gaining approval of a grant takes too much time.

Permit me to tell you a little about Colstrip and its history. The town of Colstrip was a Northern Pacific Railroad Company town from 1923 to 1959, when the Montana Power Company purchased the property. In 1968, after a nine-year hiatus, Western Energy Company, a wholly owned subsidiary of the Montana Power Company, reopened the Colstrip mine. In 1971, plans were announced to construct two 350-megawatt generating plants at Colstrip. Also, by 1971 the Colstrip surface coal mine had grown to be the third largest of its kind in America.

Colstrip's population in 1968 was approximately 100 people. In 1971 the projected 1975 population was about 3,000 people. Most of this increase occurred during a two-year period. In other words, there was almost a 3,000 percent increase in the population of Colstrip in two years. That is impact. What was done to handle that impact?

The impact-causing companies decided to work closely with county officials in handling the impact; however, we knew that the bulk of planning, construction and financing of those activities would have to be borne by our companies. Federal and state financing would not be available on a timely basis. Western Energy Company had not constructed a town before, so a joint venture of architects, engineers, and planners was hired to assist in the development.

The goal set for the joint venture and Western Energy Company was not merely one of developing a community that would mitigate the adverse socio-economic impact on Colstrip but one of planning and supervising to completion a community whose aspect and reality were positive. The companies proceeded from the standpoint that corporate responsibility militated in favor of a net improvement in the community rather than an obligation simply to neutralize adverse effects.

The first step in reaching our goal was to ask, through a written survey, the people who lived in Colstrip, or would be transferred there, what type of community they wanted. On the basis of the results of this survey, a mix of houses, apartments, and trailer lots was determined. In addition, the people indicated a preference for types of recreation and commercial facilities. The number and age of children was determined through the survey and provided to the school so that school administrators could plan accordingly.

In 1973, with the people of Colstrip's directive in hand, we began to build. Since then we have constructed a permanent town for 2,500 people, where only 100 lived in 1968. This town has a new water system. There is now adequate, quality water. People can even water their lawns. The town is green all summer for the first time in its history.

The town has a new sewage system and new streets with concrete curbs and gutters. Houses, apartments, and trailer parks were constructed, and the area around them landscaped. Temporary classrooms were constructed and given to the school district. These classrooms handled the temporary student surge during mining and power plant facilities construction. Today, permanent school facilities are being readied for fall enrollment. A motel was constructed and leased to a local family. A commercial center was constructed and spaces therein leased to merchants who provided the following services: a hardware/drug store, a medical/dental facility, a laundromat, an arts and crafts shop, an insurance office, the Post Office, and a beauty salon. Space is also provided for a dry goods store and a restaurant. Land was sold to other businessmen, who constructed a grocery store, a bank, and a gas station.

Colstrip today has a full business community where none existed before. A park system is integrated through the town so no family is far from park facilities. These facilities include five tennis courts, three outdoor basketball courts, five large tot lots, a swimming pool, two softball fields, a little league field, two Babe Ruth fields, and a community center with handball and raquetball courts, weight-lifting facilities, a gymnasium, a meeting room, a kitchen. There is a new library building and fifty-two acres of open parkland. One-half mile from town the third largest reservoir in southeastern Montana was constructed to provide water for the two 350-megawatt power plants. That reservoir has been stocked with pike, bass, and crappies and will provide an opportunity for fishing as well as swimming, canoeing, sailing, and, in the winter, skating.

At Colstrip, impact problems were eliminated or mitigated because financing was made available by the impact-causing companies when it was needed. Plans were prepared by these same companies when they were needed and implemented by these companies in time to handle the impact. Almost no government programs were available in time to handle impact problems.

Federal, state, and local governments and impact-causing companies must work together during the initial stages of planning an energy development project to insure that financing is available before the impact occurs. When dealing with impact, red tape has to be cut and action expedited, or the impacted community will have drowned in impact problems before financial relief arrives.

There are certainly issues in the development of coal in the West. There are also problems in handling socio-economic impact. Solutions have been found to many of these problems at Colstrip. Socio-economic impact has been handled. Colstrip is a stable community people are proud to call their hometown. I think back to the Pattingale River. Since that day when Dad and I first saw the intruder on our river, the road up Pattingale has been improved and a campground constructed. Now many families enjoy fishing on the stretch of river Dad and I thought was our domain. Would you suggest that that river be protected so only Dad and I could fish there? Of course not. Would anyone suggest that the largest energy reserve in America be left in the ground because there are a few problems? Responsible development can handle the socio-economic problems as well as the environmental problems.

VICTORIA EVANS, Department of the Interior: I am wondering, what is the percentage of the project cost to the energy company?

WHITE: What percentage of the cost of the town was the company's cost? With interest and everything, we have approximately $16 million in the community itself, and of that there has been a little less than a million dollars, quite a little less than a million dollars, supplied by outside financing.

EVANS: But what percentage is that of the total project cost, including the two new units at Colstrip?

WHITE: The total investment in Colstrip presently is about $250 million. So it is pretty minimal, one half of 1 percent, I think.

BRUCE KARAS, environmental lawyer: I have been going to meetings on the northern Great Plains for three or four years, and I don't think I have ever been to a meeting quite like this. The reason, it seems to me, is that the whole meeting is focused around essentially a technical problem that is susceptible of solution and many of the ideas that have been given are good solutions to it. Namely, how do you get water, sewage, recreational facilities, what have you, to a new population coming to the northern Great Plains? But that doesn't address any of the basic questions that are affecting the northern Great Plains, except for that one, which is an important one, but only one of probably a dozen.

If you had any ranchers or farmers from the area, you would have had an entirely different discussion. For example, the Wheatland project and the Colstrip project, which have been discussed, are under ferocious attack by both farmers and ranchers on a host of different issues, none of which have been mentioned. They are questions of, for example, whether you should be developing western coal at this rate at all. That has not even been raised, and that is an issue that this administration is re-examining. It is a basic question that is by no means decided.

For example, if tough strip mine legislation is passed, if a requirement is imposed of having to use scrubbers on western coal, if issues like that are decided in a particular way, there is not going to be anywhere near the level of development of western coal which is now assumed.

Now, the ranchers and farmers and people like the environmentalists may be all wrong. I am not saying that they are right. That is a subject that would be interesting to discuss. But it isn't your discussion, and it seems to me you are ending up with a discussion about a technical fix of a very specific problem. You are not going to the question of export-only policies, for example, which is a basic question in both Wyoming and Montana. Nobody has talked about air quality, water. Maybe the ranchers are concerned that the water they use is going to be taken from them. All of those basic questions have been sort of pushed aside, and you are going to discuss whether you are going to get recreational facilities. It just seems to me that you end up with a discussion that isn't the real world of what is happening out there.

HOOVER: I certainly agree that those are real concerns. I think there is some concern about that within the panel. I think Ted, in particular, wanted to make some comments. Why don't you go ahead and make your comments, Ted?

CLACK: I think that there is a slight error here. If I am not mistaken, it seems to me the water system and the sewage system and some of the community facilities were supported by county funds and not Western Energy or Montana Power Company or the consortium's funds.

WHITE: The sewer system, as I indicated, received an EPA grant to build a portion of the sewage system. That was $75,000. The only other funding that we received was from the Bureau of Outdoor Recreation, and that was matched by our own company, and it was $526,000. We put in $526,000, and they put in $526,000. To date those are the only funds that have been put into Colstrip. Now, there are funds that have been promised to Colstrip from the coal tax, but so far we haven't any of those funds in Colstrip.

HIBBARD: Is Montana Power a cooperative?

WHITE: No, it is an industrial utility.

HIBBARD: So you are not owned by the ranchers?

WHITE: Oh, there are probably some of the ranchers who own stock in our company.

HIBBARD: How about your board? Has your board got ranchers on it?

WHITE: Certainly.

HIBBARD: So in a sense the ranchers were doing it to themselves?

CLACK: The vast majority of the control of the stock in the Montana Power Company is held in New York City and in Connecticut. Sure, there are token ranchers; there always are token whatevers. But the power company is not controlled directly by the people in Rosebud County by any means.

HIBBARD: How about the Wyoming Power Company? Isn't that cooperative?

VALEU: Yes, Basin Electric Power Cooperative is a consumer-owned entity, and I might add that the Board of Directors of our cooperative in all cases are ranchers and farmers.

ENZI: John, I would like to make one point if I can. I think it is very important, based on the comment made by the gentleman just previously. I think you have to understand that the philosophical question of energy uses, the type of reserves that will be used where and when, will probably go on for the next century. The reality is that in the Great Plains states today, communities are being impacted, and we are here to try to point out that the real world is one in which those real people are trying to deal with those problems. While that is going on I am sure that here in Washington we will continue to hear for the next decade the continual discussion as to the pros and cons of whether or not coal should be developed. But we are trying to give you a perspective of how the people are dealing with the real problem right now.

WHITE: I might just specifically answer your question. There are 38,444 shareholders in the Montana Power Company. Now, 12,921 are Montanans, or thirty-four out of every hundred. The balance are spread all over the United States. Some 23,893 are in the western part of the United States. California has the most shareholders other than Montana. The State of Washington has the next, the State of Illinois the next, and the State of Florida next.

CLACK: In terms of voting, in terms of clout, Martin, how is it distributed?

WHITE: I think you would have to say in terms of voting that is how it is distributed. I am not sure how many dollars are in Montana.

Theodore H. Clack, Jr.

Project Manager for an Evaluation of Deinstitutionalization, Office of Budget and Program Planning, State of Montana

I have taken the liberty of assuming that I have been asked to this prestigious assembly as a sort of populist representative. I intend that my remarks will have a populist tinge. The other members of this panel have and will be addressing more specific topics, so I will keep mine in a general vein.

First, I want to briefly touch on what I regard to be the central issues--geometric growth and resource consumption. The nationa seems intent on pursuing a course of mindless and irrevocable consumption of non-renewable natural resources. Limits to growth have been demonstrated to exist for all natural systems, but we do not seem to be convinced of them. Until these limits have been acknowledged and rationally addressed, more specific problem solution strategies will be ineffective in the long run, in my opinion.

The nation has little time left to address this problem and will have fewer options available the longer it delays the task. Limits to growth have implications not only for the availability of energy and materials and for economic growth, but also for food, water, and the sources of those essentials. Until limits to growth are defined and understood, resource allocation decisions must necessarily be made in the dark, with uncertain knowledge of their real effect and real cost.

Now, I'll move on to some more immediate issues. People should recognize that the planetary context of energy and resource availability makes these issues even more keenly felt, at least in states like Montana. First, the nation has not yet defined what level of energy consumption it truly needs. We must differentiate between need and demand. Right now we are operating only from the basis of demand, with all the egregious waste that implies. Remember that per capita energy consumption in a number of countries, some with standards of living considered higher than ours, when you consider all the people, is far less than our per capita energy consumption. Now, Dr. Lave mentioned Sweden last night, at 49 percent. West Germany consumes, per capita, 46 percent of the energy that we consume per capita. The United Kingdom, which isn't doing too well, nonetheless consumes 46 percent of what we do, per capita; Japan, 28 percent; Denmark, 48 percent; Switzerland, 31 percent; New

Zealand, 25 percent; and the world as a whole consumes 17 percent of the energy we consume per capita.

The question arises, why should we or anyone else sacrifice that which is dear and irreplaceable to feed a glutton who knows only what he wants and not what he needs?

Second, there are constitutional issues. For example, Montana's new constitution, which we adopted in 1972, explicitly guarantees citizens the right to a clean and healthful environment for this and future generations. One section of interest reads, "The legislature shall provide adequate remedies for the protection of the environmental life support system from degradation, and shall provide adequate remedies to prevent unreasonable depletion and degradation of natural resources." Further, the Constitution of the United States vests in the states those powers that are not clearly reserved for the federal government.

What we understand to be current federal and industrial plans for energy resources in our area is in clear conflict with both constitutions, particularly Montana's. It is difficult to consider the wholesale strip mining of coal--that is an inflammatory word, but I will keep it anyway-- and its conversion to electricity or syngas by using inefficient, dirty, outmoded, or questionable technologies to be compatible with Montana's constitution.

Third, there is an ethical conflict implicit in the imposition of the unreasoned demands of the majority upon a minority. The national insistence on the development of coal reserves at the expense of the hopes, values, and needs of Montanans and other westerners is an excellent example of this. We note with interest, as other minorities surely have, that those desiring Montana coal, in order to reduce the conflict engendered by this form of oppression, have begun to characterize Montanans in pejorative terms, such as blue-eyed Arabs, greedy, unreasonable, selfish, irresponsible, and unpatriotic "expletive deleted's."

The nation should recognize what it is doing in this regard. I suggest that the central issue here, about which these and other issues revolve, is whether the nation truly does stand for equal justice, equal rights, and self-determination for all citizens.

Now, I'll move on to the area of prevention or mitigation of impacts. This issue seems less pressing of late, in part because of the work of local people in Montana and Wyoming. The federal government and industry seem to recognize that this responsibility is primarily theirs. It is their insistence that energy and material reserves be developed not that of the locals in local communities. And certainly locals in their communities lack the authority and the resources to handle all but the most local and immediate of impacts. However, I am somewhat concerned that the federal government seems more disposed to loan funds to impacted areas than to make grants. Again, it should be remembered that it is forces beyond the control of the locals who have created the impacts and not the locals themselves.

I would like to make one more point in this regard. It must be recognized that some impacts are not remediable and that a massive infusion of money is neither an acceptable nor an effective solution for some problems.

Finally, what will be left when the miners leave? Montana and much of the mountain West is dotted with ghost towns and areas devastated by various forms of development. One would hope that we had learned from our recent past, but the nation's stance and that of some industries indicates that we haven't. We propose wholesale resource development with no real knowledge of the related benefits, less of the related costs, and only a vague guess at the expected duration of that development.

Coal development in Montana, if not carefully controlled and sited can prove incompatible with the major sector of Montana's economy--agriculture. If we give up this vital sector of our economy in pursuit of energy development, what will we be able to count on in twenty, thirty, or forty years? If we build a large--by our standards--urban complex centered around fossil fuel industries, what will sustain those complexes when the coal or the market for coal runs out? How do local businessmen in small communities cope with the rapid rise and equally rapid decline in population that attends huge, capital-intensive, industrial installations?

As to much-touted tax revenues, last year nearly $136 million worth of minerals were extracted from Silver Bow County. What were the tax revenues for this? Zero dollars and zero cents. Fortunately, our taxes on coal are based on gross proceeds, as Martin White can attest. This is not the case for all forms of industrial development in Montana, however.

Let me quickly identify some impacts that have been associated with coal development in Montana. I will skip over the ones that have already been touched on--rapid population growth and decline and the impact on community facilities. Social disorganization has just briefly been touched on. A colleague of Dr. Payne, ElDean Kohrs, wrote an article discussing what he calls the Gillette Syndrome, which he characterizes by drunkenness, depression, delinquency, and divorce. Local communities, small communities, receiving a rapid influx of population with no facilities, no diversions, no whatever, are ripe for this syndrome.

Ray Gold has done considerable field work on the effects of this rapid social change on not just ranchers but also community folks who, regardless of your perception of the reality of a situation, themselves perceive the situation as untenable, unpleasant, and so forth.

There are social and economic uncertainties attendant to resource development, particularly among the farming and ranching community, and those revolve around questions such as, who are my friends; who are my enemies; whom can I trust; will I be able to buy land, which farming and ranching operations have to do to stay competitive; can I farm or ranch next to a mine or a power plant; how much land will they destroy, how long will it take to get it back into production, if ever; how much water will they take from the river or the stream or the subsurface aquifers? Downdrift of the Decker Mine, for example, private wells there have dropped considerably, twenty feet or more, as I remember. Farmers and ranchers and just plain folks in that part of the country depend on those shallow groundwaters for livestock, irrigation, and for their own personal use.

There are also rivers there. The main river in the area is the Yellowstone, and the Tongue and Powder rivers, which drain into it, which are not dammed, except for the upper reaches of the Tongue. They don't run steadily all year round. They get quite dry toward the end of the year. Industrial use of water, however, is a steady state thing, pretty much. That is part of the reason why Martin's company had to develop that holding pond, because they couldn't count on a steady supply of water taken out of the river every day of every month of every year.

There is air and water pollution to be dealt with. Dust erosion, wind erosion, coal fines from uncovered tipples, fallout from the stacks at Colstrip and other areas that we hear are going to be developed. There is definitely destruction of ranchland. Now, you can argue about whether the ranchland is productive or not. It depends on how you want to quantify land values. If you want to quantify them solely in terms of the dollars they can get you today, coal mining is probably the best alternative. There are a lot of people in Montana, and hopefully in the rest of the country, who feel that land has value other than its current market price.

Also related to energy development on a large scale in the state or in the northern Great Plains region are vastly increased demands upon land and water engendered by the growing population attracted by energy development. Attendant to that is destruction of wildlife habitats and, with the destruction of habitats, the destruction of the wildlife species, which we hold quite dear.

Let me conclude my emotional remarks with a very brief outline about what could be done, what many of us feel has not yet been done. We could rationally and comprehensively--probably with the emphasis on comprehensive--assess our energy and material resource problems, not just in the context of what we have available in this country but, since we do live in the world, what everybody has got available and what they are doing with their resources. We should examine the worldwide shift in development--industrial development, economic development in other countries, nationalism, and so forth--in the context of national and international availability of reserves and demand for those reserves. As a result of thorough research and analysis of the problem, we should then develop specific plans *after* we have got the data in our hands. I think we must consider some alternative to maintenance of the *status quo*, which I perceive as continuing our economic growth and our increase in energy and materials consumption on a geometric basis, every year. The Ford Energy Policy Project suggests that there are other alternatives available. I think the country should look at them.

Fourth, we should acknowledge our dependence upon natural and social systems and values and commit ourselves to preserving them. We do not live in a world in which the lack of water and the lack of food and the lack of air can be easily endured--we can't live on money is what it boils down to, I think. I have never tried to eat money, and I don't want to start now.

Finally, we should commit ourselves to a thorough program of public education, public involvement and input into the nation's problem solution process. It is not just a state thing, it is not just a company thing, it is a national thing.

HOUTHAKKER: Of course, it was more a declaration of faith than a reasoned argument, and as such it is beyond questioning. I did, however, toward the end of these remarks find something that maybe could be the subject of a question, namely, the desire to have rational planning of energy production and consumption on a global basis. This presumably would involve the present producers and consumers of energy. In the case of oil the main energy source in the world is, of course, OPEC. So, apparently, what is suggested here is that the energy producers of the world get together and form a rational plan in which energy will be produced where it will cause the least social cost.

If this could be done, there would be some advantage to it, perhaps. I think one doesn't have to think very hard to see that this is a very farfetched opinion. In the meantime, we in the United States and in the other consuming countries are vulnerable to exploitation by outside forces. The only defense we have against this is more domestic production, whether one likes it or not. And, therefore, it seems to me that without denying the validity of many of the things that have been said by the last discussant, it is also obvious that in the national interest we have to do something more.

I believe that there are ways of satisfying the local interests that have been advocated here. I don't understand why the county you mentioned gets no tax revenue. That surely is within its power. If it wants to be compensated for the development that it considers to be forced upon it by the outside, it can actually obtain this compensation by taxation, and I certainly would have no objection to it.

CLACK: There are a number of things I have to address. First, I would like to address the rationality of my presentation. I have to ask you the rationality of a position that maintains that there always will be resources available for us to consume while we continue to increase our consumption of those resources. A case in point is the 1972 publication of Roseman from the National Economic Resource Associates. We are taking a hypothetical fossil fuel resource, which he hypothetically assumed would last 1,000 years, supporting whatever economy is depending on it at a steady state. When he applied a 3.5 percent growth in energy consumption rate to that 1,000 year reserve, he learned that it would be depleted in 104 years. To make his case stronger, he doubled the reserve, and at 3.5 percent per year, per year, per year, the reserve is gone in an additional 20 years, 124 years. If you multiplied that reserve by a factor of ten so that you would have 10,000 years worth of reserves at a steady state, at a 3.5 percent growth in energy consumption it is gone in either 170 or 176 years.

I don't think there are any natural resource scientists anywhere in the world who would propose that we have even a 1,000-year reserve of energy, and yet we continue to increase our energy consumption and our materials consumption. As Dr. Lave mentioned last night, it was close to 5 percent last year. What is rational about that, sir?

HOUTHAKKER: I see you prefer to answer a question by another question, which I am sorry to say is not a very serious one. Of course, these projections of constant growth have nothing whatsoever to do with the subject. I know they are put forward by certain organizations that have some association with MIT and other places, and this gives them a respectability that they don't deserve for a moment. There is nothing in economics that holds that energy consumption or any other mineral consumption will continue at a constant or increasing percentage rate, and the projections you mentioned are indeed ridiculous. Why are they ridiculous? Because if it becomes more difficult to extract these resources, the price will go up, and that will generally curtail consumption. Therefore, in those minerals, and there are a few, where it becomes more difficult over time to find them, there is a built-in corrective through the price mechanism.

CLACK: Does it respond in time, sir?

HOUTHAKKER: Of course it does.

CLACK: It always responds in time?

HOUTHAKKER: Unless there are serious interferences with the market. But, generally speaking, if the government does not intervene unduly, as indeed it has, there are built-in correctives that certainly may occasionally cause transitional problems but that nevertheless keep consumption and production in line, generally, at a decreasing rate.

Having answered your question, I hope to your satisfaction, I would like to get an answer to my question.

CLACK: Which one? About why didn't the county get tax reserves?

HOUTHAKKER: That is one question. The other one is, how do you envisage rational planning of energy on a worldwide basis?

CLACK: I don't know that, sir. I am just suggesting that we could consider that because I don't think it is being done now. I don't think that the people of the world have had very much to say about it at all. Since people are the ones that are affected by the decisions, I think people should. It is an article of faith with me. I personally feel that the country has been misused in part by experts who considered that they had all the answers, and now we find ourselves in an energy crunch, with very few alternatives available to us to resolve it. Again, I admit, it is an article of faith. But then, a lot of things rely on faith.

HOUTHAKKER: What about the county?

CLACK: The county didn't get any taxes because up until about 1955 or 1957, for example, the Anaconda Company owned every single major newspaper in the state of Montana, and the laws of the state of Montana are such that it is possible for the Anaconda Company to avoid paying taxes, on the basis of net proceeds that are very amenable to careful juggling. The laws of the state have not been changed yet. That is why the county gets no taxes.

CALEB A. SMITH, Southeastern Massachusetts University: As an economist I wanted to object to the very narrow sort of economic interpretation that Professor Houthakker is using. It is very common in our profession to do this. I think perhaps we need to go back to the example that Professor Rose gave last night: if you take the value of all land in the United States today, which is determined on the basis of capitalizing the earnings that you can currently get from that land, you come up with a figure that is considerably smaller than the gross national product for one year.

What we have here is a situation that involves a conflict between two ways of looking at a problem. Given the exigencies of our present economic system, it is an inevitable result that we will have the valuation of land and of all natural resources on the basis of these discounted values with respect to the future earnings that can be obtained from the resource. This places a very disproportionate valuation on the very immediate future.

I have a little paradigm that I call the "Catch 2200 Syndrome." The Catch 2200 is, first, to take even very optimistic projections as to what the value of all the assets in the United States will be in the year 2200, projecting forward on the basis of rather outrageous growth rates. Then you use the sort of discount rates that we use to establish valuations and ask, how much can one afford to invest today in order to keep the United States from self-destructing in the year 2200? You come up with ridiculous figures between less than a billion dollars and around $75,000, depending upon the discount rate you use.

What we have here is an inherent contradiction within the framework that we are trying to argue about. Implicitly, you and the ranchers are arguing in terms of something that doesn't conform to these discount rate calculations that we make, and, implicitly, the economists and the businessmen put their faith in this sort of a discount rate. What we have to do is seek a solution for dealing with these conflicting points of view.

Angela Russell

Consultant and Educator, Montana State University

I appreciate this opportunity to make some comments concerning the Indian coal resources in Montana and what I see as some of the major issues, the problems, and possibly some of the resolutions for its development as an energy resource.

When I was initially contacted to be a panel participant, I was asked to give a somewhat objective viewpoint of the socio-economic factors fostering or impeding development of our coal. However, a detached objectivity is not a real possibility, since the coal issue has affected all of us, some more directly than others, and will continue to affect us. Therefore, my analysis has to be somewhat subjective also.

The Crow tribe, of which I am a member, and the Northern Cheyenne tribe, our neighbors, are owners of vast reserves of coal underlying their reservations. Estimates of coal reserves run anywhere from 10 to 30 billion tons, possibly more. Whatever the exact amount, it is a considerable reserve that has attracted major coal and oil companies to both reservations.

Beginning in the midsixties, before either of the two tribes knew very much about the potential of their coal, large acreages of reservation lands were already under lease or exploration permits to major coal companies. Areas under lease or exploration range from one-tenth of the Crow reservation to over half of the Northern Cheyenne reservation. It is reasonable to understand why the tribes entered these lease arrangements with numerous multinational corporations, such as Amax, Peabody, Shell, and others.

The prospect of financial returns, even at seventeen and one half cents per ton, would help alleviate some of the poverty conditions existent on both reservations--extremely high unemployment rates, per capita incomes below the poverty level, and all the other corresponding social problems usually associated with poverty, a situation which is all too familiar to most Indian reservations in this country. Not only would money be available for increased governmental services and projects, but there was a prospect of numerous jobs.

However, despite the financial benefits, the tribe soon learned of the associated impacts that would result from coal development at the levels projected by the companies. A massive influx of an outside labor force, strip mining of land that was subject to only the vaguest of reclamation standards, heavy depletion of water resources to which tribal rights have yet to be determined by the courts, and on and on.

A paramount issue also was the financial return to the tribes for such massive development. There appeared to be a real disparity when, on the one hand, the Crow tribe would receive an annual rental of a dollar per acre, while a white surface owner purportedly received from $3 million to $10 million for a couple thousand acres of land he sold to the same company. And again, while the tribe would receive seventeen and one half cents per ton of coal, the state of Montana would receive 30 percent through their new severance tax.

With this backdrop it was only a matter of time for the tribes to get organized to regain control of their land and lives from this development plan. In 1973 the Northern Cheyenne filed a petition with the secretary of the interior requesting the cancellation of their permits and leases, on the grounds that the Bureau of Indian Affairs, as trustee of tribal assets, had violated its own rules and regulations by the issuance of the leases and permits.

The Crow in 1975 filed a similar suit in Federal District Court against the secretary of the interior and the commissioner of Indian affairs to nullify four exploration permits and mining leases. In the Crow case the secretary of the interior recently ruled that exploration permits and mining leases be reduced to 2,560 acres to conform with required regulations. You might be interested to know that some of these acreages prior to the regulation enforcement ranged all the way up to 80,000 acres.

The secretary had previously made a similar decision on the Northern Cheyenne petition. At present, no new negotiations for coal mining have been signed by either tribe. The Crow, in 1972, entered into a standard leasing agreement with Westmoreland Resources in an area north of the reservation proper, referred to as the ceded area, which was opened for homesteading in the early 1900s; mineral rights were restored to the tribe by a congressional action in the 1950s. The renegotiated contract in late 1974 was heralded as the best contract between an Indian tribe and a major coal company.

The most tangible victory resulting from the renegotiated lease was the increase of the royalty rate for the tribe from seventeen and one half cents per ton to an escalating rate of the higher of twenty-five cents, or 6 percent, to forty cents, or 8 percent, of the selling price of coal, shortly after which period the state of Montana imposed their new coal severance tax.

For the Crow there appear to be many conflicting issues at stake in reaching a decision on further coal development. One has to take into consideration also the history of our previous experiences in coal development and the element of mistrust which thus exists. Needless to say, there is also the very real fear of the unknown.

One must first of all understand that the Crow are a traditional Plains tribe who still retain and practice many of the vestiges of Crow culture and tradition--native religion, tribal clanship systems, the language. As a sovereign entity the Crow maintain their right to self-government except as preempted by Congress. Although foreign concepts such as private property ownership and the profit motive of capitalism were introduced to the Crow with the establishment of reservations and white settlement, tribalism and one's identity as a Crow are of continuous importance. Many Crow believe that the very essence of their being is tied to the land. Without land they will no longer be Crow. Despite this belief, the reservation land base continues to diminish. Why? For similar reasons as have already been outlined in the poverty condition in which most Indian tribes find themselves.

To continue, the Crow life-style requires money, too, because of the extended family and clanship activities, which often require feasts and giveaways at important stages of one's life. An interesting aspect of this continual giving is that oftentimes the goods and services that are recycled are extended to reach numerous individuals.

Two major issues, therefore, appear to exist and will hold various degrees of conflict and compatibility with each other. One issue is tribal identity and its tie to the land, and the other is the tribal economic needs, collectively and individually.

What might then be some of the problems resulting from development in light of these two issues? With coal development, will not large acreages of land be out of the tribe's direct use and control? Given the fragile ecosystem of this part of the country, will reclamation work, especially in alluvial valley floors? How will coal mining and possible power plant development affect the land, water, air, and human and animal life? With skilled and semiskilled labor requirements, how much of the labor force will come from outside, and where will they live? If outsiders move on the reservation, will the Crow eventually become a minority on their own land? With an increased per capita income, will social problems on the reservation such as alcoholism, already of an extremely high incidence, intensify? What new types of exploitation will come to a people who have known only survival income for so long? Will development give energy self-sufficiency for the tribe, or is this energy to be provided for the rest of the country?

What are the alternatives for other types of resource development that may yield financial security to the tribe? And, most importantly, once the mining has ended, will the Crow as he is today still be here?

The Crow well recognize that change is inevitable. It is a process of life. The questions, however, are the rate of change and who determines that. Potential problems from coal development can well be numerous and never-ending if considerations are not given to how best to minimize them while there is still time to do so. Recognizing all the while that trade-offs have to be made, do the disadvantages of the financial return to the tribe and the prospect of new jobs justify the trade-offs?

So, the final question is, how do we make it work? What are some of the conditions under which coal development might proceed on the Crow reservation? One has to recognize, too, that there is a real possibility that the tribe could enter any number of arrangements to develop their coal tomorrow, given the fluid political process by which decisions are made and the very real pressures exerted, on the one hand, by the companies who want to develop and, on the other, by the tribe, which desperately needs money to maintain its rights as a tribe and upon which many members look for employment, credit, and numerous other services.

But in the midst of all this uncertainty are numerous tribal members who believe that if we are going to develop our coal, then let us do it right, which, they believe, takes considerable expertise, planning, negotiating, and time. First, it is essential that basic tribal regulating policies and mechanisms for their implementation be completed and functional for coal development. Part of this process will need to include clarification of tribal rights and the federal government's role and responsibilities as trustee.

Second, it stands to reason that if the Crow develop more of their coal, they should receive the maximum benefits from such development, in profits, in employment, and in usage.

Third, the tribe has to play a more participatory role in the coal-mining venture itself, for example, as partners in a joint venture arrangement, and assume some of the responsibilities and benefits from such. No longer are the Crow content with the argument that they do not have the capital investment for a project, since the coal is their investment.

And last, the coal development needs to be done in such a manner as not to lose sight of our basic values as Crow people, so that in the long run the Crow will have a bigger and better reservation. There are definite trade-offs that must be made. Development of some of our coal will change our life-style and isolation, but it may well provide the business capital to enter other areas that will not pose the conflicts that coal mining does and provide us with economic security. It was not an understatement when Crow Tribal Chairman Patrick Stands-Over-Bull declared at a public hearing in March of 1975 on the draft programatic environmental statement for projected coal development on the Crow reservation, that "the projected development could be a blessing or a curse."

ROSS FORNEY, Colorado School of Mines: I am certainly sympathetic to Ms. Russell's position. I think that all of us here feel that the Crow should have a major input into the mining and so forth of their reservation. I do think, though, that it is ironic that the Crow reservation sits on top of the lowest-sulfur coal in the country. On the other hand, we have an interest in complying with the EPA standards, which are requiring large quantities of low-sulfur coal. If the utilities in this country were not confronted with having to use compliance coal, there would be very little, if any, interest in developing the Crow reservation coal at the present time.

RUSSELL: As I said, it was only in the 1960s that we suddenly looked around and found ourselves surrounded by energy companies that had taken up large tracts of land, one-tenth of our reservation, as a matter of fact. Part of that time we weren't thinking about any type of energy development. We do need new kinds of employment coming in, we need a better economy. But, here again, you know, our value system comes in, and we have to really assess where we want to be. I think at this point, too, the federal government hasn't made any definite declaration of a comprehensive energy policy.

Robert L. Valeu

Basin Electric Power Cooperative, Bismarck, North Dakota

I would like to preface my remarks by clarifying a certain assumption that was made on my part in preparing for this presentation. And that is that while we may presently be in a dialogue concerning priorities--sources of energy, technological means of energy--there is also the reality that energy development is going on in order to meet today's demands and the demands of the consumers in the next four to five years particularly. Those demands and the energy development in the Great Plains area create a very real problem for the people who must be faced with this so-called rapid population influx.

Given that assumption--that there will be some level of development of coal in the Great Plains, at least in the foreseeable future--the real issue becomes one of, can the communities manage effectively the socio-economic impacts under this rapid growth condition? What we have found today through the voluminous studies that have inundated the West is that there are many different approaches that have been experienced. We saw and discussed earlier the Gillette situation, wherein, basically, the free economy system, *laissez-faire*, took hold and actually dictated growth, and now we have a situation where, through the mayor's efforts, the political process is trying to throttle or cap the *laissez-faire* concept and formulate some realistic approach to dealing with the growth.

The Colstrip situation is one in which the economic process just would not respond in Rosebud County, Montana, so it was left to the specific industry going into that specific area to create an environment for the people who would live there and generate the fuel source that was necessary.

We feel that somewhere between these two extremes is a rational approach to dealing with community impact problems. We were fortunate at Basin Electric Power Cooperative to initiate what we like to refer to as an integrated approach to community growth management, the integration being between the industry coming into the area and the local community.

Wheatland, Wyoming, provided the ideal situation and has proved to be a very exciting experience and, we feel, a very profitable experience, not only for the community but also for Basin Electric Power Cooperative. One of the fortunate conditions that existed in this situation was a single industry impacting a single community, and the variables were not too erratic at the time.

I do not want to go into a detailed discussion of what is transpiring in Wheatland, Wyoming, right now. I only want to say that Basin Electric Power Cooperative and its consortium of companies are in the process of building a power plant there and did make front-end financial commitments to assure that the community of Wheatland would be prepared to accept the impact as it was expected to occur in the years 1976 and 1977. This meant comprehensive planning on the part of the community, with the technical assistance of the industry representatives, to provide the infrastructure facilities in advance of the impact. Today, this day, Wheatland, Wyoming, is probably overdeveloped in the sense that our construction work force level is approximately 200 men and we have a new elementary school under construction. We have 438 housing units under construction and actually available. We have bachelor quarter facilities; we have the whole system there waiting for the impact to occur.

What we are hoping to do now, through rather sophisticated planning and monitoring processes, is to transfer this integrated approach to community impact problems to North Dakota. There, a consortium of three industries will construct energy facilities, beginning in the fall of this year, in a county with more than just one specific town that will have to accept impact.

We feel that somewhere between the Gillette experience and the Colstrip experience is a rational approach that was implemented in Wheatland, Wyoming, in particular, and that will be refined in Bismarck, North Dakota. Someone indicated that in Wheatland, Wyoming, there was extreme opposition from the ranchers toward our project. To put that in its proper perspective, there was a group of seventeen ranch families who organized themselves so that they could express their opposition or their concerns to the development. They were then able to obtain extraregional support through the Sierra Club and similar entities to finance their effort. But I wouldn't go so far to say that there was total opposition.

Another comment that was made earlier--and I think that it is important that it be stressed--is that the local community really has the last right to say whether or not any industry may come into its community through the political process of zoning and subdevelopment division, code requirements, and that sort of thing.

I might add that one of the ways in which we feel that adequately preparing the citizens of Wheatland will be beneficial to us as a cooperative is that through the provision of housing, in which we finance approximately $10 million of housing, we will be able to effectively compete with the labor force, we will reduce turnover, we will reduce our absenteeism, and we will increase productivity in the construction stage of our facility. Increased productivity is significant when your interest on construction runs as high as $184 to $200 million during a seven-year period. So those are some very positive benefits. Additionally, the facilities being put into the community not only provide an opportunity for adequate housing and other facilities for the permanent residents of the community who were there before the construction but also enhance the suitability for the permanent work force that will develop.

What we have found is that the federal response to dealing with community problems is just not there. The system of allocation of funds through federal programs just does not fit the uniqueness of impact conditions in the community. Basically, there has to be a certain understanding as to what the community consists of in approaching the question of community impacts. The participants can be defined as the public, individuals, private industry--most notably those industries that are going to effectively increase the population--specific interest groups that form themselves to represent a particular position, and political entities, or jurisdictional bodies, that have constitutional and legal responsibilities for promoting the public welfare of the community.

Once that community is formed, either formally or informally, there has to be a certain understanding of prerequisites if a community is going to manage its growth.

First, there has to be an adequate flow of information, information not only from the public sector but most obviously from the private sector. Additionally, there has to be an understanding and a recognition of participation on the part of the communities that are going to deal with this question. Further, there has to be some definition of the public-versus-private responsibilities for dealing with social and human welfare, and there has to be some basic understanding of the geographical area because, unfortunately, communities are defined by geography, whether it be political or natural geography.

Once that structure is established in a community--and we were able to do this in Wheatland, Wyoming--then there are four steps to effecting the analysis. It is a process whereby you take the community as it exists now and you take the information from the industry, the most reasonable, most accurate available information at the time, and through the scientific modeling techniques, whether it be economic or whatever, you analyze what the potential numeric impact is going to be on the community. I emphasize the numeric impact.

The community has to establish some type of expertise to carry on this analysis. It does not exist in most communities in the western Great Plains areas. The federal government has attempted to initiate all kinds of studies that would help with this expertise, but in the final analysis the industry is the most reasonable source of providing the expertise to do the analysis because of requirements under the NEPA Act or other federal, state, or local permitting processes.

The output of an analysis has got to be accepted as being in the ball park. It cannot be absolute; it cannot be accurate. The day on which an analysis is printed, it is changed. What the output is is a representation of potential forecasts in terms of employment, school enrollments, population, and fiscal aspects of the community. Those forecasts, then, set the stage for the community to initiate the second phase or the second process of growth management. That is what we call assessment or problem solving. The community must take its values and assess what this potential forecast means in its community. Do they want it to in effect occur that way, or do they want to do something to change it? In most instances a change is the desired solution. That change is created through the third phase, which is strategies, and those are the trade-offs that the community must make, determined by all the influences in the community.

The trade-offs are then established solutions that are represented in the form of mitigation, an action process by the community to implement specific strategies given all of the expertise in terms of managing the mitigation, funding it, financing it, planning for it, and timing it. Once that is completed, then the only way a determination can be made as to whether or not that action by the community has been beneficial in terms of the judgment of the community is through some refined process of community monitoring. That monitoring step is the fourth step that initiates the continual dynamic process of growth management, not only during the period of rapid growth but also after the community has experienced rapid growth and during the period of long-term growth.

The federal government has a role in each one of the steps, a role that has not been, in my opinion, adequately defined to date. In the analysis stage there has been no clarification under the NEPA Act as to what an analysis really means. There has been some clarification in the sense of the environmental concerns but not in the sense of the socio-economic aspects. Therefore, what happens under the NEPA Act is that analyses are completed and they become representative of an end rather than a means for the community. In the assessment stage the federal government has not really identified a role for itself, either through technical assistance or by providing grant mechanisms to allow the community to affect that process.

In the strategy stage in Wheatland, Wyoming, our experience, as I mentioned earlier, was that we spent exhaustive hours and tremendous amounts of money to try and ferret out federal programs that would be applicable to the impacts in Wheatland. The only one that we found was a combination of Farmers Home Administration and HUD. Unfortunately, the attempt to put two federal programs together to resolve senior citizens' housing problems in the community took twenty-four months to effect. In the monitoring stage, again, we did not have a role.

So my recommendations are, first, that there is a process whereby the community can deal with the question of growth management, a very viable process. Second, I think that the federal government, while it must rightly discuss the much broader and philosophical question of where we go in this country, also has a responsibility to deal with the realities of day to day life, and that means establishing a viable role for itself in community problems. The same applies for the state government. Third, I think that we can establish an appropriate definition of a role between the private sector and the public sector to effect this process. I would just like to say to you it is happening, it is occurring, and it is not all gloom and doom.

SPEAKER: You mentioned that your company hopes to realize increases in construction worker productivity due to the better living conditions in the town. Do you have any quantification or any feel for what this will be? I have heard of this before and have looked into it and have never been able to find an expert to quantify it.

VALEU: It is very hard to quantify because most productivity and efficiency is measured from an engineering point of view. It is hard to go out and ask a worker, did you work harder today because you have nice living conditions and your family is very satisfied and you don't have the frustrations of all types of things?

We have attempted to analyze it quantitatively. We won't know whether or not that works until after construction. But there are some ways to attempt to measure it. We feel that we will probably save in the neighborhood of $24 to $35 million on a project that is estimated to cost $1.4 billion.

WHITE: At the Western Energy Company mine we have had a less than 2-percent annual turnover because of dissatisfaction with living conditions. We have never had a strike there. I think both of those things have a value. It is hard to put a dollar amount on it, but they certainly have a value.

COLEMAN: One of the questions you raised about government strikes leads me to a philosophical response. Government is stuck in history. It is inertial. It doesn't forecast; it doesn't operate in a value consensus that allows it to take chances with future-oriented distributions of money, and I think that is the nature of the problem.

The industry does have to live to a certain extent in future orientation. Therefore, the kind of systems results orientation that you did in problem solving is something that I think is perfectly legitimate for business. I would also like to see that application when business up and leaves. That is, when we have declining industries--say, as an example, the impact of the shifts of the textile and shoes from the New England states down to the South--we should have that kind of a system for measuring and dealing with change too, as it relates to that kind of an industrial change situation.

VALEU: Industry's involvement and either its acceleration or deceleration of development are predicated to a great extent on federal policy. Therefore, I think that the federal government has to awaken to a much more real situation and it has to get involved in the public sector needs of a community.

We are running a very serious risk here, quite frankly, from a philosophical standpoint, wherein private industry is supposedly being taxed through the appropriate mechanisms to fund these public problems. But what is happening is that the bureaucratic process gets bogged down, and those tax revenues never get reallocated or back to a specific community. So you are dealing with a site-specific problem.

What happened in Wheatland, Wyoming, was that Basin Electric was taxed twice. We were paying all the appropriate *ad valorem* taxes, corporate taxes, whatever, and at the same time we had to establish a contingent liability of somewhere around $19 million with no assurance that we would be able to recapture that liability for public sector needs.

Another problem with it is that if the private sector finances public sector facilities, it is very costly to the community. An example is schools. General obligation bonds have a much higher rating and are much less costly to the community than if the private sector goes out in the open market and obtains the financing at 4, 5, or 6 percent higher than that.

COLEMAN: But the question is that you can move faster than the government can. Take our friend there with the Community Development Administration. She has been tied up in court. Why? Because of value problems, problems of literally being able to take a resource base and capitalize on it so a problem can be met in time. A public entity keeps running into those problems.

So the trade-off is the question of having to pay a higher interest rate and solving the problem or not solving the problem and living with all of the social and health dislocation that would attend that.

VALEU: I think the real solution is the ability of the federal government to react more aptly, if you want to call it "reaction." One entity that has the capacity to do that in the federal government are the Title 5 commissions, which have been able to demonstrate a very unique ability to provide direct revenue and technical assistance to these communities. But they are the only entity at this point.

DISCUSSION

HOOVER: I would like to open the floor to general questions, and there are three general topics we might discuss. I wonder if we could focus questions along one train before we move on to another.

One was this general topic of solution management at the local level: Who is involved? What are the roles of the "feds," the local community, and industry?

The second topic is regional economics: What does coal development mean to the region, particularly in terms of competition with other sectors?

The third topic is the general topic of limits to growth, either from a national or world perspective, or options for the region in the sense of exporting coal versus exporting the product.

JOHN MAYBERRY, U.S. Geological Survey, Denver: Most of the speakers have touched on the problem the communities and regions have of importation of nonlocal talent for developing the mines and the coal and attendant processes. There are several areas in Colorado where mine companies have come in and have made a very large point out of hiring local people to help them. They have given them schooling and direct employment in the local job base.

In the Navajo area of northwestern New Mexico and northeastern Arizona, the Navajo have written into their contracts for coal leasing the idea that the mine companies will hire local people wherever they can before they will bring in outside talent.

My question is, can't the regional commissions, the local political organizations, the people who own the land, the Crow, the other Indian tribes, make an impact into this area where they wouldn't have to have such worry over the importation of nonlocal talent?

STUDER: I would like to respond in part to that. I think I pointed out that most of this region does not have an unemployment problem. So even if you take the workers from the service employment and give them a higher-paying employment in construction, somebody has to take those service jobs. So we don't have a surplus labor market. I think Martin White pointed out that in their case they had used an awful lot of local people, as many as they possibly could have. We just don't have that much of a surplus of labor, though, and when we are talking about multiple mines, multiple power plants, there is absolutely no way we can fill it from the local market.

HOOVER: Are there other questions directed at local social impact management?

SPEAKER: I would like to hear Ms. Russell speak on the subject of what the Crows are doing to provide and upgrade personnel so that they can take management positions within the coal industry that is developing on the reservation.

RUSSELL: I would like to add that I have been down to some of the coal-mining areas down on the Navajo and Hopi reservations and was very impressed by the fact that there seemed to be sizable numbers of Navajo working there. I think there was only one Hopi individual. And as I understand it, it was written explicitly into the contracts that certain levels of percentages be employed from the Navajo tribe.

I certainly think this is an option for the Crows. I think I made it quite clear that we got into this whole business with our eyes closed, and there are certainly a lot of things we have learned since then; this is very much one of the concerns.

Also taken into consideration is the fact that our labor force isn't all that large either, but the employment demands are certainly there. There are many people unemployed, and in the wintertime up to 50 percent of the labor force can be unemployed.

HOOVER: Are there any questions having to do with the region's economy?

WHITE: I guess the first speaker brought up the word "colonialization," and other speakers brought up "nationalism" and "internationalism" and "regionalism." I think this is a real problem in the country today, the fact that we are pitting regions against regions over a number of issues. We want to talk about doing national planning. I don't think there is enough wisdom in the country to be able to do a national and international energy planning that will tell us whether or not we need to build a surface mine and power plant at Colstrip. I think it can give us some broad ideas.

In the area of regional economics, Mr. Clack said that he had some inflammatory remarks, and I, as a state representative from Texas, will respond to his comments.

I think that perhaps the straw that broke the camel's back in the area of regionalism was perhaps Montana's imposition of the 30-percent tax. I think what we are seeing today is a lot of states looking for tax mechanisms to shift tax burdens outside of their boundaries to other states in the form of severance taxes. I know in Texas we are currently looking at a refinery tax to shift the burden outside the state. I think we are seeing states playing off against each other. I know that the Rand Corporation found that the severance tax in Montana was substantially in excess of the social and economic impacts of coal development in the state of Montana.

I have no quarrel at all with charging taxes in a state to compensate for the impacts those states feel, but I question whether it is good judgment on the part of a state to charge a tax that is in excess of those costs. It has been found that Wyoming, Montana, and North Dakota have a sufficient share of the market; they could probably form a tax cartel and tax at a substantially higher level than they are doing now and still be able to bring in those revenues. I think they could perhaps get away with it on an economic basis, but I question whether or not such policies are in the benefit of the nation.

You are saying, we need to look at what is good for the country. Should we develop western coal? Should we develop eastern coal? We should do some planning, but I am saying that at the same time, we have regional policies that directly conflict with that attitude.

CLACK: Let me start first with an allusion to what I said earlier, and that is that insofar as we don't really know how much energy we need to use right now as opposed to what we want right now, it may be a tragic error to deplete what limited amount we have left. There are lots of uses for coal besides generating heat for boilers, hydrocarbon feedstocks, and so forth.

There is a great feeling in Montana of having been exploited. Whether or not anyone else would regard it as exploitation, there are many Montanans that do. The domination of the state by a few corporate interests for many years is a good case in point. I think there is a feeling in the state that if they are going to get that coal, which is going to disrupt us, they are going to pay us for it.

Now, it is not quite as cold blooded as that, in that right now, regardless of what the Rand Corporation says, no one knows how much money it is going to cost Colstrip--well, I guess Martin White does, because Colstrip is essentially a company town--right now nobody knows for sure how much it is going to cost Forsyth, which is about thirty miles up the road, and the little towns around Forsyth, Miles City and Circle, where Burlington Northern is perhaps planning a syngas and synthetic fertilizer plant using coal, and so on. No one knows exactly how much that is going to cost. One early estimate from Montana's former Department of Community Affairs was that in three years, just local government costs, given an influx of 30,000 people, would be $93.6 million, in the first year.

Another case in point is that the coal taxes that we have gotten are allocated to a number of places, not just the community impacted, but also to the state general fund, highway fund, alternative energy resource development, and so on. So it doesn't all go directly to the impacted community.

I think right now that the only thing I can say is that nobody knows for sure that we are the blue-eyed Arabs. No one knows exactly what it is going to cost, including us. Until such time when it is demonstrated that we are ripping the country off, I think the country should lay back and get the data.

WHITE: Texas being the blue-eyed Arab of the 1930s and 1940s and, up until the earlier part of this decade, being the state that was exploited by the FPC and other groups in the area of natural gas, I share a lot of those sympathies. I still contend that making pessimistic estimates is not in the interest of optimizing the national good, however we want to optimize that.

CLACK: But again, the question is, what is the national good? I can't resolve that. I think no one here can resolve that. I think we need to look at that.

JAY JACOBSON, Battelle Northwest: Another question I had in regard to local economics is, do any of the panelists have any idea whether secondary industry will be attracted by the availability of electric power or coal to the region? I have never seen anything on this.

PAYNE: In Wheatland, I know, we are actively encouraging that. This is something that we anticipate happening seven or eight years from now. But we are in the process of developing an industrial park and trying to attract industry. That is not to say it would happen naturally, but we are not content to just allow it to happen naturally, we actively are seeking some kind of tapering off like that.

JACOBSON: Has anybody indicated an interest in what type of industry?

PAYNE: It is a little early for us to say that. Right now, what we have done is obtained the funds for the development of the industrial park, and somewhere down the line here a ways we will have an attractive setting and then see who bites.

SUMMARY REPORT TO THE PLENARY SESSION

L. John Hoover

Our group discussed the northern Great Plains, and discussants on the panel not only covered Colstrip and Gillette but also situations in Wheatland, Wyoming, and in Mercer County, North Dakota.

I can't really say that there was a consensus, but there were general areas of concern. These are in the following areas: (1) rapid growth, how to deal with it, and who the participants in growth management are to be; (2) the general area of risk and sharing of the risk; (3) the concerns of the Indians; and (4) the general topic of regional economics, limits to growth, and the need for a national energy policy.

In regard to the first topic of rapid growth, there was a general feeling that dealing with rapid growth required a strong commitment and involvement of the community. It required community action. In particular, there was a concern that the federal government was and is too slow to deal with the problems of rapid growth, that our federal government is established primarily to deal with urban problems, and that many of the characteristics of the rapid-growth communities do not fit the federal funding requirements.

For instance, in some cases, decisions for grants by various agencies of the federal government are based on census information, what has happened between 1960 and 1970. This can give a very false impression, for instance, about a Gillette, where the growth has been significant only since 1970.

For some programs, sewage treatment plants, for example, there is a tendency for the federal government to be cautious in giving funding grants because the facility may encourage growth, whereas in fact in the rapid growth area, the facility addition, whether it is a sewer or a waste treatment plant, is needed to deal with the growth that is there now and that is a reality.

In addition to community action, there was an expression of the need for industry's involvement with the community in a cooperative mode because industry can act fairly quickly.

The problems of growth are dynamic, not static. There is no need for static assessments. In fact, there was a general concern expressed by a number of participants that they have been studied to death. One could form a boomtown for the people who are researching Gillette; they don't need people coming out there and telling them what their problems are. What they need is technical assistance. These people need a say in the new programs that are set up to assist them and any new studies the federal government undertakes.

There was some expression that there is definitely a role for the federal government, as yet undefined, that must be defined. One particular participant expressed his belief that the Title 5 commissions could be of great value in dealing with rapid growth, that they are responsive, they are helpful, and they can be timely.

On the second topic of risk, there was a concern expressed that the risk of financing service needs should be shared. The country, not just the states and the localities where coal is mined, should be investing in those areas and helping to finance public service additions. This results from a general concern about a possible change in our resource use policy in the future. The western states may someday be in the position that perhaps the midwestern and the eastern states now find themselves in: ten or fifteen years from now we may make a change such that we will not continue to use western resources. Then these states will have to bear the burden of paying for unused facilities.

With regard to Indians the general concern expressed was one of their not wanting to be exploited. Assistance is needed--technical expertise and capital. The conflict is one of the change in the culture versus the possible improvement in the quality of life. There may be an improvement in the quality of life (jobs, schools, et cetera), but it may result in a major disruption of the Indian culture. The question is, "Is coal a blessing or a curse?"

The topic of employment of Indians in mines was discussed. Some particular tribes have taken a very aggressive stance and have required that as a part of the contract to open the mine a general agreement about employment of members of the local tribe and education must be reached.

At the regional level, a concern expressed is, when does energy development no longer become a stimulus to the economy but become the economy of the northern Great Plains? This area is a ranching-agricultural economy, with coal presently making a small contribution.

Finally, the last topic, at what I would say was perhaps a global level, was concern for limits to growth, the general expression that there is a need to define our energy policy in this country, there is a need to address conservation and its role before we go very quickly and develop our resources, particularly our resources in the West.

INQUIRIES FROM THE PANEL

HOUTHAKKER: I just want to make one comment, which is only indirectly related to the panel itself. Incidentally, I thought the discussion was very useful in focusing on one particular aspect. I would have liked to have seen other aspects discussed, too. The aspect discussed most was, what can communities do and not do about the impact of some major energy development? That is a very important subject, which I am sure doesn't get anywhere near enough attention from people who live outside the area. I think there are many other questions that I would like to have seen discussed, for example, the whole question of reclamation. But that was not what the panel members were particularly interested in.

Now, I think maybe the other panel members who were not present at the discussion, as I was, may want to ask specific questions. I want to come back to one remark just made by Mr. Hoover about the national energy policy that everybody is asking for. I am a little skeptical about this idea. I am sure it would be nice to have a document laying out a national energy policy that we could all quote. But I think the past history of these things is not all that encouraging.

We do have a national transportation policy, which is the preamble of the Transportation Act of 1940, and I think it is a terrible policy. It has caused great inefficiency in our transportation system in part because it is worded in such a way that it will appeal to a large public and actually can be interpreted in ways that were never intended. I am afraid that a national energy policy might be just as bad.

Actually, what a national energy policy should be like isn't all that obscure, and we don't really need to enact anything like that. The one conclusion I have come to, and maybe others have, is that the one thing we don't need is a policy based on one particular thing, be it nuclear energy, or conservation, or coal, or whatever. We need something of all of these. I think that is about as far as we can go in deriving a national energy policy. The only thing we have to recognize is that there is a place for everything and that ultimately we have to count the social cost of all the various forms of energy, including the social cost of conservation. Beyond that, we cannot expect too much.

I found the most heartening result of the panel discussion, which I attended, to be that some of the communities described, and undoubtedly there are others, have apparently done quite well in dealing with the great stresses caused by energy development. They have grappled with all these difficulties more or less on their own, for the reasons that Mr. Hoover has just described, apparently with some success. Now, I believe that this suggests that one shouldn't rely too much on the federal government because the federal government in the end is likely to be more harmful than helpful in these matters.

ROSE: Did you consider how much coal could be taken from that area by the present or revamped transportation system by some particular year?

HOOVER: No, we did not address broad regional questions such as what is achievable for the whole region. The basic assumption, I think, of this group and the experience of the discussants were such that they lent themselves to answering the question, if we are going to have growth and development, how can we deal with the problems? That was really the central question that we addressed.

ALLEN: I had supposed that the problem of growth centering around new mining exploration and development was not a new one for this area. Is it the size or is it the modern viewpoint of what should be done to contain growth that leads you to your present inquiry?

HOOVER: I will answer that from my own perspective and then call on perhaps two of our discussants. I think its size is of a great concern out there. If one thinks about a typical synthetic natural gas plant, he is talking about a construction employment of perhaps 2,000 people and an operating force of 800 people. Many of the communities out there are of the size of 1,000 to 2,000 people, so the direct labor requirements themselves are substantial, not even thinking about the indirect or secondary development that may come along with that.

Is Mike Enzi here, the mayor of Gillette? Mike, would you like to comment?

ENZI: Growth may not be new to the West, but growth of these proportions and the expectations of the people who are coming into the areas are creating a lot of problems. If you bring someone in from Houston he may not expect a lot of rainfall, but he will definitely expect the types of cultural activities that he had at the place where he was before and will need them in order to have satisfaction even among his family. So we are seeing a lot of new types of problems with the kind of growth that we are having now, probably unprecedented growth, at least in our area.

WEINBERG: I am a bit puzzled because, again from my nuclear background, wasn't the growth in the Hanford area even larger than what you are speaking of here? Hanford, you know, sprung from a barely incorporated town to 75,000 and then leveled off at--well, Richland leveled off at something like 20,000 or 25,000.

ENZI: Once again, you get into a problem of people's background experience. I am not even familiar with Hanford, and the people that come to Gillette are not familiar with any of the other growth towns either. They know what kind of a town they left, and they therefore expect the town where they are going to be in exactly that same condition, regardless of how rapidly the additional infrastructure needs to be put in.

WEINBERG: Well, of course, Richland was a company town, and most of the inhabitants of Richland, I think, are very pleased to live there.

HOOVER: I think perhaps there might be some real differences here in the way in which capital was obtained and was available at that time versus what it is possible to do under these situations.

WEINBERG: It was government, of course.

BOYD: Most of my questions have been answered in advance. However, I would like to point out that I have been around mining camps all my life. I was born in one, and I lived in it until I came here to Washington. We never had any question of how we handled the infrastructure of the camp in which we worked. In one large mining camp that I operated, we had to go to the federal government for help, and we got into more trouble than we asked for. We would have been a lot better off if we had financed--in fact, eventually we had to finance it ourselves.

So it seems to me that we are violating the very principle we talk about when we start talking about bringing in help from the outside. This is the job of the management itself in relationship to the people who are involved in the local area. It has to be done in that way so that you can--what the economists like to call--internalize the costs of the infrastructure to produce that operation. This has been an age-old operation in the mining industry. I don't know why it should change. We can always help, perhaps, but I think we would be better off to leave the cost of doing business to management.

HOOVER: Perhaps Robert Valeu, in relation to that point, would like to talk a little bit about his experience with Basin Electric, which is precisely a cooperative community-industry experience.

VALEU: In reference to the last comment that was made, there is no question of the ability of management to internalize some of the cost relative to what we no longer call a construction camp but a bachelor quarter facility. Those bachelor facilities are more in line with the new expectations that construction workers illustrate as they work on power plant facilities in particular. What we have found is that the decision-making process in the public sector has involved itself in the decision to provide the types of facilities that either the state government or the federal government feels are suitable in the local community.

A prime example is the state of Wyoming, which has initiated an Industrial Siting Act, which actually at the state level puts the primary responsibility for meeting the infrastructure requirements of a community on industry, regardless of what the local community feels is adequate or inadequate.

HOOVER: Martin White is here too from Western Energy Company. Martin, are you in the audience? Would you like to say a few words about your experience at Colstrip, which is virtually a company town?

WHITE: Colstrip is probably more similar to the camp that you speak of. It was developed in 1973 as a mining camp. But at that time the maximum number of employees was around 100, and we had probably 2,000 employees during the height of construction of our project there. We were only one industry, and we could handle that problem. But if, as is expected, over the hill, another company, Mobil Oil, say, comes in and decides to develop a project and make an addition to Colstrip, and another company wants to make another addition--pretty soon it gets too complicated for industry, and I think at that point there has to be some government entity that involves itself in coordinating these activities. It is not quite as simple as it probably was before to just dig in and handle the problems.

PLENARY SESSION DISCUSSION

HIBBARD: I want to correct an impression that Jim Boyd might have given you. White Pine, Michigan, is no mining camp. It is a perfectly reasonable, normal town, isn't it, Jim?

BOYD: We made it that way.

HIBBARD: That is right. These are not mining camps. You go to Pea Ridge in the Ozarks, and you will find the same thing. There is a town that Bethlehem Steel and St. Jo put together, and the town is a perfectly normal town. I am just a little concerned that you are writing off the efforts that the mining companies made as inadequate or have a vision of a mining camp with three tents, a saloon, and a local sheriff. These are very nice, adequate places.
What is the population of White Pine, Jim? I have forgotten.

BOYD: Oh, it is about 2,000 now, I think.

HIBBARD: But it has its own theatre and its own everything.

DAVID WHITE: One of the things everybody talked about was the ability to make decisions, and the problem in making decisions is that we need to develop some mechanism to do this. We have also spent a lot of time talking about a national energy policy and how nice it would be to have one. As somebody pointed out, we have a national energy policy today. Perhaps we don't like it, but the fact remains that we have some sort of a policy. I heard the comment that somebody was glad to see President Carter elected simply because we would have a Democrat in the White House and Democrats in Congress and perhaps rather than adversaries in a partisan atmosphere there would be some agreement between the executive and legislative branches.

I was at a conference last week, and the problems of how large the United States is and developing a national energy policy that would recognize regional needs and needs of various consumer groups and industry groups were brought out. The thought came to me about how much a lot of people think of the idea of appropriate technologies. If you have an indigenous resource, if you have something available to you, it should be used, to generate whatever needs are at a local area. I got to thinking about how ironic it is that we think that small is good in technology but, when we start thinking about an energy policy, we want a national energy policy that will solve all of our problems.

In policy we think big is good, but on technology we are coming around to the position that small is good. I find that to be a tremendous irony. I speak from a state perspective and feel that the best thing that a national energy policy can do for us is to provide guidance to the state, not explicit programs.

I would like views from the panelists on whether a national energy policy can recognize all of the diversity that exists in the United States and make a policy that is not only appropriate for the state of Texas but also fair to the people in Montana and the people in Ohio.

HEIBERG: The obvious point is that when you have something happen in state A that affects state B and state C, it is very hard to have state A take into account the needs, desires, and demands of states B and C unless you have someone who can work across the state lines. That is one point.

The other point is that in our panel we discussed the need for a national policy not because we wanted more policy or more regulations. In fact, if I were to say there is a consensus, it would be that we all feel we are better off with less policy and regulation. However, what we don't like is unanswered questions, or questions that have not yet been wrestled to the ground, or, even worse, questions that have answer A today and answer B tomorrow. That is the context in which my panel wrestled with that one.

CALEB SMITH: On this matter of getting definite answers and decisions that stick, I can see the great advantages and in a way, the necessity of this. On the other hand, we are always in the situation of getting new knowledge, and when we get new knowledge and find out that a particular decision was definitely wrong, I think we have got to be in a position to retract that decision. It may involve indemnification of people who have gone ahead on the basis of the wrong decision that was made by the governmental authority at an earlier time, but I don't think we can just say we are going to stick with the decision even though new knowledge makes it now certain that is is a wrong decision.

SINGER: I would like to get back to land. One of the speakers in the northern Great Plains discussion, who put forth, as he called it, a populist point of view, asked, rhetorically I am sure, how much land is going to be torn up to give us the coal we need? He didn't bother to answer the question, so I would like to do it for him.

To those of us who are familiar with eastern coal and its really quite thin seams, the thickness of the western seams is really fabulous, ten to fifty times as great as seams that we would consider to be commercial in the East. This means that the amount of land that is torn up becomes relatively small. If one considers that all of the coal produced in the United States were to be produced in Montana or Wyoming, the amount of land torn up every year would be something less than twenty square miles.

I wondered why this speaker who objected so much to the tearing up of this land, which is then later restored, didn't object to or wasn't so vocal about the fact that some of this land is torn up to make highways, interstate highways, and is torn up for other various purposes. I don't see that one is socially more desirable or useful than another.

I think that one needs to consider these things in perspective, and perhaps the small amount of land that needs to be disturbed to get the coal we need is worth reflecting on. In fact, of course, we would not produce all of our coal in Montana or Wyoming. I am sure the East would like to produce some coal too, so that whatever land is disturbed will be a great deal less.

CAROLYN ALDERSON, Burnie, Montana: There is a question of the amount of land that is going to be torn up, a small amount in relation to the total national acreage, and a question of having an efficient mining camp in Colstrip or Gillette. Another assumption that is made is that there is nothing or nobody out there doing anything and that it is simply there for the taking.

I live not very far from Gillette and even closer to Colstrip, in the same county, in fact. There are lots of people there. They aren't many in numbers, but they are doing something. There is a productive industry going on. It doesn't produce the amount of dollars, but there is something else going on, and that is that people do eat. There has to be some consideration of the population and the people, the few of us that still live there.

HOOVER: I think this point of conflict with agriculture is central. It may not be the conflict of the amount of land but the conflict with another sector of the economy and the conflict with individuals that become crucial.

In the West the Indians' concern and their land ethic is a key issue. There is perhaps a different respect and feeling for the land than we Anglos have at this point in time. We have to be very careful about that ethic and about protecting it.

DAVID WOLCOTT, Energy Research and Development Administration: The more I hear representatives of the utility industry expressing concern about the delays inherent in litigation processes and the more

I hear about environmental groups with perhaps superfluous lawsuits, the more I wonder if the time is not coming in which environmental mediation services, either private or on a government level, will be appropriate means of settling out-of-court types of disputes.

ALLEN: Could I make a brief comment on that? I think this is an interesting idea, but you don't do any mediating until both sides have gotten their firm bargaining position. If you deny the *pro bono publico* environmentalist fellow a right to intervene in court he has no bargaining position, and I don't think he is going to accept you.

WEINBERG: Any other comments or questions?

GREGORY GOULD, Fuel Engineering Company of New York: I have been listening to the discussants here last night and today, and I recall back in September of 1973 I was addressing a group very similar to this. It was sponsored by the Illinois Capital Development Board, and they had originally invited me to come talk on energy conservation. I told them I would like to come and talk, but I felt that maybe we ought to be talking about the availability of energy. And in my address I pointed out that we were developing a very definite dependence on foreign sources of supply for energy needs, that the Arabs seemed to be far more informed on this than we were, and that we could expect some very adverse effects of it.

About a month later an oil embargo was underway. Then we began to learn something about "what happens if."

Now, I think all this discussion is great. I think it is necessary. But I also begin to wonder, do we really have enough sense of urgency about what we are up to? I think one of the subjects that we should be discussing is some kind of an emergency plan that we would dedicate ourselves to in providing sufficient energy for basic needs. I don't think we really have to worry about the Arabs pulling another embargo. I think that is out of the question. I have been over there; I have talked to some of the people, and I have a very different feel for their attitudes toward us than I used to. But it is evident to me--well, there is a consensus here. We are going to burn more coal. We are haggling over how, where, when, and how much. But we are going to burn more coal. All of this talk may turn out to be academic tomorrow if it blows up again in the Middle East, and I think the possibilities for that are tremendous. Then what are you going to do? Then the time for talk has ceased, and you are going to have to do something. That might be tomorrow.

I would like to hear the panel's response to those thoughts.

HOUTHAKKER: I think the point just raised is a very important one. I am also very concerned about our present vulnerability. We are not vulnerable to an embargo. The embargo in 1973, contrary to much popular opinion, was really a complete failure and did not succeed in cutting off our oil supply to any significant extent. It was, however, of great psychological impact because it put us in an absolute panic, of which the gasoline lines were, of course, the most spectacular manifestation. So we are not vulnerable to another embargo because, by now, most OPEC members are importing as much as they can afford, and they all need us to put spare parts into their fighter bombers and whatnot.

So the embargo is not the real risk. The real risk is a continuing increase in prices, and the possibility for that certainly exists because we are as vulnerable to that as we ever were. Indeed, we are more vulnerable, and I think to the extent that we have to think about energy policy, we have to think about the fact that by increasing forever our demand for imported oil, we are on a rising supply curve and will have to pay more and more for each additional barrel.

WEINBERG: I think the time has come to close the discussion. I was pleased, however, that the last speaker did say that there was a general consensus that we must burn more coal. Again, going back to my nuclear background, I have recently heard the arguments about nuclear energy characterized as being a religious war, the essence of the matter being that, as far as nuclear energy is concerned, there does not seem to be that consensus. So I hope you do not leave this auditorium crying "Oh woe, oh woe," as our good friend Professor Rose suggested last night, because at least the coal controversy has not yet risen or sunk--however you wish to say it--to the level of a religious war. Therefore, we trust that it can, one way or another, be resolved without the War of the Roses or the Thirty Years' War.

DAY III

INTRODUCTION TO DAY III

Walter R. Hibbard, Jr.,
Cochairman

We have heard speakers discuss several key questions: How fast should coal be developed? Where should coal be developed? What will be the environmental impact and costs? The general consensus seemed to be that we ought to use as much coal as we can, wherever we can get it, provided we can meet the constraints of the environmental health and safety issues.

Frank Zarb seemed to think that for the next ten years the die had already been cast, that the decisions had been made in the board rooms with regard to how much coal we are going to have, and that any change in policy or discussions at this Forum would have only a limited impact on that amount.

Richard Newcomb indicated that he thought that coal could be cleaned to a degree that would be environmentally acceptable, and he projected a price of $1.50 per million Btu, which is something on the order of $40.00 a ton. This is one solution to the burning of coal, although it is not the total solution.

We have heard three case histories. The first relates to the Ohio River Valley, where 140 million tons of coal are mined, and 75 million are shipped by water out of the area. The evidence is that there is plenty of water to handle more coal shipments, but there are three locks, two on the Ohio, one on the Mississippi, that might be bottlenecks with respect to barge shipments.

We heard about Kaiparowits in southern Utah, a large, 5,000-megawatt electric plant for which planning was started in 1963. After delays due to water and environmental questions it was canceled in 1976 after its costs had escalated from $500 million to several billion dollars.

We then heard about the northern Great Plains, Colstrip and Gillette, and the whole spectrum of questions of social, economic, political, and quality-of-life issues. How can one sustain the rapid growth of very small towns with populations in the range of several hundred to a couple thousand? What are the consequences if these growths are not properly handled? The general feeling was that assistance from the federal government is too slow and that the planning and implementation of rapid town growth can be best handled by the industry involved. It is particularly effective if it is one town and one industry.

We have had a series of seven workshops. The ones that I observed were quite vigorous, and I am looking forward to the reports from those workshops.

COAL TECHNOLOGIES

Thomas V. Falkie
Senior Vice-President, Natural Resources, Berwind Corporation, Former Director, Bureau of Mines

I am going to talk mostly in terms of mining, coal preparation, and transportation technology, hoping that George Hill and the others will pick up on utilization, conversion, and combustion.

There are three basic problems connected with the mining of coal. All are related and one cannot be considered without the others: (1) health and safety problems, (2) productivity problems or, if you will, production and economics problems, and (3) problems connected with environment. Problems are associated with each and all of these. In health and safety we have a steady trend of improvement with which no good engineer or technologist is satisfied. There are still major problems to be solved in such things as dust suppression and control, mine lighting, noise suppression and control, roof and ground control, gas suppression and control, and a whole host of others that I could list here.

In the environmental area, in underground mining, there are problems connected with mine subsidence and mine drainage control and treatment. In surface mining, a topic that seems to be of interest to many of the participants here, the challenge is to build the reclamation into the mining system; and there is probably some particular set of challenges with regard to reclaiming land in arid and semiarid climates.

Let us talk for a minute about productivity and production. I think we can classify some eras of productivity in coal mining. Until the 1930s, the middle and late 1930s, most of the coal that was mined in this country was mined by hand-loading methods. Mechanization began in the 1930s, and the 1940s was a period when mechanization was done in earnest, much of it with track-mounted equipment. The underground-mining productivity rose from 4 tons to 6 tons per man-day. In the 1950s and the advent of trackless equipment, the productivity rose from 6 to 11 tons per man-day. The 1960s was the age of the continuous miner, and the productivity rose from 11 to 16 tons per man-day. The 1970s is the era of productivity decline, and the most recent number I have seen from the Bureau of Mines is that the productivity last year was something of the order of 8.5 tons per man-day. In other words, there was a drop from 16 tons per man-day to 8.5 tons per man-day in seven or eight years. The 1980s will be covered under future remarks.

The fact is that the design or theoretical capacities of existing mining systems--continuous mining, conventional mining, longwall mining--are much greater than the attained capacities. In fact, the maximum attained capacities, the records that we have been able to attain with these systems, are much greater than the average that we are attaining.

There are reasons for loss of productivity in underground mining. I might add that there is a downward trend in productivity in surface coal mining as well. These reasons are sometimes stated too emotionally and are often not stated completely.

First, in the period of productivity decline there was a great influx of inexperienced workers, engineers, and management into the coal industry. Second, the work rules have changed. There is a job-bidding procedure and some other work rules that have had an effect on productivity. Third, it has been an era of labor unrest, wildcat strikes. While you are looking at it strictly from the engineering standpoint, you say, well, when they are out on strike you don't count that in the productivity calculation. The fact is that the turmoil before and after and the start-up and shutdown problems do in fact contribute to a decline in productivity. There is no question that the Coal Mine Health and Safety Act of 1969 and other government regulations have had a negative impact on productivity.

There are also some contractual changes between the bituminous coal operators and the United Mine Workers, such as the adding of extra people on work crews, that have a negative effect on productivity. There has been a general decline in growth rate of productivity in our country. Some people like to call this a worker attitude problem. I don't like the term, but nevertheless that is how it is identified.

There is no question that there have been more difficult mining conditions in recent years and more marginal mines in an era of expansion following an era of contraction in the coal business. Also, because of this, because of the fact that there are many new mines in the seventies, there is more development work as opposed to production work in mining. To sum it up, there are more people in mines now doing the same thing to produce similar or less tonnage. So there are many, many reasons other than just the health and safety law that have contributed to a decline in productivity.

I think it is fair to point out that when we think about productivity, we should also think in terms of the cost, the price in dollars per ton.

There are technological factors limiting productivity and there are many common problems to be solved: more difficult working conditions; deeper, thinner seams; underground mining of thick seams; equipment reliability problems; problems of whether we should be in fact satisfied with an average underground recovery of coal of something like 50 or 60 percent.

There are a lot of goals that we can set for ourselves. Let me just cite a few of them for developing and improving underground mining systems. We must increase the average production per shift from both conventional and continuous mining methods by increasing the reliability of existing equipment and improving productivity substantially by automating and combining or "remote controlling" as many tasks as possible. We must accelerate the use of longwall mining to increase the percentage of coal recovered and to mine efficiently and safely coal deposits at greater depths and under difficult strata conditions.

We must recover the methane from coal seams prior to mining to speed production by eliminating methane problems, to make mines safer, and in some cases to make use of methane as an energy source in itself. We must develop mining systems that can safely and economically recover 80 percent or more of the thick and steeply pitching seams of coal, especially in the western coal fields. We must advance the technology to provide adequate protection of the surface environment from underground mining, that is, from subsidence and water contamination.

We must reduce the time required to open new mines and bring them into full production. We must conduct critical proof of concept tests that will lead to the development of the new mining systems of the future and identify health and safety problems connected with these new mining systems and eliminate them from the systems. We must eliminate or reduce disasters caused by fires and explosions in underground coal mines by improving detection, suppression, and extinguishment technology.

On the health side, we must continue to improve the technology to protect miners from exposure to respirable dust, noise, toxic gases, and, in the case of other types of mining, radiation hazards.

We must prevent coal bumps and accidental falls of roof, rib, and face by improving artificial support, hazard detection, and mine-opening design. We must develop and improve the technology to improve the probability of a miner surviving a disaster. That is, we must improve rescue methods and mine-reopening techniques.

Of industrial-type hazards, we must identify and provide technology to correct hazards in electrical systems, mechanical systems, illumination, communication, haulage, materials handling, and other areas. We must improve the business of waste management, waste not only from mining but from the processing step as well. And, in my judgment, an area of great potential is human-factors-technology development for underground and surface coal mining.

What has been accomplished and what is in the hopper? What does the future look like? In underground mining and the health and safety aspects there has been quite a bit of development over the last seven or eight years, development in environmental monitoring, measuring gases, and continuous dosimeter-type alarms. In this kind of thing the technology has come along very, very rapidly. In respirable dust control, although we are not satisfied with it, there certainly have been major technological advances in the last eight or ten years.

The basic technology has been developed for draining and recovering methane. Communications systems, the art of communicating underground and between the ground and the surface, are improving at a very rapid rate. Again, I could cite a whole litany of other accomplishments in this area.

In the area of underground mining connected with productivity--I don't like to talk about it separately from health and safety--I think that what we will see in the next ten years are more high-speed development systems in terms of using tunneling-type borers and other boring-type equipment to open mines faster, to shorten the lead times in which mines can be open.

I think we are going to see automated longwall systems and remote control longwall systems. I would like to make a distinction because when you throw up the term "remote control," it conjures up removing people from the mines. I don't think we are going to see complete underground mines with no people, perhaps not in my lifetime, but I do think we are going to see much in the way of remote control in underground and in surface mining in the next ten years.

We are going to see, perhaps, an automated, remote control, continuous miner, a programable, continuous miner; no men will have to work right at the face itself. We are going to see further improvements in automated roof support systems. One of the things that the mining industry is very enthusiastic about now is the miner-bolter concept, the concept of combining steps in underground mining. This is something that the Bureau of Mines is doing in conjunction with the industry, and in my way of thinking it is going to be a success within the next five to ten years. We are going to see great improvements in coal haulage systems from continuous face to preparation plant. Probably the biggest single part or step of underground mining that prevents us from attaining the theoretical production capacity of any of our systems is removing the coal from where it is mined at the face and taking it out so that it can be transported to the surface. I think we are going to see many advances in things such as bridge conveyors, extensible belt conveyors, rail haulage automation, and improvements in shuttle car haulage that will, as I said, make this process more continuous.

There is quite a bit of work going on right now in the business of mining coal by hydraulic methods, slurry systems--not only the mining of the coal but the transporting of the coal. There will be rapid strides made in the next five to ten years, although I personally do not look for a great percentage of coal to be mined by slurry systems by 1985.

In a new concept area it is highly likely that such things as borehole mining of coal from the surface and pitching seams in the West and perhaps in anthracite will find limited use. I think there is going to be some underground augering done.

In the surface mining area there is also a great potential, and I think that here it is not a case of our finding dramatically new methods, or even new equipment, for that matter, but it is a case of putting together existing pieces of technology into more systematized approaches, building the land reclamation into the mining system. We will have, for example, trials on hauling spoil or hauling overburden by conveyor belts, where this is rarely done now. This has potential for being advantageous both in the mountainous mining of the East and in the area mining of the West.

I think there is going to be further automation of the dig/haul/dump/spread cycle in surface mining, and there is going to be a lot of work done and accomplishment made in terms of making the equipment more reliable.

We are going to see some advances made over the next ten years or so in the development of special mining systems for underground mining of thick seams, especially in the West.

In the environmental area of underground mining, subsidence correction and prevention will get increased attention, and we must find ways of designing mining systems and operating mining systems--better ways to minimize the results of surface subsidence.

There has been a lot done on waste bank fire control and on acid mine drainage. I don't look for any radical breakthroughs in these areas. I think the fundamental technology is there, but there is always room for improvement.

There is, in my judgment, a lot of work that can be done--in coal preparation and in the business of waste disposal, not only recovering carbon values from waste but also minimizing the environmental and health safety problems connected with storing wastes, especially fine coal wastes.

The general trends then in technology in coal mining are more automation and remote control, combination of unit processes, and leveling-out of size of equipment in surface mining. I would predict that we are not going to see too many Big Muskie-size drag lines built. I think it is just a case of putting too many eggs in one basket, and you will see a leveling-out, not only of the excavating equipment but also of the haulage equipment. There may be some increase in size of support equipment, like bulldozers, and this kind of thing. You will see greater equipment reliability and integration of excavation and reclamation in surface mining.

There is going to be another trend, and here I would like to go out on a limb a little bit and make a prediction, especially in underground mining. I hope that the labor unrest will calm down over the next ten years. But I think that you are going to see a limitation on the size of underground coal mines because of the labor problem. I think that people are going to be reluctant to start huge mines and subject them to the possibilities of problems due to wildcat strikes and so forth. We will see whether that prediction comes true or not.

What we will see, and have been seeing since around 1970--and I can't prove this--is a greater ratio of technically trained people to capital dollar invested. What I am trying to say is that I think we will see that we will build more engineering into mining systems.

The government has a role, a definite role. Right now most of the research and development dealing with extraction of mining is force-fed by the government. It is dominant, even though the government is spending only somewhere between $100 million and $150 million, and I say "only" in relative terms. There is no question that the government is the dominant force. Very few mining companies have their own formal research and development organizations. The total research and development done by industry is not adequate, but industry does more research and development than what it gets credit for. Industry is also getting more involved in government programs through cost sharing and cooperative projects. That is a good sign.

There has been some discussion in this Forum on coal preparation. You prepare coal to do several things--to remove sulfur, to remove ash, to dewater coal--and coal preparation is concerned with waste control, recovering carbon values. To an increasing extent--I would predict over the next ten years--attention will be paid to the business of trace element control. I think we are going to see no real, dramatic, major breakthrough in coal preparation in the next few years. However, there will be more automation in preparation plants. There will be better control systems. There will be a lot of striving toward higher recoveries of carbon values. The average recoveries now in coal preparation vary all over the map from 90 or 95 percent to 40 or 50 percent.

We will see some technology development, some unique froth flotation methods, pyrite flotation, for example. We will see perhaps high intensity magnetic separation for the pyritic sulfur. We will see a great deal of work being done in the removal of organic sulfur by chemical means, and we will see a lot of progress in the business of dewatering coal, not only for recovery purposes, but for waste control purposes.

I am not going to say much about transportation. I will say a few more words about it when I give the report. I think that most of the problems on rail, barge, and slurry transportations are not technological but rather economical or political.

In conclusion, we can no longer plan, design, build, or operate mines by seat-of-the-pants technology. That era is gone. We must find ways to build more engineering and science into the mining systems. This has been going on for years, but it needs to be improved. The fact is that there still are some people who are what I would call running up mines by the seat of their pants. I would contend that not only is there a need for a change here but also you are almost forced by government regulations to build more engineering into the system.

The technology in the area of mining, preparation, transportation, and perhaps even exploration will be evolutionary rather than revolutionary. We had a lot of talk about whether we should have a moonlike approach. I don't think that will work because we are dealing with a problem that has real-world economics connected to it, and the actual costs of mining and preparing coal are translated into costs that the consumer must pay when he buys power or pays for steel.

I do feel that the government and private sectors have definite roles to play. I think that the government sector role is limited and, at the present time, not at the level of what it should be. There is a need, in my judgment, for more centralized mining research and development on the part of industry. There may be a mechanism in place through Bituminous Coal Research. I am not boosting them, but I do feel that the industries can do some things collectively, without too many problems connected with the antitrust business, such as are being done in South Africa, Poland, England, and Germany.

Incidentally, I am not one of those who gets ecstatic over what is being done overseas in comparison to here. I think our technology is better than anywhere else, and our health and safety records and productivity records are still better than anywhere else. Nevertheless, they do have some good things, and one good thing that most foreign countries have going for them is collective research on the part of their mining companies. Some of them, obviously, are government controlled.

The government research and development programs, which really were boosted by the Coal Mine Health and Safety Act of 1969, are beginning to pay off at a very rapid rate, and accelerated results are coming and will come in the next five to ten years.

I have one comment, perhaps a biased comment, that I would like to make about the governmental effort. There is a proposal right now to fragment the government research and development effort into so-called productivity research, health and safety research, and environmental research. While we all recognize the need for reorganization in energy, I personally think that will set back the government's mining research program. The fact of the matter is that mines are designed around health and safety problems, and you cannot separate these three factors. All underground mines are designed around two things: ventilation and ground support. When you develop a mine system to get adequate ventilation and ground support, you also must consider the productivity problem. If you design a system that is going to cause too much subsidence, you don't have a good system at all. So all these problems are interrelated. What we really need is an integrated systems approach.

Technology is not the limiting factor in doubling coal production in the next ten years. But I don't think that anybody will be satisfied with the productivity or the health and safety and environmental progress if we do not continue to improve this technology.

THE KEY ROLE OF COAL

George R. Hill

*Director,
Fossil Fuel Power Plants Department,
Electric Power Research Institute*

I appreciate the opportunity to talk with you about planning to meet the energy needs of the United States. I would like to discuss and analyze the problems that have been preventing achievement of national goals in energy utilization.

My secretary made an interesting Freudian slip as she typed up my dictated notes. My statement as prepared reads: "Analysis of problems preventing achievement of national coals in energy utilization." To meet our *goals*, we must use coal.

What I would like to do is to help us recognize where we have been and where we are and then chart a course to where we need to go in terms of utilizing coal to solve our energy problems. I would like to begin by laying a background with which many of you are already acquainted; however, I would judge from previous discussion that there is merit in an overview.

Until just a few years ago we were operating in this country as if we had an infinite supply of natural gas, oil, and coal. These resources were assumed to be of zero value in the ground. Now we recognize that the resources available at up to two to three times present cost of production are indeed finite. We should not have allowed ourselves to misuse gas and oil. We finally realize that we must optimize the use of each of our fossil fuel resources, making them environmentally acceptable with maximum efficiency.

A graphic method is used to predict the production or lifetime of a single oil field. It consists of the superposition of two areas. In a rectangle the total producible oil in a particular oil field is given. The total recoverable oil would be in the area included in the rectangle. A bell-shaped curve represents the production pattern. It will be low at first, build up to a maximum, and then taper off as the rate of production goes down. The time and magnitude of peak oil production can be varied quite a bit. But the area under the bell-shaped curve must equal the total producible oil and hence equals the area of the rectangle.

Logically, the total world production pattern of oil will follow a similar pattern. Three alternative production patterns are possible. If we exercise no controls on oil consumption whatsoever, as we have done in the past, the consumption of fossil fuel could increase exponentially with a catastrophic decline as the resource runs out. If we institute reasonable controls and emphasize conservation, we can do a great deal to postpone the time of serious energy shortage--but not enough to solve our problems. We could have an early peak with a tailing-off that would allow us to extend the availability over a longer period. If we have a very serious effort at conserving and if we are willing to change our life-styles, then of course we do extend our resources much further.

Now, let us look at the nature of growth in demand--a phenomenon which causes concern. We have an almost insatiable demand that is reflected in an exponential increase in energy consumption with time. It is inevitable that any exponential demand curve will intersect the supply curve for the product, which generally increases linearly. With natural gas, that happened during the past decade. The ratio of proved reserves of natural gas to annual consumption peaked out in about 1968. There are many places in the country now (last winter provided a dramatic illustration) where insufficient gas is available to meet the demand. Worldwide, the demand rate for petroleum is still at the beginning of the exponential increase curve. Probably before the year 2000, all major nations will be competing severely for petroleum, and I believe we will be in a very serious situation.

Table 1 will help orient us to how much coal we have (which we are concerned with in this particular symposium). These data are from Parent and Linden.* The publication has analyzed various World Energy Conference reports and other sources of information and estimates that we have close to 600 billion tons of recoverable coal reserves. The total resource of coal, i.e., the coal available at very high cost of development and production, is of the order of ten times the present economically producible reserve of a trillion tons.

You are all familiar with the widespread distribution of coal reserves in the country. The Appalachian and midcontinent area coal reserves are familiar to all. The lignite in Texas (formerly considered spotty deposits) has increased in amount until it has almost become a continuum. The lignite is essentially distributed clear across the state. The intensive core drilling in the state, which has proved up the reserve is principally due to the electric utility interest. The vast majority of the coal in the United States is in the lignite,

---

*J. S. Parent and H. R. Linden. Survey of...production...reserves...resources of fossil fuels and uranium. *IGT*, January 1977.

TABLE 1 Comparison of 1974 Coal Resource Data of Averitt and WEC, 10^9 metric tons

Region	Original Resource Estimates of Averitt[a]		WEC, 1974 Current Resource Estimates		
	Identified	Total Resources	Reserves Recoverable	Reserves Total	Total Resources
North America	1,560 (1724)	4,173 (3992)	187.3	372.6	3,033.3
South and Central America	18 (27)	27 (36)	2.8	9.2	32.9
Europe[b]	562 (272)	753 (726)	131	365	653
Asia, including USSR	6,350 (3629)	9,979 (9979)	234.1	613.7	6,821.7
Africa	72.6 (81.6)	218 (227)	15.6	30.3	58.8
Oceania	54.4 (63.5)	118 (118)	24.5	74.7	199.7
United States	1,474 (1908)	2,971 (3947)	181.8	363.6	2,924.5
Canada	86.6	1,202	5.5	9.0	108.8
USSR	5,897		136.6	273.2	5,713.6

SOURCE: J. S. Parent and H. R. Linden. Survey of...production...reserves...resources of fossil fuels and uranium. *IGT*, January 1977. Figures in parentheses are Averitt's revised estimates.
[a] In beds at least 1 ft (30.5 cm) thick and generally less than 4,000 ft (1,219 m) below the surface
[b] The sum of the individual WEC data for European countries is greater than the total figure given by WEC. The former is arbitrarily used in this table.

subbituminous, and bituminous fields in the Far West, where coal occurs in multiple and thick seams.

CONVERSION DRIVING FORCES

Now let us look at the driving forces that are going to determine in large measure how we convert coal to oil and to gas and how we use it most wisely. The first of these is the change in environmental perception that has occurred in our time. We have the basic requirement to protect the health and welfare of the people of the country. There were, and still are, some effluents from automobiles and from industrial plants of various kinds that need to be reduced.

A higher level of control than just to safeguard health is one we are currently striving for, viz., that of achieving zero environmental degradation; it may cost us more than we can afford. The present environmental control law (even before the one that is now under discussion in Congress) has been interpreted as requiring plants to use the best available technology to reduce effluents regardless of cost to the consumer. This means that many of the control devices that will be required in industrial plants of various kinds can exceed the health and safety requirements as we move toward the goal of zero degradation of the environment.

The extension of this, of course, is the achievement of zero growth in population and industrial productive capacity. This proposal would freeze things as they are and then try to improve what remains. The public may be a bit reluctant to accept that solution without more reasons than have been given at the present time.

The second driving force in planning coal use development is to increase the efficiency of utilization of the resource. We really have a responsibility to safeguard future generations in terms of saving some fossil energy supplies for them. Unfortunately, most of our efforts to improve the environment have been counterproductive to increased conservation. Almost every equipment addition required to meet EPA regulations has been at the cost of increased consumption. The increases are of the order of 25 percent to 40 percent more fuel used. Some apparent compensation has been achieved by reduction in size and weight of plant (automobiles, for example), but the apparent gains in efficiency are more than counterbalanced by the losses. We need to strike a balance between how much improvement we are willing to pay for and what it will cost in terms of lack of availability of those resources for future generations.

The third driving force that we need to consider is national security. Petroleum imports, prior to the OPEC embargo and the price hike, represented some 30 percent of the oil consumed in the United States. By last year that figure had risen from 30 to about 45 percent. The bulk of the increase, of course, has come from the Middle East and North Africa. The imbalance of oil payments of $25 billion to $35 billion a year, our dependence on that oil for our very survival, and the physical location and political instability of those areas are causes for great concern. We must take a good, hard look at whether we ought to have a technology in place, or at least available, that would allow us to survive if we were to be denied that oil.

The problem facing the government, facing you, facing all of us, is to solve these problem equations simultaneously and to come up with the best, or at least reasonable, solutions. At this meeting, we should suggest solutions that hopefully maximize health and safety, clean up the environment, optimize the utilization of our resources, and protect our national security. This is an extremely tough job. No one has yet written a computer program that begins to solve those simultaneous equations.

COAL UTILIZATION OPTIONS AND PROCESS IMPROVEMENTS

Let us now look at the coal utilization methods that I am asked to address directly. First of all, the direct utilization options that will help meet environmental regulations include coal cleaning, mentioned by Dr. Falkie, former director of the Bureau of Mines.

Table 2 tabulates the permissible levels of sulfur oxides, nitrogen oxides, and particulates that can be emitted from power plants--per 10^6 Btu of fuel. Present day maxima are for 1.2 pounds of SO_2 per million Btu of coal fired in a power plant, and 0.7 pounds of NO_x. These are achievable with clean coal and with slightly modified power plant operations. Suspended particles are controlled to 0.1 pound per million Btu. For fine particulates, there is no national standard at the present time; however, one state has a standard of 0.02 pound per 10^6 Btu.

TABLE 2 Electric Utility Emission Standards: Gas Cleaning--New Source Performance Standards (lb/10^6 Btu)

	SO_x	NO_x	Suspended Particles	Fine Particulates
1975	1.2	0.70	0.10	N/A
1980	0.8	0.40	0.05	0.02
1985	0.6	0.15	0.05	0.02

Water cleaning: eleven constituents in seven different streams in the plants must be controlled to prescribed levels.

Thermal discharge: for any plant started after 1974, there must be no thermal discharge from main condensers.

Reduction to lower levels will be expected in new sets of performance standards in 1980. NO_x is dropping to 0.4 pounds per 10^6 Btu. Suspended particles may be cut in half, and there may be a fine particulate standard imposed nationally. These have been discussed by EPA but are not yet in place. In 1985, some regulation levels could be established that would require every power plant, regardless of how low the sulfur content of coal it burns, to have a flue gas desulfurization unit. The projected NO_x level appears impossible to achieve through control of combustion and would require an, as yet, undeveloped technology for decomposition of the NO_2 and NO in the stack gases. Suspended particles are at a level that we believe we can achieve with the new technologies, some of which I will talk about in a moment. This includes the fine particulate matter.

Figure 1 illustrates the coal that would become available for direct firing (meeting current regulations) if we are successful in developing the magnetic separation techniques and the chemical coal cleaning processes currently being developed. Three systems are being tested now.

Each bar in the figure represents the coal that could be produced annually from mines currently in operation or under development in various regions of the country. The top portion represents western coal, the central bar is the coal from the central part of the United States, and the bottom part is eastern coal. The improvement in coal quantity, if you go to a physical cleaning operation using current technology,

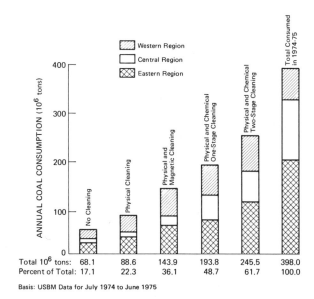

FIGURE 1 Potential impact of coal cleaning on annual coal consumption by utilities in compliance with NSPS (1.2 pounds of SO_2 per 10^6 Btu).

would be about 30 to 40 percent above what we now have available. These numbers were obtained by using Bureau of Mines data, principally. Extrapolating from the Bureau of Mines data by utilizing the data that have been published on the degree of sulfur reduction that can be attained in magnetic cleaning results in an increase of 160 million tons of coal that could be produced each year.

By utilizing new chemical cleaning techniques, including a two-stage chemical cleaning operation, it is conceivable that you could increase the quantity of coal to over 250 million tons per year from the present level of 60 million to 70 million tons. That would be coal cleaned to a quality that would meet the present, mind you, 1.2 pounds of SO_2 requirement. When you drop down to 0.816, you reduce those columns very severely--almost to zero. That is why I draw the conclusion that unless we come to an adjustment in our SO_2 attitude, we will in fact require the expense of putting a flue gas desulfurization system on every coal-fired power plant. The increased costs of these systems may run $100 a kilowatt in a plant whose base cost is $600 to $800 a kilowatt to construct (plus another $50 a kilowatt to dispose of the sludge that is produced by the flue gas desulfurization system).

Second-generation, regenerable scrubber systems reduce the sludge disposal problem. Several regenerable scrubber options have been proposed to the Electric Power Research Institute for support. The institute is cooperating with ERDA, with EPA, and with individual utilities to develop operable, reliable scrubber systems that will meet the requirements. We recognize that the population of the United States has come to expect available electricity at all times. Adding to the complexity of the power plant reduces the availability of electricity from the plant. The overall availability is determined by multiplying the fractional probability that each step in the system will operate without interruption. If you put two more units in a power plant whose individual reliability might be 70 percent, you reduce the overall reliability of that system to below 50 percent. There is a lot at stake in developing these regenerable, more complicated, flue gas desulfurization processes. They will be more costly but will produce marketable sulfur and/or sulfuric acid as products rather than the present sludges that are very costly to dispose of or to store.

Let us now look at what is being done to improve particulate removal. Electrostatic precipitators, operating in their present mode, do not do an effective job of removing the very fine particles (in the 0.1- to 1-μ particle size range). This is the size particle that is inspired as we breathe and that can remain in the lungs. These particles can cause congestion. Unfortunately, most present electrostatic precipitators do an inadequate job of removing particles of that size. They do well in removing the large particles but the small ones tend to go on through the system and up the stack.

Soon to be marketed is a high-intensity ionizer that has been developed from a concept in the mind of a university professor to modular physical testing at TVA. The ionizer is a device designed to add a high-intensity electric charge to particles entering the electrostatic

precipitator. The high-intensity electric field charges the small
particles when they go into the electrostatic precipitator so that they
are attracted much more efficiently to the electrodes. This can result
in a 70-percent increase in efficiency of particulate removal. This
system reduces by about half the cost of rebuilding and enlarging the
precipitator--the alternative method of achieving the same goal.

Another development we are excited about is the superimposition of
an alternating-current field on the charged particles to clump the particles together and reduce even more the escape of fine particulates.
I believe we will be able to achieve the particle removal that is
needed in order to protect the health of the population.

COST-BENEFIT RATIO

The EPA forcing function is going to prove expensive. In my opinion,
we need to strike a balance between how much we are going to demand
in terms of quality of effluent and how much it will cost the public,
who ultimately pays the bill. Estimates of the increase in electric
bills range as high as an additional $400 per family per year in the
Southwest if we have to put on all of the devices that may be required.
I am not sure how much the public is going to be willing to pay in extra
costs that are over and above the increased cost of the power due to
increased fuel costs. I hope that meetings such as this will help
us reach a balanced judgment in terms of what we are going to do.

COAL GASIFICATION

Basically, in gasification, one reacts the coal with air or with oxygen,
and with steam, to produce a mixture of carbon monoxide and hydrogen.
If air is used, the product gas ends up diluted with nitrogen and is
a low-Btu gas. If oxygen is used, the concentrated CO and hydrogen
(after the gas stream is cleaned) can be recombined to produce methane
(SNG) or chemicals. At the Sasol plant in South Africa, motor fuel components and chemicals are being produced commercially from such coal-derived gas. The gas is an excellent fuel to replace natural gas
for industrial or power plant use.

One unfortunate thing about the conversion processes is that it costs
25 to 30 percent of the energy originally present in the coal to carry
out the conversion reactions. It also costs a billion dollars for each
plant to do the job. It is imperative that we move ahead with these
gasification programs to produce industrial gas and, more importantly,
synthetic natural gas (SNG). (The additional methanation step required
for production of SNG from the primary CO-plus-H_2 mixture has been
demonstrated in three or four devices developed through OCR, AGA, and
ERDA contracts recently.) We are in pretty good shape, as far as the
technology is concerned, for producing an industrial gas, a gas for
boilers, a gas that furnishes the hydrogen for upgrading of liquids,

and a gas that makes hydrogen for synthesizing ammonia. All of these can be done now. We have new technologies being developed that may decrease the cost of the synthesis gas generation to perhaps 25 percent less than the cost of gas from Lurgi, Koppers-Totzek, Winkler, and Wellman Galusha gasifiers that are commercially available now.

COAL LIQUEFACTION

In liquefaction of coal, one can dissolve the coal in a coal-derived solvent, add hydrogen to it, and produce a large quantity (at a thermal efficiency of 70 to 80 percent) of petroleumlike liquid. The oil can be hydro-treated in typical petroleum refinery operations to produce the equivalent of gasoline, diesel fuel, chemicals, and boiler fuel, which we desperately need to supplement our reducing supply of naturally occurring petroleum.

The Solvent Refined Coal (SRC) process takes the first step toward producing a liquid. SRC processing produces a solid that has the mineral matter and the unreacted coal separated from the organic constituents by filtration. The inorganic sulfur and ash constituents are essentially removed. The product is a high-heating-value substitute for coal.

Carrying the SRC process one step further, by hydrogenating a little more, the solid is converted to a liquid. The place in the system where you separate the mineral matter is fairly critical. With SRC and in the H-Coal or Exxon Donor Solvent (EDS) process for making coal into liquid, you filter or separate, by centrifugation or distillation or some other technique, the residual solid matter after you have generated the liquid. In the case of the EDS process, the oil is distilled from the solid and then is catalytically hydrogenated. This avoids the problem of contamination of the catalyst used for the hydrogenation stage. In addition to the three options mentioned (which are being developed with partial funding by ERDA), Gulf Oil and other petroleum companies also have programs under development. The Electric Power Research Institute is cofunding three of the processes with a $40 million commitment to one, and about $10 million in the others to encourage the development of the alternative sources of liquid fuels. We will need continual supplies of clean fuels that can be burned in locations where coal itself cannot be burned and in the existing oil- and gas-fired turbines and boilers, which cannot be retrofitted to use coal. When you burn coal in a boiler, a much larger chamber is required than that for oil or gas firing. This is due, in part, to the fly ash and the slag that are produced. One cannot convert a boiler that has been designed to burn oil or gas only to a coal-firing mode--thus the need for coal conversion.

Another approach to coal conversion is the process called hydrocarbonization. The COED process developed by FMC under Office of Coal Research sponsorship ten years ago is one such process. The Coalcon process is an alternative designed to maximize the yield of gas and oil, leaving as residue a smokeless char, which could be burned directly for power generation. Such processes are difficult to commercialize.

Oil companies are primarily interested in the production of a liquid. The natural gas industry needs synthetic natural gas. The utility industry wants a fuel that it can use as simply and cheaply as possible. None of them have an interest per se in the simultaneous production of all three fuels to raise the capital to meet all the environmental complexities that would be required to build gas, oil, and power complexes. Here, an incentive program is clearly needed.

IMPEDIMENTS TO PROGRESS

What are the reasons for lack of success to date in moving into these coal conversions options? There are really four of them. The first of these is the marketplace itself. An oil company, if it were to build a coal-to-oil plant (and we don't yet have the technology proven at the commerical size) would be investing its stockholders' money in a plant to produce an economically uncompetitive product. A drop in the OPEC oil prices could wipe it out. There is just no way that an oil company, on its own, is going to invest money that way. Therefore, we have the marketplace problem that has to be solved before we can move ahead with the synthetic fuels program.

The second difficulty is the unfortunate phenomenon that if the work is done in national laboratories, the Congress-devised requirements to protect the public's interest preclude a particular company from exclusive commercialization rights. A process developed in a national laboratory or in an energy research center must be equally available to all industrial developers. As a consequence, many great things done in these laboratories never reach the marketplace. The rare exceptions are where a process is developed in a field where there is no industrial counterpart. We did develop nuclear reactors; we did develop the moon landing capability and the space shuttle through our national laboratory efforts, but in each instance there was no preexisting component in the economy with operations and investments in the field. In the liquefaction and gasification areas, there already are companies that are in the business of distributing oil and gas.

A third basic factor that accounts for lack of success is the wish by Congress to protect the taxpayer from "overpaying" for a product. This has been reflected in the historic FPC regulation that held natural gas prices to levels below replacement cost to protect consumers.

The fourth problem is the public's inadequate perception of the energy demand. The public wants convenience and availability of energy at low cost. The basic problem is that each decision maker, as ecumenical as he feels he is, is principally concerned with a limited aspect of the world. Our dilemma reminds me of the reactions of four blind men who were taken to a zoo. Each was given the privilege of feeling an elephant and then describing it. One, as you recall, grabbed hold of the tail and said an elephant was like a rope; another got hold of the leg and said an elephant was like a tree; another one grabbed the trunk and said it was like a snake; the fourth felt the side of the elephant and said it was like a wall.

I get the feeling that depending on our point of view, each of us is like one of these blind men. It is going to be difficult to open all our eyes to see what a complicated elephant the energy animal is. And we had better learn to manage its entirety, or the nation is going to be in serious trouble.

There is another aspect of this, of which we need to be aware. Building new technology, even with everything going for it, requires at least ten years before there can be a demonstration plant. That plant will have to be on line for a couple of years to prove its operability and reliability. This is a long induction period.

Furthermore, the time it required to change from dependence on one energy source to another is much longer than ten years. For example, from 1850 until 1910, wood was the prime fuel. It took sixty years before wood was essentially replaced by coal. And then from 1920 to about 1960, coal use declined as oil and gas filled an increasing fraction of the demand. The time required for that transition was about forty years.

Now we must reverse the trend and get coal back into the picture and decrease the quantity of oil and gas use consumed. We must restrict our use of oil and gas to smaller units, not large industrial plants. We need gas to heat our homes and the like. Coal can be used in environmentally acceptable ways in large industrial and power plants.

Another fact we must recognize is that the rate of penetration of the market by a proven technology is also very slow. For example, it has taken between twenty-five and thirty years to get a penetration of our electrical energy production system by nuclear power of only 10 percent. Even when a technology is fully developed, the penetration rate in the marketplace takes decades.

It is going to be difficult to make much change in our present mode of operation certainly in the next ten or fifteen years. That does not mean we must not work hard at it. On the contrary, recognizing the long time for change requires that we get at it now--with enthusiasm and increased vigor.

DISCUSSION

BOYD: I don't think that I could add much to what these two speakers have put out in such complete detail, but I think the crucial question is what, in each of their minds, needs to be done to accelerate the process of bringing technology into use, such as George Hill and Tom Falkie indicated. What is it we are not doing in the matter of government policy or in the way we operate that we should be doing in order to accelerate the process of adapting technology to our daily needs?

HILL: In my opinion the element that is missing is the element that brought us to success in World War II in the synthetic rubber industry, which was nonexistent prior to our being cut off from supplies by the Japanese. At that time companies that had the technical capability to do a commercial level job were contracted with by the government to produce a product. If we in the synthetic oil industry were to say to a given oil company or to all oil companies, you shall have on line by a certain date 10 percent or 5 percent of your oil from coal, I think that that kind of a challenge--with the support necessary to support the infrastructure, the pilot plants, and the like--would in fact accelerate tremendously the real development of a synthetic fuels industry.

The same thing would go for power generation improvements. If the companies were given the challenge at the commercial level, if you please, where the rubber meets the road, I think we could achieve it in a much shorter time. I don't see the transition being accomplished otherwise.

FALKIE: I think it is primarily an economic problem. Let me try to quote some numbers from memory from the Federal Power Commission. Through September or August of last year the delivered cost per million Btu of coal to coal-fired power plants was something on the order of $0.90 or $1.00 per million Btu. The delivered cost for gas to gas-fired power plants was something of the order of, let us say, $1.75. Now, if I am too far wrong, Alex, correct me.

But oil-fired power plants were something of the order of two dollars per million Btu, roughly twice for oil as for coal. Now, a power plant manager today, when he goes out to buy coal, does not go out to buy coal at two dollars per million Btu. He goes out to buy coal at one dollar per million Btu. Somewhere, we have got to make up our minds in the overall energy program of the country, or in our philosophies toward energy, that a good part of that gap is in essence going overseas for what we are paying for foreign oil and that we will ultimately be paying more for coal. And if we pay more for coal, the economics will improve the situation to where some of this technology will be adapted, especially by the smaller operators.

ALLEN: Can I make a brief comment about the utility buyer of coal? I think that the thing that the utility buyer of coal is doing is trying to find coal he can burn and then he looks at price. This gets back to one of the three equations. We have moving targets for a health objective, which is real, but not the scientific approach that we should have to determine a reasonably soluble set of three simultaneous equations.

But let me emphasize that no utility is going to buy coal at one dollar that he has to pile up and leave on the ground.

FALKIE: But will a utility buy coal at two dollars? I am not saying he should, by the way, but *will* he buy coal at two dollars per million Btu that he *can* burn today?

ALLEN: He can look at the alternatives.

FALKIE: Okay, that is the point I am trying to make.

ALLEN: He also knows the risks of putting in equipment that will burn the two-dollar coal today but will not be good enough to burn it three years from now.

PLOTKIN: I believe it is important to question some of the statements made by Mr. Hill concerning the cost of environmental degradation or environmental controls. There are three points in particular I really ought to address. First, Mr. Hill used the figure of 25 to 30 to 40 percent of the original value of the resource for environmental protection. I have seen lots of studies about these kinds of cost, and I believe that you are an order of magnitude too high, sir.

Second, you use the term "zero degradation" and talk about best available controls, without any regard to cost to the consumer. I find this language to be incredible, because I think it has no basis in law. The Environmental Protection Agency standards are not set this way, and the legislation is not worded this way.

Third, you use a figure for a future NO_x new source performance standard that you say cannot be met by any technology, yet the definition of new source performance standard demands the existence of a technology before it can be set. Now, can you explain how you can do this?

HILL: On the first statement we are both right. My statement referred to the fact that if we are required to convert to oil to burn on the boilers, in order to meet new source performance standards that can't be met by burning coal directly in cleanup systems, the conversion efficiency of that operation in terms of the raw material is about 75 percent.

It is true that a flue gas desulfurization unit on an existing power plant only derates the plant by about 5 percent. So again, you are right from that point of view, but if we do have to go to the limit that is implied in some of these things, it will cost us about 25 percent of the original energy.

Regarding zero degradation, I think there has been a federal court case where EPA was taken to court by the Sierra Club for not requiring some technology that is presumably available. The court ruled that the law does require that the best available technology be used and that the cost to the company or ultimately to the consumer is not presently in the law. That was the point I was making. If I am mistaken, I will be happy to know about it.

Also, regarding the new source performance standards, and NO_x, we have now conducted tests with Babcock and Wilcox, with combustion engineering, and with modifying the combustion process for coal. The best we have been able to achieve thus far is something in excess of 300 parts per million of NO_x, which is way above the 0.05 level. So my comment was that the combustion process itself will not allow us to achieve that level, and I hope it isn't imposed, if it ever is, until there are some stack gas cleaning techniques, presently not demonstrated to my knowledge, for coal-fired systems that will allow that level to be achieved. Maybe it can't be.

PLOTKIN: In fact, EPA cannot establish a new source performance standard without a technology, so that is just a red herring.

HILL: I wish you would tell your colleagues not to print those numbers then, because we got them from EPA. I am in trouble for ever putting them on the board because my utility people say, hey, if you do that it is going to be assumed that that is what is going to happen. So I would like you to all forget that NO_x number, if you will.

REPORTS FROM THE WORKSHOPS

WORKSHOP 1: ENVIRONMENT

Clinton W. Hall

Director of the Energy Coordinating Staff, Office of Energy, Minerals, and Industry, Environmental Protection Agency

The purpose behind this environmental workshop was the fact that environmentalists and environmental regulations are often blamed (or credited, depending on one's values and viewpoint) with impeding the greater utilization of coal in meeting our national energy requirements. Indeed, it has been said that coal is the only viable near-term option for lessening this nation's dependence on foreign energy sources; except you can't mine it and you can't burn it. Certainly, environmental concerns exist for every phase of the coal cycle: extraction, transportation, conversion, combustion, and utilization. Many of these concerns stem from inadequate knowledge of environmental processes and the inability to satisfactorily predict the short- or long-term environmental consequences of various energy development activities.

In recognition of this the federal government has mounted a sizable research and development program to study the health and environmental effects of increased energy production. The Environmental Protection Agency alone is spending in excess of $100 million a year in this area. If new initiatives are approved, this figure could double in a year or two. The Energy Research and Development Administration has a similarly sized program, and many other federal agencies devote a reasonable portion of their environmental budget to the energy/environmental impact area.

We recognize that the R&D area is not going to answer all of the problems of burning more coal in a timely and environmentally acceptable manner, but it is surely an important ingredient in that recipe, and I think there is enough of a budget being devoted to it to make a difference. Those of us who are involved in spending this money periodically ask ourselves, what happens if five years from now, we discover that we spent the money on the wrong issues, if we asked the wrong questions. With this in mind, the question that was posed to the environmental work group was, how do we burn more coal in an environmentally acceptable manner? What are the environmental issues, or the perceived environmental issues or bottlenecks that impede more rapid, more efficient utilization of coal, and how can the federal R&D budget be directed toward addressing these issues? The following list of issues and research requirements resulted.

ENVIRONMENTAL ISSUES IMPEDING INCREASED COAL UTILIZATION AND AREAS OF RESEARCH REQUIRED TO ADDRESS THESE ISSUES

1. *Issue:* Conversion of coal to a clean-burning fuel. Cleaning of coal for conventional or advanced combustion processes and technologies for gasification or liquefaction of coal all leave residuals or byproducts having unknown or unquantified environmental consequences.

    Research Requirements: Process or technology emission/effluent characterization and determination of environmental transport, transformation, fate, and ecological effects of representative classes of compounds.

2. *Issue:* Meeting existing air quality standards. Existing air quality standards are exceeded by "natural" conditions in some areas, or the available margin for additional incremental degradation is sufficiently small to preclude any additional power-generating capacity utilizing presently available technology.

    Research Requirements: Reassess stringency of existing air quality standards relative to public health requirements. Assess the feasibility of regional versus national standards. Develop more effective and efficient emissions control technology.

3. *Issue:* Water availability in the West. The amount of water available for energy-related use in the West is controversial. While existing western water law is based on the sequence of historical appropriations, overappropriation of available supplies, especially in drought years, could have severe consequences to many sectors of the economy as well as the environment.

    Research Requirements: In-stream flow and quality requirements for fish and wildlife maintenance must be determined. The hydrological and groundwater quality consequences of mining coal aquifiers must be assessed. Techniques for efficient water storage must be developed.

4. *Issue:* Visibility reduction/atmospheric transport. Fine particulates (submicron) emitted from coal combustion cause light scattering and reduce visibility. Visibility in the open West is a cherished aesthetic value as well as vital to much of the tourist economy.

Research Requirements: Development of technology for fine particulate removal prior to emission.

5. *Issue:* Solid waste disposal. Even if the clean combustion of coal can be achieved, leaching and runoff from surface disposal of wastes generated during fuel upgrading activities or combustion gas cleanup result in another entire set of environmental problems.

Research Requirements: Development of regeneration capability for advanced technology wastes and assessment of impacts of regeneration products on existing markets.

6. *Issue:* Trace elements and heavy metals. Trace elements and heavy metals are ecosystem stresses common to most if not all phases of the coal cycle. The effects of these contaminants on specific organisms is generally well known or relatively easily attainable.

Research Requirements: Development of the ability to extrapolate acute or chronic organism impacts to changes in ecosystem structure, function, or diversity and to predict ecosystem resiliency once the stress has been removed.

7. *Issue:* Reclamation. Reclamation, more than any other activity, embodies the issue of environmental aesthetics.

Research Requirements: Development of reclamation techniques that will ensure biological productivity after maintenance (irrigation and fertilization) have ceased and techniques for limiting off-site (stream pollution) effects of strip mining both during mining and after abandonment.

Other issues identified but which the group did not have the time to discuss included the following:

8. *Issue:* Ecological implications of polycyclic organic emissions.
9. *Issue:* The role of "technology assessments" in identifying or addressing ecological requirements.
10. *Issue:* Ecological impacts associated with energy source material transportation.
11. *Issue:* Quantification of the ecological impacts of existing facilities.

WORKSHOP 2:   LABOR
_____

Thomas N. Bethell

Director, Brophy Associates, Inc. (Coal Research); Former Director of Research, United Mine Workers

Trying to discuss the labor situation in coal today is like trying to put socks on an octopus: trying to do it at all is probably an exercise in futility, and you can be reasonably sure that before the job is complete, you will have to go back and start over.

The basic problem is further compounded by attempting to summarize a workshop discussion that explored a broad range of issues over a three-hour period. Necessarily, some oversimplifying must be done, at the risk of mis-stating a position and/or ignoring entirely some of the subjects that were discussed.

In general, the participants concerned themselves with some very short-term and some more intermediate-term problems. The short-term problems can best be summarized by the question: "What's happening in the United Mine Workers today, and what does it mean?"

The United Mine Workers still represents nearly three fourths of the men and women working in American coal mines and accounts for more than 60 percent of the coal produced in the country. This fact alone suggests enormous economic and political clout at a time of increasing reliance on coal. But that power remains hamstrung by the internal problems of the union. These are extremely complex and deep-rooted problems, and there is a great risk involved in oversimplifying them. The reform movement within the UMWA first challenged the old-guard leadership in 1969 and toppled it in 1972. But the victory in 1972 was narrow and has never been consolidated. Although Arnold Miller carried out his campaign pledges, opening up the union to democratic reforms, eliminating corruption, and providing rank-and-file miners with the opportunity to ratify their contracts, he failed as an administrator and was unable to build any kind of working coalition leadership. His problem has been greatly compounded during the past two years by a succession of crippling wildcat strikes.

Now, Miller stands for re-election in June. He is opposed, on the one hand, by Lee Roy Patterson, a member of the union's International Executive Board, who supported W. A. Boyle in 1969 and 1972 and has the support of many district and local officials--an important point because it is they who get out the vote. On the other hand, Miller is opposed by Harry Patrick, his Secretary-Treasurer, who ran with him on the reform slate in 1972. Patrick remained loyal to Miller for four years but broke with him openly in 1976, mainly over the issue of Miller's apparent inability to administer the union and his reluctance to confront day-to-day problems.

The conventional wisdom at this point is that Miller and Patrick will split the reform vote, permitting Patterson to win. If that happens, the presidency of the United Mine Workers will be occupied by a traditional wages-and-hours union leader who can be expected to take a very conventional approach to collective bargaining. Some coal operators openly favor Patterson on the theory that he will restore discipline to the ranks of the miners. It is understandable that they would feel this way, given the general deterioration of labor relations in the industry.

But it may be wishful thinking. First, the vote may not split along the anticipated lines, for reasons that were discussed at length in our workshop but that are too complex to summarize here. More importantly, the theory that the election of an apparent strong-man--Patterson--will get the trains running on time may prove to be a false hope. The roots of the labor problem today are too deep to be reached by a single election.

That brings us to the various subjects that occupied much of last night's discussion. Again, apologies are offered to the participants for unavoidable oversimplification of the various issues discussed, which are summarized in the following questions:

1. "Will the upcoming negotiations help the situation?" Here the depressingly realistic answer seems to be: "Probably not." It appears that these negotiations will be conventional, i.e., the industry will come to the bargaining table trying to nail down iron-clad contract language absolutely forbidding strikes for any reason during the term of the contract, with harsh disciplinary measures provided for any militant miners who don't get the message. The union will go to the bargaining table most probably led by a brand-new president troubled by the need to restore order on the one hand and to demonstrate his manhood on the other. This could be a formula for disaster.

2. "Will the union continue to lose its grip?" At the moment, the trend seems bound to continue. First, it is almost inconceivable that any of the three candidates can win by a wide margin. This imposes on the winner the obligation to mend his fences with the supporters of the losers. Otherwise, he cannot govern effectively, and the dismal experience of the Miller administration will be repeated. At the moment, it is not clear that any of the candidates will be able to build such a coalition. If that happens, the internal but very public bickering will continue, and the union will be unable to organize effectively. It will also be unable to present its positions cogently and effectively on Capitol Hill. Thus, as is the case today, it will not be an effective voice for a rational coal-based energy policy in the United States. This, as far as I am concerned, is a major tragedy, because it is the members of the UMWA and the communities in which they live that are most deeply affected by the lack of direction and commitment in our "non-coal-policy" of today.

The net effect of all this could well be that the United Mine Workers by 1980 will have lost its foothold west of the Mississippi and could be losing it in many parts of the East. Multiemployer bargaining, a mainstay of what little stability there has been in coal during the past quarter-century, could become a thing of the past. This kind of fragmentation could produce even more turmoil than we have today.

3. "Are there any alternatives to this doomsday scenario?" Yes, there are, but they require a highly unlikely infusion of visionary thinking into coal labor relations. Even before that, they require a much better understanding of the root causes of today's problems.

The labor force in coal has changed drastically in the past ten years and continues to do so. Ten years ago it was stable and comparatively old, with the median age around 46. The reason was that the industry had long been in decline and was not recruiting new manpower. With the turnaround since then, the median age has plunged. It was 34 when we went to the bargaining table in 1974. It is about 30 today. A third of the working members of the UMWA today are in their early twenties. Parenthetically, these miners were not around during the bitter battle between

Tony Boyle and Jock Yablonski in 1969, and many of them were not around in 1972--thus the conventional wisdom that assumes Miller and Patrick will split the 1972 reform vote ignores a large bloc of voters who were not part of anybody's constituency five years ago. Thus the risk of assuming anything about this year's election or its immediate consequences.

Oversimplifying again, today's miner is young, better educated than his father, more economically and geographically independent and mobile, and almost totally unresponsive to the old school of labor relations in coal, which was very much a one-way street, in which the word of the boss could not be challenged and men would do just about anything to hold on to a job. Those days are gone, and none too soon. But the industry has been slow to recognize this fact, and one reason for harboring some pessimism about the upcoming negotiations this fall is that a great many coal company presidents seem to feel that the solution is to impose a heavy dose of authoritarianism via the contract language. It won't work.

Similarly, there is a growing tendency on the part of many industry people and even more so on the part of many securities analysts to write off the UMWA as a lost cause. This has the effect, first, of accelerating the shift from the labor-intensive East to the relatively problem-free West, but it doesn't solve anything. The great coal resources of the United States lie in the Midwest and the East adjacent to the overwhelming preponderance of the nation's industrial plant and population, and any rational plan for coal development would be based in the East. Furthermore, this tendency to want to walk away from today's labor relations problems is uncomfortably consistent with the coal industry's historical tendency to wreak sustained havoc and then get out of town. It is not something we should be willing to accept, if there are alternatives.

In my view, and I think in the view of many of the people who participated in last night's workshop, there are alternatives, beginning with a better understanding of the complexities of today's problems. Coal mining, like it or not, is still the most dangerous occupation in the world--and incidentally this applies increasingly to surface mining as well as underground mining. All the propaganda of the National Coal Association notwithstanding, the coal miner is still exposed to extreme hazards: extreme temperatures; extreme quantities of respirable dust, water, and mud; electrical shock; fatigue; spinal problems; and a whole host of less obvious threats to his well-being. Both management and labor should be launching a massive program of training--not just job training but training in more elusive skills like people-to-people communication. On a more basic level, both sides should be exploring the possibilities of different working conditions. The four-day week is one. The wildcat strike in a way is nothing more than a sloppy, unpredictable approach to institutionalizing a four-day week, but there is a valid point in asking why anybody should have to do that kind of work five or six days a week. Nobody in this room would do it. Why should anybody else? Yet, when you mention a simple proposal like this in any convocation of coal people, you can be assured of a hostile reaction. That, to me, is the greatest tragedy of the labor problem in coal today--that there is so little willingness to experiment with new solutions. Until that attitude changes, today's problems are not going to go away. They are, in fact, going to get worse.

We are all the losers. Because it seems obvious to me that a disciplined, demanding, questioning, creative labor force could be just what is needed to provide the impetus for intelligent use of our coal resources in this country. The first step in that direction could come from greater public involvement in the labor relations process--involvement which is admittedly difficult but, theoretically, at least, not impossible.

Apologies, again, to the workshop participants for unavoidably oversimplifying a long and wide-ranging discussion. It is clearly presumptuous to have tried to summarize such a complex subject in this length. But the coal industry was built on presumptuousness, so there's no good reason to change now.

HOUTHAKKER: The report you have given us on the labor situation, the union situation in particular, is very depressing, and I have no reason to quarrel with this assessment. I would, however, ask if there aren't other aspects to the labor situation that may also have come up in the panel discussion or on which you could comment otherwise. The question of how much labor is actually available to do underground and surface-mining work is, I think, really more fundamental than what happens in the union itself. I wonder if you could comment on that, whether there is any apparent decline in the willingness to work underground, whether from that point of view there is an advantage to surface mining.

HIBBARD: I am sorry, I can't answer that question. It is not my report, and I wasn't there.

HARRY PERRY, consulting economist: I attended the workshop, and the matter was discussed. I think we concluded that history has shown that although people have been concerned about being able to get underground workers, in truth there has not been a labor shortage either for underground workers or strip miners. There appears to be no problem at all for the western strip type of workers. They can be recruited and trained relatively easily from other types of jobs for which there is a large pool of labor. The industry will have to take the opportunity to train underground workers, something that they never had to do before. But there are training programs already available. More are being planned, and I don't believe our panel felt that there would be any labor shortage.

CARLOS STERN, University of Connecticut: In some circumstances when there are labor problems, one way to get around them is to develop an incentive for the worker, such as profit sharing or even some kind of cooperative ownership. This might be especially true in the condition that Dr. Falkie pointed out, where the capital investment is increasing but the productivity is dropping because the workers themselves are unwilling, for one reason or another, to put out as much as the employer expects. It would seem to me that some kind of profit sharing might, if nothing else, lubricate the process whereby worker incentive combines with that of the company.

I am wondering why your talk focused so heavily on the technological side, when it would seem here, as in so many other areas of energy policy, the human side was so intricately mixed in, the technology improvements were inhibited by the lack of adequate consideration to the human aspects of the same issue.

FALKIE: I feel very strongly that the human aspects of technology and the interaction between the operators and management is a very critical point. I won't get into that. I think that Tom Bethell alluded to the trying of different approaches, and one of the things that he and I discussed earlier was autonomous work programs, where a worker could get more involved in deciding the kind of work that he is doing. Getting involved in the decision making certainly should be tried, perhaps more than the degree to which they have tried it here.

Counterbalanced by that, in my judgment, is that you would not get a favorable reaction from the union for an incentives-type approach. I think the union would go more for the guaranteed wage, dollars-per-day, type of approach. While some people--I am not going to speak for Tom Bethell on that particular subject--think that the incentive approach may offer a lot of potential, I think the union would turn it down.

I feel that if the operators went in in December and proposed an incentive-type arrangement, it would get turned down by the union.

STERN: What about the 25 percent of the mines that are not unionized?

FALKIE: Some of them have programs--not too many of them, but some of them do. And some of the ones that do are working very, very successfully.

HIBBARD: Any other questions?

JOHN DOUMELE, Moore McCormack Resources: I was at that session also, and one of the main points that Mr. Bethell made was that certainly for this year, but also for the three years that the upcoming contract will cover, there will not be the energy to expend on innovations in contract terms, that it will be spent all on internal reorganization and power building. That is the tragedy he referred to, that there will not be any efforts made on incentives or anything else that can improve productivity.

WORKSHOP 3: DISPUTE RESOLUTION/RULES OF REASON

Milton R. Wessel

Attorney and Visiting Professor, New York University School of Law

The national energy controversy, which is the subject of this National Academy of Sciences Forum, is one of a class of modern disputes characterized by several similar factors. First, and most important, it is a public interest dispute. That is, the absentee and largely uninformed public has a vital stake in the outcome of the issues being debated by the disputing parties. Second, it is a socio-scientific dispute. That is, the critical concerns are social, determining how we will live on this planet, and depend in major part on scientific judgments and analyses. Third, it involves extremely complex fact and opinion considerations in several highly technical disciplines, much of which is far beyond the competence of the lay person and aspects of which are beyond even the scientific expert skilled in a particular specialty. Fourth, there is no absolute answer, no right or wrong, no good or bad. Any decision must be based upon a balancing of a large number of pluses and minuses and of risks and benefits. Different people with different values will come inevitably to different decisions.

In a democratic society, in which ultimately the public must make the final decision, the combination of these factors means that credibility of the dispute resolution process is essential. Our dispute resolution workshop was agreed--as we all certainly must be--that our society has not yet developed such a process, which the public accepts as achieving optimum resolution of these issues. As one of the speakers at yesterday's plenary session said, "There is a crisis in dispute resolving." Cornelius B. Kennedy, a discussant at last night's workshop and chairman of the American Bar Association's Administrative Law Section, emphasized the historical inadequacy of legal and administrative processes, taking Dean Roscoe Pound's 1906 speech in this regard as his text. In present-day socio-scientific disputes, these concerns are intensified: Existing mechanisms are far too long and costly; large segments of the population are dissatisfied with the outcome of any case that they don't really understand and that, rightly or wrongly, they believe is determined by wealth, entrenched position, or activism, rather than merit. There is too much confrontation, with mistrust and hatred turning us against each other and putting too many people at the throats of others.

Not surprisingly, also, the workshop was unanimous that society needs to explore new approaches to resolution. It must find among those who are concerned with the present process, the leadership that will develop the necessary trust and confidence.

A number of such new approaches were explored. Dr. Amos Jordon, executive director of resource programs at the Georgetown University Center for Strategic and International Studies, described the National Coal Policy Project, which is bringing together opposing industrial,

environmental, and other interests in an effort to reach accommodation--
a sort of "environmental conciliation" procedure. The workshop was agreed
that the earlier Forum discussion, rejecting mediation, was premature.

Richard Fleming, chairperson of the newly organized Chemical Industry
Institute of Toxicology, described that institute's research effort to
develop and publish scientific data in which the public can have confidence and on which it can rely. James N. O'Connor, assistant general
counsel, Dow Chemical, United States, described two socio-scientific
public interest cases in which the "sporting" approach of so much modern
American litigation has been replaced by a "rule of reason," with resulting accommodations acceptable to all parties, and in the public interest.

In the discussion that followed, it further appeared that any new
approaches to dispute resolution require the following as a minimum:

First, vehicles must be created so that accommodation can be sought
among the presently enormous number of scattered and splintered views.
If we are to achieve nonconfrontation resolutions, we must first have a
workable number of units and spokespeople who can discuss, understand,
and compromise. This was the thrust of the experience related by Dr.
Jordan.

Second, the so-called "establishment" must accept the primary leadership role in reducing confrontation and introducing reason and logic into
the resolution process. Presently, there is extremism at many points on
the opinion spectrum. Whether or not extremism is ever justified--and
remember, it was just last year that England as well as the United States
celebrated the bicentennial of our own revolution--it may be at least
understandable on the part of the deprived and disenfranchised when they
have no other means of getting attention. The thrust of the experiences
related by Messrs. Fleming and O'Connor is that the leaders of science,
industry, and society must respond to the extremism of "outsiders" with
data, reason, and understanding. There can be no excuse for the all-too-
common response of equal or even greater extremism--such as the characterization in this Forum of one participant's environmental opponent's
successful litigation conduct as "undemocratic," a characterization not
dissimilar to the charge of "communist" a quarter-century ago; the refusal
of one of the utilities, at least peripherally involved in the Kaiparowits
case, to respond to the Sierra Club's several written requests for information, justified by some of the participants here as not the utility's
responsibility and giving rise to the comment of an environmental leader,
"We are sustained by the arrogance of our adversaries"; or the act of
violence, however minor, which occurred on the floor of this great assembly hall at last month's Forum on Recombinant DNA in bitter reaction to
the vigorous and outspoken but nonviolent and orderly demonstration of
those opposing further research. All of this has contributed to the
mistrust of industry, and science, and the establishment by a major segment of our society to the point where a good part of the public simply
will not believe the unanimous assertions of the oil companies and industry regarding even this energy problem.

Third, as was pointed out by Robert White, director of the Academy
Forum, the scientific community must intensify its efforts to enable the

public to understand, or at least to evaluate and judge, scientific information so that public consensus or accommodation is possible. Reference in this connection was made to the usefulness of standards, professional accountability, certification of procedures, codes of ethics, and other specific measures in the effort to enable the lay public to deal with esoteric technical matters. The concerned lay person who traveled here all the way from Montana because of concern about the quality of life in her home community must be as satisfied as the most technically qualified energy expert. The public must be treated with dignity and respect not patronizingly or with public relations gimmickry.

And, finally, the workshop was agreed that there was to be a major effort by all to understand the views of any opposition and to accommodate to it wherever possible. The "scientific method" of consensus may be no more possible in many of these areas than it is between Russia and the United States, but accommodation *is* possible and may spell survival: People on one side or the other of a dispute are not usually "bad"--they simply have different perceptions and values, resulting from different environments and backgrounds. It is the unique spouse who can get through the first weeks, months, and years of living together without such understanding and accommodation. Society is no different.

Not long ago, our cities were wracked by riots, caused by the disaffection of major segments of our society. The socio-scientific issues discussed in these Forums have not yet generated the hatred or despair of the civil rights or Vietnam debates; but, unchecked, one day they might. It is up to those who pretend to leadership to see that they do not and that our nation continues in democratic fashion to search for and to find solutions to even the most troublesome and difficult problems, without confrontation, force, or violence, in the interest of all.

ALLEN: I think I have just one question. You ended on a note of hope, that the present mechanism for resolving these very complex public issues is not working, but you think there is some way of doing it. Do I understand that you feel that public discussion and airing of both points of view, as we are doing here, are the best mechanisms?

WESSEL: I think that they are helpful.

ALLEN: I agree.

WESSEL: Discussion moves things along, but there is no single thing that is going to resolve the issues. I wish there was time to describe the paper that was delivered by Mr. O'Connor, who was assistant general counsel of the Dow Chemical Company. He made reference to and described some cases in which they specifically applied the kind of response that I have referred to successfully for them and successfully for the so-called environmental opposition who was looking for satisfaction.

ALLEN: Let me ask you one further question. I thought the most significant thing you said was that we can't reach consensus on these very complex and multifaceted issues but that we can reach accommodation. Second, that accommodation is a negotiating problem and you can reach accommodation only if the spokesman, and therefore the negotiators for each of the separate interests, are believed by the rank and file. What belief do we have that the public, who gets pretty excited about these things, will permit a responsible negotiator to give something from the standard position, be it the utility side, or the establishment side, or the environmentalist side, or the consumer side, or whatever. Is that a role that a negotiator can in fact accept and perform to the satisfaction of his public?

WESSEL: It is a tremendously difficult problem. Dr. Jordan spoke particularly about the concerns of those who were in the environmental caucus in this Coal Policy Project. If they go too far or they don't maintain the proper posture, they will lose their constituencies. Yes, it is a terribly difficult problem, but it is no different from the problem than any labor leader has had over the years. It is a different magnitude in one sense, because there are many more people involved, but any labor negotiator faces that all the time. I live in New York, and Mike Quinn used to do what he did every New Year's Day or the day before it because he had to keep his people. He didn't want to strike. He knew very well a strike couldn't happen. It wasn't until my classmate, John Lindsay, had to take him up on it, that we had a strike. But Lindsay didn't understand Quinn.

ROGER F. SWANSON, Veterans Administration: I might be entering into the area where angels fear to tread, but Peter Drucker quite a while ago mentioned that each corporation is a socio-economic-type organization. We read recently that the corporations in Japan have possibly suitably overcome some of these problems because they have taken on some social as well as economic types of responsibilities. The people and the concerned citizens we saw up here in the case analyses are near stranded; the corporations that get in there have to help people get established. We can't always go to the federal government.

The other point is that it appears that the Sierra Club, which I have great respect for, is equally responsible for balancing the environment and the quality-of-living equation. This is a hard thing. You were talking about balancing these three simultaneous equations. This is part of it. They not only have got to sit down and say, how do we save the forests or the air? They also have to ask, how are we going to heat George Hill's home? The thing that scares me right now is that you people are talking about things on a very sophisticated level that are going to be cast in concrete soon.

One of the things that bothers me is that you haven't gotten into the health care area. What are we going to do with these 500-bed hospitals? What are we going to do with the 200-bed hospitals? One half of a hospital down in Tappahanock, Virginia, is heated by electricity? What are you going to do for it?

WESSEL: Let me duck the topic of workshop 7, which is health, and only say that it is a problem that has to be dealt with. But in an effort to give you a brief response--first of all, there is no doubt in my mind, and I hope not in many others, that corporations do have a social responsibility. More and more of them are recognizing it. There has been great change in the last twenty-five years. There will be much greater change in the next twenty-five years. But I doubt that there are many responsible people in the country who don't regard the large corporate structure--with 200,000, 300,000, 400,000 people depending directly on it and perhaps millions depending indirectly on it in many different respects--as having some social concern as a part of its process.

    Second, you suggest that the Sierra Club must do a balancing act. Yes, we have to all be responsible. We all have to somehow try to understand the other side of the concern and not go too far. But after all, we have different value systems. The Sierra Club is not a chemical company, and it is not a coal company. It is an environmental group. Its job is to present that position in a responsible way in some kind of a forum or mechanism where the other side can also be presented and where there is a vehicle for accommodating or deciding between the two. I don't want the Sierra Club deciding for me whether I do or don't have gasoline or what size car I have. I am willing to let the country make that decision in some way, although I do also respect the Sierra Club.

FALKIE: Perhaps I am the only coal operator here now. This last question touched me because the first major decision I made with Berwind Corporation was to enter into a real estate development in southern West Virginia, a development in real estate, which we know nothing about and which we may lose money on. We feel that there is a need for housing and that we should start doing something with some of the land that we own. It was the very first decision I made as a senior vice-president, and, lo and behold, the environmentalists are blocking it because they do not want any housing developed near the New River Gorge Bridge, or the New River Gorge. It is a prime piece of land. If we held onto it for several years we could undoubtedly make a lot of money on it. That is an example of what kind of a decision-making process we have.

WORKSHOP 4: TECHNOLOGY/RESEARCH AND DEVELOPMENT

Thomas V. Falkie

Senior Vice-President, Natural Resources, Berwind Corporation; Former Director, Bureau of Mines

This particular workshop on technology decided that there was very little said in previous discussions about technology per se and that we should not philosophize too much but should talk about specifics in the areas of coal preparation, transportation of coal, underground mining and surface mining, and exploration. We didn't get to exploration.

We talked about coal preparation. One interesting fact, other than the various technical things we spoke about, was that a greater percentage of eastern and midwestern coal will be beneficiated and that coal preparation plants are being built at a very rapid rate in these areas. We talked about the fact that you must consider the water in the transportation of coal and the various ways of taking advantage of that technologically. The government role was discussed extensively, and most participants felt that the government had to play a somewhat larger role in the area of coal preparation.

We touched on the problem of trace elements in waste from coal preparation plants, and we discussed some misunderstanding on the part of the public as to the nature of low-sulfur coal reserves in the East and mentioned that some of these eastern low-sulfur coal reserves are actually low-volatility, metallurgical coal, which is a prime commodity in itself.

There was a general discussion on R&D and what the government should *not* do, and it was an interesting interaction. As we went through mining and preparation everybody said that the government should be doing more research in these areas. But the group also more or less agreed that, in general, government should get into R&D only in areas in which the technology is very difficult or far removed from commercialization and where the market potential has political uncertainties connected with it.

We talked about the economics, the cost in dollars per ton. We asked, what could you afford? What does it cost to prepare coal? Where is it going to go? How is the balance going to be maintained? What would the public be willing to pay for coal in relation to oil?

On the subject of transportation, we asked ourselves, is there a railroad or a coal transportation problem? There was a discussion about whether railroads can meet increased production if it comes. There wasn't complete agreement on this. The yeas said that coal-bearing railroads are generally the best fix financially, except Conrail, and, given modest lead times, they can respond. The nays said that governmental policy uncertainties, single tracks in the West, systems coordination problems, and possible environmental impacts of increased traffic could hamper railroad expansion. We got into some technology problems, but, generally speaking, people felt that technology was not the limiting factor; rather, it was economics.

There was an interesting discussion on the effects of the inability of railroads to perform on need and what effect this would have on importing coal into the Northeast--in particular, coal from South Africa and Poland, low-sulfur coal.

The bottom line on the railroads is that the railroad situation in terms of ability to transport larger amounts of coal in the future is in-between, not as bleak as some skeptics say, and not as good as some railroads say.

We talked about barging, and, again, technology is not particularly a problem, although the lock system is something to look out for. We talked about pipelining, and again, no real major technological problems exist, although we need to do more R&D in such things as coal slurries. This led to some discussion, which we limited, on the political and water problems of the pipelining of coal.

As to underground coal mining technology, there was considerable discussion on health and safety aspects: why captive coal mines have historically been safer than noncaptive and are mining metallurgical coal at a higher price and, in fact, have a historical record of taking a more progressive management attitude toward the human, nontechnical points. Generally speaking, there is need for more automation, more remote control.

There was considerable discussion on productivity, somewhat along the lines of what I gave in the earlier paper. There was debate on what productivity in coal mining would be ten years from now but not agreement. There was general agreement that we will be lucky if we get back to the sixteen tons per man-day that we had ten years ago.

There was considerable discussion on how to get the technology accepted and agreement that the Bureau of Mines does a good job on technology transfer, but even when things are actually available for commercialization, they are not being widely used fast enough. Industry should participate more in R&D, and more involvement is needed to enhance this.

There was considerable discussion on training. The bottom line is that there is growing interest in training and a realization that education and training may be the most effective tool in improving health and safety and productivity. There was agreement that training approaches and techniques will be radically different ten years from now from what they are today.

There was some discussion about the use of diesels in underground mining, which I won't go into. The next ten years, in general, will probably see more effective management, better mine design, better control of the methane problem, more monitoring--which not only will improve health and safety but will give management more control over the system-- improved ventilation, dust control, lighting, and so forth.

On surface coal mining we didn't feel that we had an adequate discussion. We ran out of time. But there were discussions on the timing of reclamation, the fact that it depends on the geographical areas. There was discussion about whether it is really necessary to reclaim to original contour or even original use. We can reclaim faster by building land reclamation into the mining systems. There was a lot of discussion on steep slopes, high walls, arid lands, and aesthetics, and, interestingly enough, some in the group said that we could not get away with mining all the mountaintops in southern West Virginia even though the flatland is needed for other developments there.

Economics is a problem in some cases. Some people felt that you can reclaim just about anything if the price is right. I don't think that is quite a fair way to put it, but it was discussed, nevertheless.

A general conclusion was that the technology of extraction and preparation will depend on what happens in desulfurization technology and policy. That is East versus West; the East versus West problem was discussed. More government R&D is needed in coal preparation and waste disposal. There was unanimous agreement that more government R&D in extraction technology is needed. We must continue to give technological emphasis to health-safety problems. Not only are they technical problems, but they are crucial social problems as well, as, for example, in the case of black lung. Some people felt that you can save lives by increasing productivity, and we pursued that discussion briefly.

Again, training perhaps has the most potential for enhancing health, safety, and productivity, and the group felt that the primary responsibility for doing this training was with the coal operators.

More demonstration is needed for technology transfer purposes, and we discussed, from an engineering and managerial standpoint, the negative effects of fragmenting the government R&D program in coal extraction research.

BOYD: I have one question. Tom has pointed out that the transfer of technology developed in the federal establishment into industry is not as fast as it should be. In order to carry out mining research, you have to have the laboratory, which is the mine itself. Therefore, the federal establishment, if it makes a contribution to research, but in cooperation with industry, must do it there.

As I have been around and talked to the bureau staffs, I get the feeling that plenty of money can be spent on research but you have got to have people to supervise that reserach and carry it out and they have got to be able to go to the laboratory. Is this an inhibiting factor? Is this the approach that the government uses in establishing its R&D in relation with industry that inhibits the transfer of technology?

FALKIE: Yes, it is an inhibiting factor. The government increased its dollars being spent on extraction research three- or fourfold in the last seven or eight years and has essentially fewer people than when it started out. Also, I think the government has the responsibility to maintain a stable in-house program. This is difficult under those circumstances. But there is more cooperation from the industry, more participation in cost sharing and cooperative projects in this area of technology. Not only does that catalyze the industry into doing more research, but it also parlays the government's money and, hopefully, promotes faster technology transfer.

HIBBARD: Are there any other questions?

AL SCHULER, U.S. Geological Survey: Dr. Falkie, in his earlier remarks and just a moment ago, referred to the improved technology in mining. In view of the extensive advertising one sees for longwall systems and the existence of a number of demonstrations of longwall operations in Appalachia, I wondered whether he might be able to give us a statement of what percentage of our Appalachian mining might be accomplished by longwall mining in the next ten years. The inherent interest in this question is the relationship of improved recoverability to our reserves of both old coal and low-sulfur coal.

FALKIE: We discussed that fairly extensively, and I think everybody agreed that there is no question that longwall mining will have to be used more in Appalachia, especially as you go to the deeper and the thinner seams.
   Now, there was a hot debate as to what the percentage is going to be. It is definitely going to be higher than what it is now, not only for recovery purposes but for efficiency purposes, just to allow you to mine these deeper and thinner seams.

SCHULER: Would you say that it would be a substantial improvement?

FALKIE: I don't know what "substantial" is. What are we mining--3 to 5 percent now--by longwall? If we double that percentage in ten years, I will be surprised. I don't think it is going to be order of magnitudes greater in the ten-year period.

WORKSHOP 5: COAL CONVERSION

Kenneth C. Hoffman

Associate Chairman for Energy Programs and Head, National Center for Analysis of Energy Systems, Brookhaven National Laboratory

The role of coal in the United States energy system is generally viewed as being limited by our ability to use it in markets in an environmentally acceptable way. Thus, coal conversion technology is central to the increased use of this abundant and versatile energy source in the existing infrastructure for the delivery and use of liquid and gaseous fuels as well as electricity.
   The coal conversion workshop addressed the technological aspects of the conversion of coal to other energy forms. While these technological aspects determine the feasibility of conversion, the need for, and competitiveness of, these processes are determined by economic, social, and

political factors. These aspects of coal conversion were also discussed. The processes considered for the delivery of converted energy forms included combustion for heat and power, gasification, liquefaction, and combinations of these processes.

There seemingly is quite a broad range of options with respect to technologies, scale of conversion, and locations. Many novel and imaginative techniques can be proposed to overcome specific problems. Upon close examination, however, the technical realities of handling large volumes of coal and contacting it with the required amounts of oxygen, steam, solvent, etc., become apparent. Many conversion processes have been reduced to commercial practice. The workshop felt that research and development must be continued to improve and refine these processes and to reduce their cost, but there does not appear to be any breakthrough imminent that will significantly change the economics of coal-derived synthetics relative to conventional sources of oil. The economic benefits of research and development in reducing costs by, say, 20 to 30 percent clearly justify the planned levels of research and development activity.

Some concern was expressed that the development of environmental regulation and standards must better reflect the technical options that exist. Regulation requiring the best available technology could force, for example, the use of such combinations of control strategies as the deep cleaning of coal and flue gas desulfurization. Under such circumstances the rate payers and consumers would not receive a level of additional protection commensurate with the increased cost. Improved institutional mechanisms may be required to balance the costs and benefits of controls with due consideration of the technological capability and the likely response and its effects on other goals of energy policy, including economics and security.

A review of the economics of converting coal to liquid and/or gas indicates that costs in the range of $3.50 to $5.00 per million Btu of delivered product may be anticipated. These costs are, of course, above the current market prices for oil and gas, and, thus, the processes are not commercially viable under normal circumstances. Under some special circumstances of location and the lack of availability of alternative supplies, some current low- and high-Btu gasification processes do become viable, and several are now being planned. The workshop observed that at some point in time, the world price of oil and gas would increase to the point that the coal conversion processes could be viable. The timing of that event depends on many uncertain parameters, such as OPEC production ceilings, new discoveries, etc., but is inevitable. There was deep concern that when the event occurred, the domestic and international situation could deteriorate quite rapidly, and we must lay the basis of an industrial infrastructure that would allow coal-derived synthetics to assume the role of marginal supplier of liquid and gas as soon as possible. This implies a commitment to develop a synthetic capability and to protect the markets for these synthetics over the period of time when they can be undercut by monopolistic pricing practices. The technological aspects of coal conversion were discussed with respect to several individual markets.

UTILITY MARKET

The primary utility markets that were discussed include base load and peak electricity. Combustion processes are well developed for base load service, and current practice involves the use of flue gas desulfurization for sulfur control. The workshop did note that while this control technology was available commercially, additional development work is needed to improve the economics, reliability, and operability of these chemical systems.

Fluidized bed combustion appears to be a quite promising alternative for utility application, as does low-Btu gasification for use in combined cycles. The major development problems in the latter concept involve the removal of particulates from hot gases and the development of turbines capable of high inlet temperatures and of handling gases with some particulates present.

There was considerable interest in the conversion of electric generating plants from oil and gas to coal. This could release significant amounts of oil and gas for other markets. There are difficulties in such conversions of a design nature. The tube pitch in oil- and gas-fired boilers is considerably tighter than that needed in coal-fired boilers to avoid slagging and fouling problems. Conversion to coal therefore requires complete replacement of the boiler and ducts, which account for about 40 percent of the cost of the plant.

Fuel cells were discussed as a promising option for peaking service.

INDUSTRIAL MARKET

The industrial market consists of the requirement for process and direct heat, often from a clean fuel. Low-Btu gasification technologies are available to this market at a cost of between three and four dollars per million Btu. Small fluidized bed combustion devices are also applicable, but there is a point at which the delivery and handling of small amounts of coal becomes uneconomical. At this point, synthetic liquids and gases become more attractive coal conversion options.

There was some expression of support in the workshop for the use of processes based on the pyrolysis or carbonization of coal. In these processes there is a natural mix of special liquids, medium-Btu gas, and char-produced that can be used in certain industries. Indeed, it is current practice in the steel industry to use such by-products from coking and steel-making operations. It was judged to be quite difficult, however, to market these by-products beyond the industrial sector.

RESIDENTIAL, COMMERCIAL, AND TRANSPORTATION MARKETS

There exists an extensive infrastructure for the delivery of liquid and gaseous fuels to these markets that must be utilized. The methanation step in the conversion of low-Btu gas to a high-Btu product has been proven, and several facilities are now under construction. The cost of the high-Btu gas is estimated in the $3.50 to $4.50 per million Btu range.

There are liquefaction processes that are available now, most notably the Fisher-Tropsch SASOL Plant in South Africa. Other promising concepts based on solvent techniques are under development. Current cost estimates for liquids produced by these technologies range from twenty-five to thirty dollars per barrel.

ROSE: Perhaps I should have addressed this question to Milton Wessel, but you brought it up, Ken.

You talk about consumer protection. Do we get what we think we get? Do we trust who said something or other? And it raises the ancient question: *Quis custodia ipsos custodaes*? or "Who watches these watchers?" Did you have any suggestions as to some organization that decides these questions?

HOFFMAN: I was really calling for balance and the strengthening of institutions that would make for an effective balance in this debate. There have been discussions of a science court and operations like that. I am not too enthusiastic about that, and I judge from your reaction that you aren't either.

ROSE: No. It is a question of who holds the scales.

HOFFMAN: You are getting too close to the restructuring of government, I think, for me to feel comfortable in making a recommendation of that sort.

CHARLES A. ZRAKET, Mitre Corporation: I have two questions. None of the speakers have mentioned the possibility of *in situ* gasification of coal. Why?

HOFFMAN: With regard to *in situ* technology, that is, of course, one of the processes that was discussed, and there is active work going on in searching for appropriate bodies of coal on which to do such processes and methods of controlling the process. There is a good deal of experience in the Soviet Union.

HILL: I think that *in situ* processing, which really is a way of making low-Btu gas, does have a great deal of merit if you can find a coal seam near a site where that product will be demanded and can locate your plant so you can get a good lifetime production. We don't know, yet, about deep seams; but, for thin seams, it does have a great deal of merit. However, we are quite a ways from answering all the critical questions before industry will be ready to invest much money in it.

ZRAKET: What effect will ownership patterns have in both the coal mining industry and the coal conversion industry on the kinds and rate of technology that will be introduced?

HOFFMAN: You are referring to the size of the leases that one can engage in?

ZRAKET: Who will own these conversion plants? Who will run them?

HOFFMAN: Many of the activities now are sponsored by the pipeline industry. I guess there is a whole range of options there.

HILL: This is one of the key problems. The electric utilities are fighting like mad against having to become chemical plant operators and having a gasification and liquefaction plant that they have to run. They don't have inherently that capability. There is great reluctance on the part of the public to trust an Exxon Corporation, for example, with that responsibility, although that is where the expertise lies in the existing industrial component. It looks to me like we are going to have to almost create an industrial component that is not now existent, and that is not going to be an easy thing to do.

WORKSHOP 6: HABITAT

Raymond L. Gold

Director, Institute for Social Research, and Professor of Sociology, University of Montana

The workshop's panel members from rural areas of Appalachia and the West, which are underlain by extensive coal reserves, described the areas and indicated some of the more noteworthy features of the residents' lifestyles. Southwest Virginia was described in such terms as the following: "It is an area very largely controlled by coal companies who have been treating it as though it were a Third World type of colony." The natural resource, coal, is extracted largely to the detriment of locals and for the benefit of the coal companies and other outsiders. Slipshod strip mining has defaced the hills and adversely affected air and water. The blasting methods used also create structural problems in many nearby residents' houses. Poor and coal-dusted housing, poor diets, abused roads, and easily intimidated people who feel they have no future are constant reminders that the colonization process is well established.

The Appalachia experience with strip mining of coal was represented as a case of economic development that has been done at great social and personal cost to the many locals who have become dependent on this coal industry for their livelihoods. It is a case of how *not* to do right by either the people or the land. The coal companies have operated under the assumption that they need not be guided in their actions by other than economic and technological considerations. This operational mode has resulted in the oppressive social scene noted above.

Social impact at Gillette, Wyoming, reported earlier was described again by its mayor, Mr. Enzi. The community is growing rapidly owing to actual and expected development of the county's enormous coal reserves. (Gillette is the only incorporated town in Campbell County.) Growth problems abound. Much technical assistance is needed in order to anticipate and deal with questions and issues related to the rapid growth. The town's feeling is that it cannot depend on federal help, so it is seeking ways of meeting these problems with essentially local resources. The lessons of similarly impacted communities are being examined for what they may do to help Gillette avoid or minimize development-related impacts.

The people who live in the vicinity of Birney, a tiny ranching community in southeastern Montana, have deep social and cultural roots and a land ethic that is strikingly like that of nearby reservation Indians: both have a reverence for the land on which they live and work and feel it their duty to see that the land is in viable condition when passed on to their children. Like their Crow and Northern Cheyenne Indian neighbors, the white ranchers at Birney value their isolated, rural way of life and the opportunities it affords to rear children in physically and emotionally secure circumstances. Both whites and Indians in this part of the state are worried about what may happen to their cherished ways of life should the prime coal under their pastures be strip-mined. Both fear cultural invasions by coal industrialists, even though both stand to make money in these events. Both feel too tied to their land to survive displacement. This is especially so in the case of the Indians, who feel that there cannot be a substitute for what they regard as their sovereign territory. Both have a western, rural antipathy toward land use and other formal social planning, an attitude which will have to change if their interests are to be protected by local zoning.

The whites, whose land is underlain by coal that is largely owned by the federal government, are disinclined to go along with public agencies that are asking them to participate in planning for coal development, which these ranchers regard as planning for their own demise. They strongly resent hearing agency and other people (such as those at this meeting) talk about the northern Great Plains as if it were a blank check. They want to have an important "say" in coal-related decisions affecting their area, including decisions *not* to develop it at all or to do so only as a last resort. They feel that their present contributions of food to the nation are of long-standing importance and that coal ought to be taken from places that are already industrialized or of much less agricultural value.

The Indians own their substantial coal reserves but on both reservations are very uncertain that the cultural risks are worth the economic benefits they may realize. Both tribes are poor and have much chronic unemployment with attendant social problems. Both are still in the throes of trying to figure out how the risks of development might be reduced to culturally acceptable size.

The Birney people (with very few exceptions) feel that these risks are simply too great, particularly since the coal development would involve strip mining and all the well-known *people* pollutions such mining causes in rural areas. They emphatically believe that coal strip mining and western agriculture cannot accommodate each other.

In the discussions of the foregoing descriptions some of the more noteworthy points made were these:

1. Energy parks in rural areas are tantamount to totalitarian interventions: they centralize control for the good of the state and industry but to the detriment of the locals.

2. Decentralization of energy production and consumption would be much more acceptable to rural people immediately and to all Americans (and others in the world) ultimately.

3. Too many people in the United States are not actually interested in or concerned about where their technologically produced energy comes from and are unaware of the high social prices rural people in coal areas are being asked to pay for helping the energy industry cater to the wishes, desires, and demands of urbanites, collectively, albeit mistakenly, referred to as "needs."

4. Indeed, the question of what "needs" really means must be examined carefully, to sort out and reclassify those which will have to change to something less insistent in light of decreasing capabilities to cater to them.

5. Basic human needs and desires have to be distinguished from lesser ones and from ulterior motives that are essentially just self-serving. For example, industry tends to interpret and define energy needs for itself and its customers for its own purposes, not basically to serve the nation or mankind. Comparable self-serving motives exist among consumers of energy.

6. The meaning of land ownership, with attendant rights and privileges, needs closer scrutiny. Many privileges are erroneously assumed to be rights, such as the notion that one can do with one's land what one pleases, even if one thereby unilaterally circumscribes and intervenes in the lives of one's neighbors. An example is the action of a landowner who, without consulting others, opens the way to development of a strip mine on his land. This action, in the view of neighbors, constitutes an outrageous trespass on their way of life and an intolerable interference with their ability to continue in the agricultural business.

7. Those of us who have been living and/or working in areas where coal is being stripped or is apparently about to be stripped agree that the coal-related "people problems" are the really critical ones and that the people clearly most concerned are the residents of coal areas. Their lives are obviously so much more fatefully affected by coal development decisions and actions that it would be unfair, unjust, and downright wrong to fail to provide them with a "special" say in coal-related matters that directly affect their lives. How to define this special voice and use it are questions that urgently need to be addressed by all concerned--the midwestern users of electricity generated by Montana coal along with the Montana ranchers and other appropriate parties of interest.

8. In the West, coal-related social impact is now largely anticipatory;

perhaps this is the most severe impact phase becasue it is the time of greatest and most widespread uncertainty, anxiety, and distrust.

9. Why do so many western ranchers in coal areas conclude that strip mining (and therefore the operation of any mine mouth coal-processing facility) is inimical to a ranching way of life? One important reason is that they perceive (and in my judgment their perception is empirically sound) that, unless social impacts can be prevented or avoided, they will have to be suffered practically forever. Damage to social relationships, human organizations, and other aspects of quality of life is essentially irreparable. Accordingly, any effort to "mitigate" (i.e., repair or alleviate) such damage is a case of too little, too late.

PAUL RONEY: One of the problems is that the raw, earthy reality of the so-called human dimension is a far cry from the highly ionized reality that is lived in the scientific world. The object is to articulate in such a way that the reality of the scientific world coincides with the reality that is seen by the average person in this community.

I would suggest an approach in which all the information in this thing is "mastercharted." We can do this now with computer mechanisms. This would solve the problem of not having the right mechanisms to approach these multidimensional problems.

The second suggestion I make is animated, diagramatical articulation, which would help in getting this across. All this reality that is being expressed here today will never penetrate and will never be internalized by those people who most need it. I suggest that we use a few more different approaches.

FALKIE: I would like to make a quick comment about the antidevelopmental tone of the remarks of your workshop. What is happening is what I like to call a provincial opposition to development of any kind with regard to removal of things from the ground. The people in Florida are rebelling at phosphate mining. The people in Virginia are rebelling at vermiculite mining. The people in Minnesota are rebelling at copper-lead-zinc mining. I guess my bottom line--again a rhetorical question--is, where are the people in Montana going to get the phosphate that they need to conduct their farming if this kind of attitude prevails?

GOLD: This kind of rhetoric that you have just uttered we hear all the time in the coal areas. Keep in mind, I am a Greek messenger here. Don't slay me, now.

What you call provincial opposition is the operating reality of human beings. This is the way they see things; this is the way they find meaning in things; and this is the way they manage to get through the day and live in this world. That meaning-making process has to be understood and respected and not simply labeled and written off as provincial opposition.

HOWARD DRAPER, consultant: As an engineer and a scientist I would naturally tend to look favorably on the most highly technological solutions, and I suppose my first reaction might have been that this is provincial opposition except for the fact that I lived for six years in Colorado and absorbed some of the attitudes the gentleman reported. These are genuine, and they are not based on fancy.

The people in Appalachia I am sure received promises and were told that they should not worry about those mud slides or about acid in the streams. Of course, these horrible effects and devastation resulted, and the companies simply evaporated and, financially, could not be reached.

I wonder if some analysis should be made, despite my obvious knowledge of the efficiency of strip mining, as to which undervalue structures we could do without. Obviously, underground mining or processing can be done in almost all the cases you mentioned. If the country accepted a value system that the Montana and the Wyoming ranchers' land shouldn't be stripped, to what extent could electricity be supplied by very large nuclear power plants or very large coal power plants? Is this an impossibility, or must such "provincial" opposition be overridden?

HIBBARD: You will find the same opposition to the nuclear plants, I am afraid.

WORKSHOP 7: HEALTH AND GEOCHEMISTRY

Bobby G. Wixson

Associate Professor of Environmental Health, University of Missouri, Rolla

Albert L. Page

Professor, Department of Soil Science and Agricultural Engineering, University of California, Riverside

This Forum presented a pertinent opportunity for a recently formed National Research Council Subcommittee panel to discuss objectives of a study underway concerned with health and geochemistry. Our existent energy crisis, coupled with the determination of the United States to achieve independence from imported energy supplies, has projected the expanded utilization of coal, which is our most abundant fossil fuel resource. As noted at the Forum, coal is known to be an impure material which varies with composition, energy values, sulfur, ash, and toxic trace element composition dependent upon its type and geographic locality.

There are, however, significant environmental and health effects associated with increased coal production. Problems that are associated with underground and surface mining include leaching, water contamination, and possible burning; washing, transportation, and open storage present other problems, not only in the release of sulfur dioxide but also with various trace or fugitive elements (e.g., mercury, arsenic, lead, and molybdenum). For these reasons, attention must be focused on the trace element geochemistry of coal development related to health.

To improve understanding of the potential dangers presented by the future increased U.S. coal production and the inadequate knowledge of the fugitive trace elements emitted through the process of coal resource development, a Panel on the Trace Element Geochemistry of Coal Resource Development Related to Health (PECH) has been assembled to help solve these problems.

The panel functions under the guidance of the NAS/NRC Subcommittee on the Geochemical Environment in Relation to Health and Disease (GERHD) of the U.S. National Committee for Geochemistry. The panel was initiated in December of 1976 and members are Bobby G. Wixson (Chairman); Albert L. Page (Cochairman); Jack Beckner; Vern Swanson; Rodney Ruch; James F. Boyer, Jr.; Richard Neavel; Leonard D. Hamilton; and Pope A. Lawrence. The makeup of the panel represents a balanced blend of university, industry, and agency members concerned with various aspects of coal resource development with special emphasis on trace elements.

Since the panel has just recently started its study, the workshop presented an unusual opportunity to present the panel's planning and objectives to the scientific and interested lay public and to enter into discussion on how the panel might best fulfill its objectives. It was emphasized that the panel is looking at trace elements in coal primarily in terms of present technology as related to coal resource development. The activities the panel will be investigating in relation to coal resource development include: surface and underground mining; preparation products and waste; surface and pipeline transportation; combustion, including fly ash, bottom ash, and scrubber sludges; coking; gasification; and liquefaction. All of these activities are being investigated in relation to the environmental impact of trace elements in coal on water, land, air, and biota (including people).

The panel objectives were then presented to the workshop participants as follows:

1. To determine the current state of knowledge on the health hazards of various trace elements in coal.
2. To evaluate additional burdens of coal-derived trace elements, modes of intake, and health-related effects associated with deficiencies or excesses of trace elements.
3. To point out areas where there are "gaps" in our knowledge about the role of trace elements in geochemistry of coal resource development.
4. To compile and disseminate information and make recommendations regarding further studies or research needs.

Following the presentation of the panel objectives a good informational discussion was held between the workshop participants. Of particular interest were some of the biologically active elements considered by the panel, such as arsenic, boron, beryllium, cadmium, fluorine, mercury, lead, sodium, zinc, antimony, chlorine, molybdenum, nickel, thallium, thorium, uranium, vanadium, polonium, tellerium, and germanium. In one example it was noted that molybdenum and selenium emitted from coal-fired power plants may accumulate in soils to the extent that forage crops will absorb amounts that are toxic to animals that consume the forage. Boron was also discussed as another element which frequently occurs in many coals; is volatile upon combustion; and enters the atmosphere, soil, and water to levels potentially detrimental to man.

Routes of absorption by the human system including air, liquids, and food were discussed, and it was generally concluded that greater percentages of trace elements are absorbed from ingested air. It was also noted that there is little information available on potential health problems associated with coal gasification.

Most participants agreed that the source of coal for significant increases in consumption is expected to come largely from the western United States. On the basis of this assumption it was considered to be important to evaluate trace element contamination in relation to particular air basins, since the chemical composition of air basins varies and atmospheric chemistry of trace elements in the ecosystem is needed because their effects on biological systems vary with the form.

The condensed findings of the workshop were presented to the Forum and will be incorporated into the development of the report on the trace element geochemistry of coal resource development related to health.

ROSE: One of the other Academy panels, the CONAES study, will be much interested in hearing what you are doing, and vice versa.

COLEMAN: A statement was made earlier that the conversion from gas-fired or oil-fired boilers to coal boilers might cost as much as 40 percent, if I understand the question, of total new plant construction. I would like to see how that figure is arrived at, that 40 percent figure. If there is some misunderstanding that I have, I would like to have it cleared up.

HILL: I will be glad to respond to that. A boiler that is designed to burn gas and/or oil, and not designed to burn coal, cannot substitute coal directly into that system. You have, therefore, if you are going to burn coal in that plant, to replace that part of the system by a coal-fired unit. I have been told that in a new plant the boiler segment of the system, including the electrostatic precipitator, and so forth, represents something on the order of 40 percent of the total plant cost. That is the basis for it. It could vary quite widely, depending upon how many units, electrostatic precipitator, flue gas desulfurization, and so forth, have to be added. It might run higher than that.

SOME LONG-RANGE SPECULATIONS
ABOUT COAL

Alvin M. Weinberg and
Gregg H. Marland

*Institute for Energy Analysis,
Oak Ridge Associated Universities,
Oak Ridge, Tennessee*

We shall speculate on the role of coal in the intermediate future, say to the year 2010, and in the very distant future, beyond 2010 when replacements for fossil fuels may have to be used on a large scale.

DO WE HAVE ENOUGH COAL?

When will our coal reserve be seriously depleted? In Table 1 we summarize our coal reserve based on estimates given by P. Averitt in 1975.[1] Note that the U.S. reserves of coal, defined as 50 percent of the coal that has been accurately identified and is *currently* economically recoverable, amounts to but 5 percent of the total U.S. resource--the latter being defined as all coal in seams no thinner than 14 inches (35 centimeters) nor deeper than 6,000 feet (1,800 meters). We are thus

TABLE 1 U.S. and World Coal Resources

	United States		World	
	Billion Metric Tons	10^{15} Kilojoules	Billion Metric Tons	10^{15} Kilojoules
Resources	3,600	86,300	15,074	360,000
Ultimately available resource	941	22,600	3,940	95,000
Reserve base	394	10,300	1,650	43,000
Reserves	197	5,150	824	21,500

confronted with an immense range of possibilities between what is now judged to be economically recoverable and what the total resource is. Since coal in seams as thin as 14 inches (35 centimeters) is mined in some places, and in mines as deep as 3,500 feet (1,100 meters) in others, one simply cannot assert that we shall never exploit a great part of the resource. The second entry for the United States, designated "ultimately available resource," represents Averitt's estimate of how much coal there is at 3,000 feet (900 meters) depth or less in seams 28 inches (70 centimeters) thick or thicker; and he regards 50 percent of this as "ultimately available." For the world we show the same fraction of "resources" as Averitt estimated for the United States.

Let us now examine possible bounds on the amount of coal that is actually used. These bounds may arise from limitation of overall energy demand, from environmental effects, or from competition from alternative sources of energy.

Most recent estimates of the future demand for energy have been lower than they were, say, a half-dozen years ago. Much of our basic energy policy in the United States was formulated when we took as a law of nature, more or less, that energy demand would increase at the historic rate of 4.5 percent per year *ad infinitum*. But beginning with the Ford energy project,[2] and now with increasing frequency, we are getting accustomed to much lower projections of energy growth. Thus, the ORAU Institute for Energy Analysis, in its study[3] "Economic and Environmental Implications of a U.S. Nuclear Moratorium, 1985-2010" (abbreviated USNM), projects a high aggregate U.S. energy demand in the year 2000 of 126 quads (133×10^{15} kilojoules), a low of 101 quads (106×10^{15} kilojoules). The low figure agrees with the Ford zero energy growth estimate.

At the time Ford zero energy growth was promulgated, much of the establishment reacted violently; yet, only three years later such low projections receive nods of approval and seem to be incorporated into our national energy policy. Three major factors give to such lower estimates a feel of plausibility: (1) a lower fertility rate (1.8 per woman) and therefore a population around 250×10^6 instead of 350×10^6 by the turn of the century; (2) a trend toward conservation induced both by government fiat and by rising energy prices; and (3) a lower rate of increase of labor productivity and therefore of gross national product (GNP). Though there is no constant relation between GNP and energy, the slower rise in GNP would lead to a more moderate rise in the use of energy.

Nevertheless, even with these more modest rates of energy growth, the per capita demand in the United States is, according to our estimates at the Institute for Energy Analysis, going to rise--from 340×10^6 kilojoules per person per year to around 400×10^6 kilojoules per person per year in the low scenario and 480×10^6 kilojoules per person per year in the high scenario. Moreover, we believe it likely that the fraction of energy that goes through electricity will also increase--from 28 percent at present to about 50 percent. This amounts to about 200×10^6 kilojoules per person per year going into electricity.

Let us set aside for the moment the question of whether this

electricity will be generated primarily by nuclear energy or by coal
and speculate on the pressure put on our coal resources if all of this
electricity were produced from coal. This amounts to, say, 50×10^{15}
kilojoules per year from coal, which corresponds to about 2×10^9 tons
of coal per year. Now returning to Table 1, we see that at this rate
we would use our reserves in 100 years and our ultimately available
resource in about 500 years. These times do not consider the opposing
factors that other energy sources will supplement coal for electric
generation and that some coal will be used to make liquid and gaseous
fuels. Nor do we take into account a likely shift toward coal as a
source of industrial heat, a shift that would be encouraged if a reliable fluidized-bed boiler is developed.

These estimated times to depletion are reassuringly long, but we cannot be complacent. We have assumed that our asymptotic energy system
levels off at 400×10^6 kilojoules per person per year and 250×10^6
population for a total of 100×10^{15} kilojoules. With equal plausibility
we could consider the demand to level off at 180×10^{15} kilojoules--say
600×10^6 kilojoules per person per year and 300×10^6 population. In
that case we might imagine coal supplying 90×10^{15} kilojoules per year,
and our reserves and ultimately available resource would last, say, 55
and 240 years, respectively.

WORLD DEMAND FOR COAL

These figures take little account of the rest of the world. R. M. Rotty
has projected what the future demand for energy in a developing world
might come to.[4] He divides the world into six regions and projects the
energy growth of each independently, with the presently underdeveloped
countries expanding both their populations and their per capita energy
expenditures much faster than the United States. He assumes a yearly
per capita growth rate in these countries of about 3.5 percent per year
and a population growth of 2 percent, which gives an overall energy
growth of 5.5 percent per year (compared with 8 percent during the past
decade). This leads to an average per capita expenditure in these countries about one sixth that of the present U.S. expenditure--say, 50×10^6
kilojoules per person per year by the year 2025. Rotty's estimated
total world energy demand reaches about $1{,}300 \times 10^{15}$ kilojoules by 2025,
distributed among countries as shown in Figure 1. If we assume 50 percent of this is supplied by coal, the *world* reserve and ultimately
available resource would then last roughly 30 years and 140 years. But
this is by no means an upper limit. F. Niehaus has proposed a scenario
based on 10^{10} people using an average of 200×10^6 kilojoules per person
per year for a world total of $2{,}000 \times 10^{15}$ kilojoules per year.[5] If 50
percent of this is met by coal, the depletion times are, respectively,
20 years and 90 years. Moreover, since the world's coal seems to be
confined largely to the United States, the USSR, and China, such world
scenarios would undoubtedly put much more pressure on U.S. coal reserves
than we would estimate simply from the projections of U.S. demand alone.

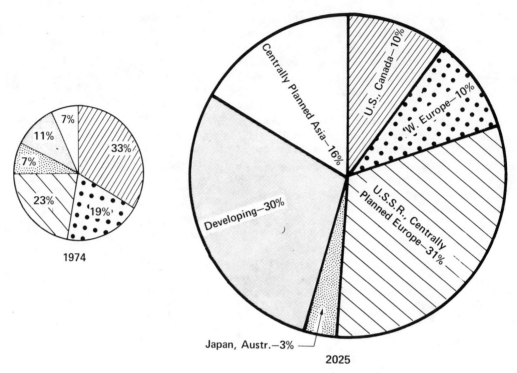

FIGURE 1 Global energy consumption by world segments.

Indeed, there may be an inconsistency in much of our energy "futurology": we usually *assume* that eventually the underdeveloped countries will use energy at some fair fraction of the per capita energy demand of the developed world, yet we don't really face squarely the question of where they will get their fossil fuel. Should, say, one fourth of the world's energy be derived from U.S. coal--this proportion is the same as the U.S. coal reserve compared to the world reserve--and if the world's energy demand reaches $1,300 \times 10^{15}$ kilojoules, the United States would have to supply some 13×10^9 tons of coal per year. This seems like an astronomical amount of coal.

Obviously, such numerology cannot be taken fully seriously. Nevertheless, we would claim that the world's coal resource of 15×10^{12} tons, which seems so enormous compared to the present use of coal at the rate of 3×10^9 tons per year (1973), may be a far smaller usable resource than the raw estimates suggest.

ULTIMATE CLIMATIC LIMITS ON THE USE OF COAL

The atmosphere contains about 710×10^9 tons of carbon in the form of CO_2. This corresponds to a concentration of 330 parts per million (ppm).

There is some evidence that about one half the CO_2 injected into the atmosphere by the burning of fossil fuel remains there. In Table 2 we give the concentration of CO_2 in the atmosphere if each of the coal reservoirs as well as the world's oil and gas are burned and one half the released CO_2 remains in the atmosphere.

The carbon dioxide burden in the atmosphere has been increasing at a rate of about 1 ppm per year, at least for the 20 years since accurate monitoring of CO_2 began (Figure 2). To be sure, during this period tropical forests have been cleared, and there is controversy as to the extent to which the decay and burning of this wood have contributed to the present CO_2 burden. At the recent ERDA-sponsored workshop on the global effects of CO_2 in the atmosphere,[6] it was concluded that we do not really know whether the biosphere is a net source or a net sink of CO_2. On the other hand, W. Broecker presented evidence that if clearing of the biosphere is a source of CO_2, the amount of CO_2 thereby added to the atmosphere must be small in comparison to that contributed by burning fossil fuel. Moreover, since the biosphere contains only one tenth as much carbon as does the world's coal resource, even if the entire biosphere were removed (an entirely implausible event), it could ultimately contribute only 10 percent as much CO_2 as would the burning of all our fossil fuel. We would conclude that the CO_2 increase is probably related to the burning of fossil fuel and that in the long run we may encounter extremely large increases of CO_2 as a result of our use of fossil fuel.

Carbon dioxide absorbs energy in the infrared. Solar energy incident on the earth's surface is reradiated at wavelengths that can be trapped by CO_2. H. Flohn[7] estimates that the present yearly rise of CO_2 increases the energy output into the climate system by about 1.6 TW.

If the use of fossil fuel increases along Rotty's projections, the CO_2 concentration would *double* within the next fifty to seventy-five years. We do not really know how the climate would change if CO_2

TABLE 2 Projected Atmospheric CO_2 Concentration from Fossil Fuel Consumption if Half of the Released Carbon Remains in the Atmosphere

Fuel Burned	Carbon Retained in the Atmosphere, 10^9 tons	Atmospheric Concentration CO_2, ppm	Years at Asymptotic Level
Present concentration	710	330	
Reserve	285	460	30
Reserve base	570	600	60
Ultimately available	1,365	960	140
World resources	5,225	2,760	550
Oil	125	390	
Natural gas	95	370	

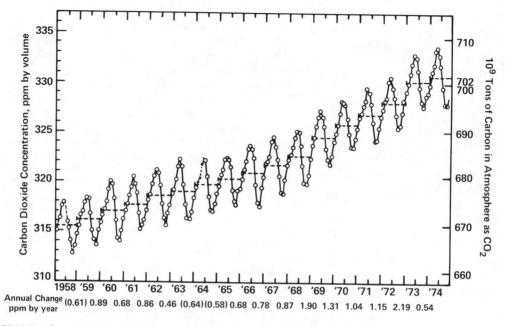

FIGURE 2 Atmospheric carbon dioxide concentration at Mauna Loa observatory (1958-71 data from Keeling et al., 1976; 1972-74 data from Keeling, private communication).

concentration *doubled*. Manabe and Wetherald,[8] with their elaborate climate model, predict a global average temperature increase on the order of 2°C and an 8°-10°C increase in temperature of the pole for a twofold increase of CO_2. To be sure, the models are crude, and they do not include the effect of clouds. What estimates have been made of the effect of clouds, by Ramanathan[9] and by Schneider et al.,[10] however, suggest that clouds may not be the potent source of negative feedback we had originally hoped for.

The *concentration* of CO_2 in the atmosphere is more strongly tied to the utilization of coal than to that of oil and natural gas simply because the coal resource is so much larger than the oil and gas resource. The total carbon content of the world's oil and gas reserve (not counting oil shale) is about 440×10^9 tons; this is but 40 percent of the reserve base of coal. Thus, even if all the oil and gas were burned, the CO_2 concentration would be increased only by 30 percent; whereas, if coal equivalent to the reserve base were burned, the CO_2 concentration would increase by 80 percent. (These figures do not include any contribution from oil shale.) Although an increase of 15 percent, i.e., 70 ppm (which could be released by the burning of oil and gas) might be dangerous, the potential contribution from coal is five to ten times greater. This suggests that although CO_2, if it is indeed a problem, will be exacerbated by our use of oil and gas, it is the burning of the much larger reservoir of carbon in coal that must be our primary worry. So to speak, the world can burn its oil and gas without worrying too much about CO_2; it can eat into its much larger reservoir of coal only at the risk of elevating the CO_2 in the atmosphere to what may be serious, or even catastrophic, levels.

THE CHOICES BEFORE US

Let us set aside for the moment the possibility of limits on our use of coal imposed by the CO_2 problem, or indeed, by other environmental constraints, and instead ask whether coal will be chosen on strictly economic grounds over other possible sources of energy. As a fuel for generating electricity, coal is now in direct competition with nuclear energy. D. Phung of the Institute for Energy Analysis has estimated the relative cost of electricity from coal and from light water reactors (LWRs), and his results, taken from USNM, are presented in Table 3. The relative prices of nuclear- and coal-generated electricity are sensitive to perceived rates of inflation and to assumed interest rates. Roughly speaking, since fuel costs for coal boilers are relatively more important than for nuclear reactors, whereas the situation is reversed with respect to capital costs, one would expect coal to suffer in relation to nuclear energy if inflation is high and to show up better if

TABLE 3 Effect of Fuel Escalation Rates above Inflation Rate on Differences between Cost of Nuclear- and Coal-Generated Electricity

Assumptions on Net Fuel Cost Increase, per year	Cost Advantage (Disadvantage) of Nuclear over Coal Electricity, %	
	Nuclear over Coal Electricity without Scrubbers	Nuclear over Coal Electricity with Scrubbers
Coal, 0%; Nuclear, 0%	9	18
Coal, 0%; Nuclear, 2%	-1	8
Coal, 0%; Nuclear, 4%	-12	-4
Coal, 2%; Nuclear, 2%	17	24
Coal, 2%; Nuclear, 4%	4	10
Coal, 2%; Nuclear, 6%	-11	-5
Coal, 4%; Nuclear, 4%	26	31

NOTE: Percent Difference = $\dfrac{\text{Coal Cost} - \text{Nuclear Cost}}{\text{Nuclear Cost}} \times 100$

Base case assumes 5 percent inflation and 11 percent operating discount rate (6 percent net discount rate).

inflation is low. Phung's analysis bears this out; on the other hand, it is clear that nuclear energy enjoys a distinct advantage only where low-sulfur coal is unavailable. In the mountain states, coal is cheaper than nuclear energy.

The relative cost of nuclear- and coal-generated electricity is sufficiently uncertain that one would be hard put to insist that, over the next thirty years, nuclear energy will enjoy as large an advantage as our analysis suggests. Nevertheless, if one takes our results at face value and projects the demand for coal and nuclear energy out to 2010, we come up with the figures shown in Table 4. Should a nuclear moratorium be imposed that allows no new reactor starts after 1985 but allows reactors already on line to continue operation to 2010, then the total amount of coal required in 2010 is about twice that required in the absence of a moratorium. Thus, in the case of a nuclear moratorium and a high demand, we would estimate the coal required to be almost 5×10^9 tons--and this does not do justice to the demand that may arise among the lesser developed countries (LDCs).

At the moment we are unclear as to when breeder reactors will be introduced in the United States. Under the circumstances the total energy that we can extract from LWRs is limited by the uranium supply; but since the latter is really not known, we can do little more than speculate as to how large their contribution to our energy system can be. Since a 1,000-MW electric LWR without recycle requires about 200 short tons (180 metric tons) of uranium per year, and its heat equivalent is 0.06×10^{15} kilojoules per year, we can compute the amount of coal we can displace with uranium in LWRs for various assumed uranium

TABLE 4 Estimates of Future U.S. Coal Requirements (from USNM), 10^9 metric tons

	1985	2000	2010
Low demand			
Nuclear option	0.64	0.71	1.04
Coal option	0.64	1.36	2.39[a]
Annual percent increase (coal option)		5.1	5.8
High demand			
Nuclear option	0.66	1.74	2.28
Coal option	0.66	2.40[a]	4.36[a]
Annual percent increase (coal option)		9.0	6.1

[a]Average tonnages for the years 2000 and 2010 have been increased 17 percent to account for western coal having about three fourths the heat content of eastern coal [26.5×10^6 kilojoules/ton (22.8×10^6 Btu/short ton) is the average heat value used up to annual production of 2.0×10^9 tons].

supplies. These are given in Table 5. We see that even with the low uranium supply, LWRs would displace about 20 billion tons of coal--i.e., about 10 percent of what Averitt estimates to be the U.S. coal reserve. If the highest uranium estimate proves correct, then some 82 billion tons, i.e., almost one-half of Averitt's reserve, could be used for purposes other than generating electricity.

No one can say whether such displacement will occur. Although, as we see the situation now, nuclear energy appears cheaper than coal in most parts of the country, it would be imprudent to decide that economic forces will cause the large displacement of coal suggested in our Table 5.

If we look at the matter in broadest terms, we find ourselves beset with a profound dilemma. The difficulties and risks of the nuclear path have been delineated often and in detail. Of these, proliferation of nuclear weapons probably poses the greatest risk, though one must always remember that power reactors and chemical plants provide a sufficient, not necessary, technical basis for proliferation. The major risk in the coal path is the possible CO_2 catastrophe. In a way this is the coal analogue of nuclear proliferation: it is global, uncertain, possibly catastrophic. Thus, we see the dimensions of the dilemma: the two energy systems upon which we are expecting to depend, at least over the medium term, are flawed to a degree that is at present essentially impossible to fully estimate and that, indeed, may never be fully possible to estimate. To those who embrace coal as a fission-free bridge to a solar future, the CO_2 question should inject a note of prudent concern: we can turn the phrase around and ask whether fission based on reactors of current design perhaps will have to serve as a coal-free bridge to a fusion, breeder, or solar future. We must also consider the possibility that both nuclear energy and coal will be judged by future generations to be fatally flawed and the question, "Can we make it on solar energy alone?" will have to be rephrased: "How can we make it on solar energy alone?"

What can we do under the circumstances? The most obvious first course is to expend much more effort in trying to understand the CO_2 question: does it or does it not pose a real threat? It is gratifying that ERDA as well as other energy research agencies throughout the world are

TABLE 5 Displacement of Coal by Light Water Reactor (from USNM)

Uranium, short tons	No. of 1,000-MW(e) LWRs	10^{15} kJ per Year	Total 10^{15} kJ (30 years)	Tons of Coal Displaced
1.8×10^6	300	18	540	20×10^9
3.6×10^6	600	36	1,080	41×10^9
7.2×10^6	1,200	72	2,160	82×10^9

finally addressing the question seriously. In the meantime we can do no better than to keep our options open--meaning nuclear, geothermal, solar, and coal options and try to clarify just what we can do should the twin swords of Damocles, one called CO_2, the other called proliferation, begin to fall upon us.

REFERENCES

1. P. Averitt. Coal resources of the United States, January 1, 1974. *Bulletin 1412*. U.S. Geological Survey, Washington, D.C.
2. Ford Foundation Energy Policy Project. *A Time To Choose*. 511 pp. Ballinger Publishing Company, Cambridge, Mass., 1974.
3. Charles E. Whittle, Edward L. Allen, Chester L. Cooper, Herbert G. MacPherson, Doan L. Phung, Alan D. Poole, William G. Pollard, Ralph M. Rotty, Ned L. Treat, and Alvin M. Weinberg. Economic and environmental implications of a U.S. nuclear moratorium, 1985-2010. *ORAU/IEA 76-4*. Institute for Energy Analysis, Oak Ridge Associated Universities, Oak Ridge, Tenn., September 1976.
4. R. M. Rotty. Present and future production of CO_2 from fossil fuels--a global appraisal. *ORAU/IEA (0) 77-15*. Institute for Energy Analysis, Oak Ridge Associated Universities, Oak Ridge, Tenn., June 1977.
5. F. Niehaus. A nonlinear eight level tandem model to calculate the future CO_2 and C-14 burden to the atmosphere. *Research Memorandum RM-76-35*. International Institute for Applied Systems Analysis, Laxenburg, Austria, 1976.
6. L. Machta. Energy Research and Development Administration Workshop, Environmental Effects of Carbon Dioxide from Fossil Fuels Combustion, Miami Beach, Florida, March 7-11, 1977 (in press).
7. H. Flohn. Man-induced changes in heat budget and possible effects on climate. In: *Global Chemical Cycles and Their Alteration by Man*, ed. W. Stumm, pp. 207-224. Dahlem Konferenzen, Berlin, 1977.
8. S. Manabe and R. T. Wetherald. The effects of doubling the CO_2 concentration on the climate of a general circulation model, *J. Atmos. Sci.*, *32*:3-15, 1975.
9. V. Ramanathan. Private communication, 1976.
10. S. H. Schneider, W. M. Washington, and R. M. Chavin. Cloudiness as a global climatic feedback mechanism: Sensitivity experiments with the NCAR GCM, *EOS, 57*(4):253-254, 1976.

DISCUSSION

HOUTHAKKER: I think that the projections put forth by Dr. Weinberg are interesting. There are, of course, many projections around. I myself also dabble in this projection game. I play with numbers. I no longer believe my own numbers; I believe other peoples' numbers even less because I don't know what assumptions go into them. This is also the problem I have with some of the numbers he mentions. I don't know to what extent his models reflect the influence of prices on both demand and supply, which is very important.

I must say that I find a world demand of 1,300 quads in the year 2025 inconceivably high. I don't know any projection that takes prices into account that gets anywhere near those numbers. I would say that 500 quads for the world as a whole in 2025 is a more likely number.

But anyway, there is one detail which I am sure Dr. Weinberg and I could sit down together and argue about. I also am not at all convinced that the demand in the less-developed countries will grow that much more than in the rest of the world. I don't know what the evidence for that is. So far, the demand has grown relatively less. Once they start industrializing, it may grow somewhat more.

Now, let me make a general remark as to the competitive position of coal, which is one of the main things that Dr. Weinberg rightly called attention to. This, I think, is something which really cannot be decided on an aggregate basis. There is a great danger in models of this kind, to interpret them as if there is some choice that society can make. We go all out for coal, we go all out for nuclear power, hydropower, oil, gas, and what have you. That, fortunately, is not the way the world works. These decisions are decentralized and should be decentralized. To try and make decisions of this kind on a centralized basis is only to guarantee that we all make huge mistakes. If decentralized decision makers make mistakes, well, that is their business. They go bankrupt, get fired or retired, or what have you. If the government makes a mistake, then we all suffer from it. So this, I think, is the first point to recognize. I don't raise it as a criticism because I think Dr. Weinberg would agree with that. Nevertheless, it is often overlooked in the interpretation of these statements.

Now, once we start thinking what decentralization really means, then we find that it means that we have to have a price mechanism that appropriately reflects social costs. That means, in the first place, that commodities have to be valued by their scarcity value in relation to the demand they have. In the second place, that externalities--what we call environmental effects, health effects, et cetera--should be duly taken into account.

Now, undoubtedly, we are some direction away from that, although we are moving closer. I was happy to see just a few days ago that a federal judge has declared the Price Anderson Act unconstitutional. I am not a lawyer; I don't know what the legal aspects of this are. As an economist I applaud it, even though I am not opposed to nuclear power as such. I believe that the Price Anderson Act is one of the ways in which social costs and private costs are made to diverge. So the main comment I would like to make is that if we are not going to make big mistakes in this area, we have to be sure that we can reach decisions on a properly decentralized basis. That applies not just for the United States but for the world as a whole.

WEINBERG: I have no quarrel at all with Professor Houthakker in faulting these projections. To be sure, he is perfectly right, we did not consider prices. On the other hand, I guess my impression is that at least as far as world energy demands are concerned, there have been, as far as I know, rather few models that really take price into account in anywhere near the detail that price has been taken into account in some of the more recent projections of the United States' energy demand. Perhaps Professor Houthakker can give me a reference for a world energy model that does take prices into account.

HOUTHAKKER: Well, there is one model which deals only with oil. It has been published in an American Enterprise publication called the *World Price of Oil*. Another one, dealing with the five principal energy resources, coal, oil, gas, hydropower, and nuclear power will be described in the CONAES report, I hope, because I am writing the international section on this at the moment, which is based in part on a world model of this kind.

WEINBERG: The second point is that even at 500 quads one will be throwing carbon dioxide into the atmosphere. Of course, you then have to argue what fraction of that 500 quads is fossil and what fraction is not fossil. Even at 500 quads, if you take Flohn's estimate seriously, you get into trouble in, say, fifty years. But that is certainly part of the uncertainty. You remember, he said that if the carbon dioxide effect were equal to the natural fluctuation in the power driving the climate, which is only 100 to 300 quads, then at 500 quads--100 to 300 terrawatts, then 1.6, which would be at 500 quads, say three, three multiplied by 50 years is 150, and you are into the region that Flohn, at least, claims is going to cause difficulty.

With respect to decentralization, I guess I don't really understand the force of your argument, Professor Houthakker, in the following sense: of all problems, if this is a problem at all, I can think of none that has to be looked at more from a global point of view than something that could cause irreversible and possibly undesirable changes in the climate. If one does not look at the matter globally, then 50 or 75 or 100 years from now one is going to have to take whatever comes. I don't quite understand what your point was there, Professor Houthakker.

HOUTHAKKER: As I said when I made my first remarks, I didn't think it was something on which we really disagreed. I don't think it is a misinterpretation into which you yourself would be led. But I believe that people who read these projections without having occasion to think a great deal about them sometimes think of this as a choice between coal as one big option and nuclear energy as another big option with maybe one or two other big options. But that is just not the way it works. We are going to use both these options to some extent to formulate the problem properly, and I don't think you would disagree with that. It is a question of balance rather than of a choice.

WEINBERG: Let me sound kooky, Professor Houthakker. Suppose that ten years from now we have the very best climatologists in the world, the very best ecologists in the world, and the very best oceanographers in the world who meet in Geneva, let us say, or at the United Nations in New York, and they make a sober pronouncement that the world is on a catastrophic course. What would we do then, because of the carbon dioxide?

HOUTHAKKER: I would agree, if there were evidence of that kind, then maybe some form of energy, be it coal or nuclear or hydro, would have to be stopped. It is quite possible that evidence will be forthcoming in the future that might lead us to foreclose one particular option altogether. I suppose what I am saying is that at the moment I am not aware of any such evidence. Rather, what we have is something wrong with every form of energy we know about, with the possible exception of natural gas, of which there is not a great deal.

WEINBERG: Without again unfurling my nuclear banner too blatantly, I would simply like to inform the Forum, conceding that I am a lousy prophet, that ten years from now this may happen.

HIBBARD: Are there any other questions?

WILLIAM RAMSAY, Resources for the Future: Might I ask you the same question I asked this morning about possible abatement of carbon dioxide? Obviously, the amounts are very large. We are into, say, from one plant billions of cubic meters per year. However, a variety of exotic schemes are proposed: liquefying it, taking it to the bottom of the ocean. These are rather desperate measures, but if you are talking about this as a desperate problem, perhaps they deserve consideration.

WEINBERG: Well, I might say that at the Miami Beach Conference, Freeman Dyson of the Institute for Advanced Study and Gregg Marland of the Institute for Energy Analysis did try to catalogue all of the possible measures that could be taken. For those of you who aren't familiar with Marchetti's scheme--he would localize all use of fossil energy in a few places that would manufacture hydrogen for transport, and the effluent would then be directly aimed into the down-welling sea, say, near Gibralter, where there is a serious down-welling. This way you would get rid of the carbon dioxide.

None of the schemes really appealed to anybody as being all that great. The one that Freeman Dyson proposed was that everybody should go out and plant a trillion trees. A trillion trees would maybe do it. But it would do it only for about fifty years, because then these trees would decay, and you would get back the CO_2.

There are those who say you are not going to have to worry, because CO_2 is going to increase the photosynthetic activity, and it will come back to an acceptable balance. Most of the ecologists do not believe that that is likely to occur. My own feeling about it is that we will have to look very much more seriously at biomass as a source of energy than we have thus far. Biomass takes in CO_2 at the same time it gives it back, so you aren't adding to the CO_2.

I don't think that anybody at Miami Beach came up with anything that sounded at all really credible. Of course, you do have the possibility of an all-nuclear or all-solar or all-fusion scenario manufacturing hydrogen or other transportable fuels. I was a little bit embarrassed even talking about this matter in a serious way because it is so bizarre, and so far reaching, and has such implications. It is very hard to take it seriously, and yet I fear that people will have to start taking it seriously.

HARTNETT: Dr. Weinberg, you showed only the concentration of CO_2. You did not show the concentration of particulates in the atmosphere, and this of course would move us in the opposite direction. It does strike me that we have an open question as to whether we go in the direction you suggested, or whether we might go in the opposite direction, and that the only way this can be determined is by monitoring the world temperature. I understand this is being done. Do you have any indication as to what those measurements give us?

WEINBERG: Let me perhaps disagree a bit with your assertion that the particulates necessarily have a negative feedback. It depends on where the particulates are injected. If the particulates are injected high, as in the case of volcanoes, then it is true that they increase the albedo, and it cools the earth down. But if the particulates are inserted low as a result of agriculture, it is not at all clear that the particulates have that effect. This point was discussed at Miami Beach, and people rather threw up their hands and tended to say no. They did not really think that the particulates would prove a sufficient feedback to stabilize.

ROSE: There is an interesting observation on that very point. The CO_2 is globally distributed. The particulates are in the Northern Hemisphere only; some recent measurements tend to show that the two trade off in the Northern Hemisphere but in the Southern Hemisphere the temperature is going up.

WEINBERG: Murray Mitchell, one of the foremost climatologists in the United States, did give a talk on exactly this question: Is there any evidence at all that CO_2 is breaking out of the noise? The answer is, no, there is none. But Murray Mitchell's estimate was that by about the year 2000, if there is anything to this whole story, then you will begin seeing the CO_2 breaking out of the noise.

CHARLES A. ZRAKET, Mitre Corporation: Independent of the CO_2 problem, have you done any calculations that show when you get into trouble just from heating, whether it is from nuclear or fossil fuel, or other sources?

WEINBERG: Flohn's energy budget gives the answer to that. The present energy contribution from man, direct energy, is 8 terrawatts. That doesn't cumulate; you produce that. The CO_2, though, does cumulate, so that for the CO_2 you integrate it. In fifty years, assuming that we are still at 8 terrawatts, even if we don't increase, the CO_2 contribution will be 80 terrawatts even though the heat would still be only 8 terrawatts. So the conclusion that most people come to is that although you do have local effects due to the heat, the CO_2 seems to be by far the dominant effect, by a factor of twenty or more.

ZRAKET: What if we went to ten or twenty times the energy consumption we have today?

WEINBERG: All right, at ten times the energy consumption, you would be up at 80 terrawatts which starts getting to something like the 100 to 300 of Flohn's estimate. But if most of that is from CO_2, then the CO_2 would be increasing correspondingly. And the CO_2 stays there.
That really is a thing that has happened in this business. People used to say that the ocean was going to take it up and that would be the end of it. But what the oceanographers have now discovered is that the top layer of the ocean, the top 50 or so meters, saturates in CO_2, and it takes many, many hundreds of years for the carbon dioxide, which is converted into carbonate, to go into the deep ocean. That is really the new point that bothers people.

ZRAKET: Even if you went up to 80, it is really the fluctuations in that power that drives the climate. If you had around the temperate Northern Hemisphere a uniform increase, and you had an 80-quad production, you might not get the climatic effects that you get from the fluctuations of the same amounts. Would you agree?

WEINBERG: I am not a climatologist, so I am a Greek messenger, and I hope I am not shot or beheaded.

ZRAKET: Forgetting the particulate and the CO_2 and so on, I was surprised to see what you pointed out about the 4,900 quads and that you would be perhaps ninety years using up the entire reserves, which means, perhaps, in the last forty years of that you would really have trouble, inaccessibility, cost, and so on. Would this not mean that even if you didn't have particulate problems, trace element problems, and CO_2 problems, you would still be concerned in another fifty years with demand for another energy source, nuclear energy or something else?

WEINBERG: That 4,900 quads is in what is called the reserve, which is already half of the resource base. Mr. Falkie, in particular, was saying that there are these better methods of mining, and so on. So I think that there is probably going to be lots of coal, and we are, I think, going to be able to dig it up. Although, I must say that 5 billion tons of coal per year in the United States just seems like an awful, awful lot to me.

HIBBARD: Professor David Rose of MIT has asked if he could make a few comments regarding the Forum.

ROSE: In reflecting on this meeting it seems to me that the debate continues to support the delusion that the energy problem is mainly the technology of supplying it, in this instance, coal. There have been some notable exceptions, including the recent discussion over the lively topic brought up by Alvin Weinberg, with whom I very much agree. The CO_2 problem is a real one and is probably going to be.

But we lose sight of the fact that energy is a means, not an end. Energy is the servant. So the question becomes: What kind of servant is it in comparison to other servants? Compared to what? Here the "compared to" seems to be nuclear energy.

A year ago I asked, "Nuclear power compared to what?" And now one can ask, "Coal compared to what?" One cannot look at either of these things in a vacuum.

In comparing the status of the technology, of the debates, of the resources, of the difficulty of working on the problem, I have said that if the same kinds of things were said about nuclear power they would shut it down tomorrow. I would like to amend that statement by saying they would have shut it down years ago. That is, the situation with coal looks to me so difficult.

So I will cry "Oh woe, oh woe," because it does seem as though we are in a lot of trouble. Compared with the nuclear technology, proliferation included, it seems that we are in the equivalent of, say, wooden pressure vessels, or, at best, cast iron ones, or something of that kind, and still have a long way to go.

Regarding damage, I have yet to see very much attention paid to the damage to the public either in mortality or morbidity or discussion of what is going to be needed to resolve that question. But on the basis of the inadequate information that we have, which is itself a condemnation of our lack of attention to the problem, we would need something like a catastrophic accident to a large nuclear reactor somewhere every year in order to equal the misery caused by coal at present. And that we don't see.

To that one can reply that people may want to value their misery differently, and one person's misery is not another person's, so we should have free choice. That is all very well for those who have free choice and don't have to sit out there enjoying the disadvantages. Those in Appalachia do not always joyfully go about their jobs, singing songs as they work. They have many difficulties. Those who can get well enough educated leave. When I lived there, I once made the suggestion that the best thing for Appalachia would be to secede and then apply for international aid; it would do much better.

We hear a lot about environmental impacts. The best description of the environmental impact of mining that I have seen so far is not to be found here but in the first ten pages of Agricola's *De re metallica*, first published in 1556 and translated by Henry and Lou Hoover in 1912. It is much better than most of the statements I have heard today. I would recommend it as a good primer on such topics.

We have underestimated the task of coal and overestimated the state of the art. One cannot help comparing nuclear and fossil fuel. There are less than 100 elements, which do their thing in very serious and complex ways, to be sure. But there are many more than 100 compounds; there are tens of thousands of them. They act in very different ways, and each time we uncover some new coal technology, we find that there are new effluents, new effects, new things we didn't know before. And it looks to me as though it is an art that is just being commenced. So we see lack of communication--lack of understanding between sectors, lack of understanding of what the problem is--in the public sector, in the Congress, perhaps even in the National Academy of Sciences. That is not a condemnation of everybody or anybody because it is the nature of such problems that they grow up that way. People feel their way through them that way just the same as my students. If they knew what they were doing in the beginning, they wouldn't have done it that way. So you learn as you go.

One thing I don't see enough of here is discussion of questions of ethics and of charity. I mean "charity" in the biblical sense that it is going to need a very great deal of charitable nature to resolve this problem. I am not against exploring the use of coal and trying to use it wisely, but I can see coming from the experience in the nuclear debate a polarization in the debate about coal, which would be corrosive, as people discover things and say, "You didn't tell us, we didn't know, they didn't let us know, they hid it from us," et cetera. It has all the markings of another nuclear calamity, and by this I mean the kind of corrosive debate that we have had about nuclear power, which has practically brought it into total social dishevelment.

So, by all means, let us do the job. It is going to take not less effort but more. The effort to date has been, despite all the efforts of those who try so hard in this program and many others, just not enough. So the task for ERDA is not to do less but to do more and to do it better.

HIBBARD: Spoken like a true professor of nuclear engineering. I am sure there must be some questions or comments for Professor Rose.

WEINBERG: I would hope very much, Dave, that the religious war that the nuclear debate has become really will not characterize the debate on coal, and I think there are certain reasons why that won't happen. I think the most important one is that despite what you say as a professor of nuclear engineering--and I am a guy who has made my living for all these years on nuclear energy--there really is something different about nuclear energy. It is the fact of radiation. The existence of radiation on a large scale is basically a new human phenomenon; and until and unless people generally become inured or educated, I think there really will be essential differences in the nuclear debate and the coal debate.

ROSE: The open debate brings us problems that are not in the control of man, except by abstinence, probably, and that are determined by laws of nature made by God. The nuclear debate has to do with how people will misuse nuclear energy because it has to do with proliferation. Therefore, it has to do with the nature of man himself. There is the big difference. I do not draw comfort, by the way, from the fact that perhaps the coal problem is going to be easier than the nuclear problem. There is a difference, indeed, but it may not be just in the dimensions that are so often so easily put.

HIBBARD: The other big difference is that we have been using coal for 125 years, and we haven't been using uranium. As a resident of Appalachia, by the way, I am going to put my hat by the door, and I would like to take up a donation for my poor condition.

SPEAKER: There is the issue of nuclear disposal as opposed to mere nuclear waste, which presents significant hazards over a much longer time horizon than that of the coal problem. There do appear to be many more irreversibilities in the area of nuclear power abuse than in the coal area and the possibility of uncontrolled genetic effects that we really have no handle on at the present time.

ROSE: Would you like to leave an inheritance of what I think are tractable nuclear wastes or an inheritance of no coal, no gas, no petroleum, and so on, for future generations? Which is the better thing to leave to future generations? I would rather take those fossil chemicals and not burn them. It doesn't say anywhere that I know of in the books of Genesis, Exodus, Leviticus, Numbers, or Deuteronomy that those things are fuels.

WHAT MUST BE DONE?

John McCormick

Environmental Policy Center

Speaking for the environmentalists and for the Environmental Policy Center, I want to begin by saying that I recognize the pressing need for greater coal and I encourage the use of that coal in a manner that complements the environmental concerns at the same time. I am not endorsing coal's use to the degree and in the amounts necessary to return to an average annual growth of electric demand to the 6 or 7 percent level. Coal, as I see it, is a suitable transition fuel but is not a long-term energy source. We must, in sequence with our coal development, do more, spend more money, and place a top priority on the development of solar technology and other benign technologies.

The manner in which we mine coal and the responsibility that the federal government has and industry has toward the protection of the public health and welfare, either in the mining of the coal or in the reclamation of the land after the mining, will determine just how much this nation is going to increase its production of coal from present levels. We are seeing that in the northern Great Plains states. Where there is considerable interest in developing perhaps 200 to 300 million tons of new production a year, there is a growing sentiment among the public that this level of productivity is not in harmony with the present form of economic endeavor, agriculture. I think we will see as time goes on a greater resentment and perhaps some more militant public attitude against that level of development. Certainly, on-site conversion of that coal either to electricity or a synthetic fuels product will be difficult if not impossible to attain at the levels that have been projected.

Coal slurry pipeline transportation of that coal will also continue to come under heavy public opposition and may in fact be curtailed in some areas that are considered primary targets for slurrying of that coal.

As the Congress debates and puts the final touches on the strip-mining legislation, I see a real potential for curbing some of the difficulties in strip mining of the coal, at least, and that is that the mistakes of the past will be amended. The companies will be required to replace the land in a way that it can be used again. Perhaps the expensive social costs borne by those communities and municipalities downstream from abandoned strip mines, perhaps downstream from abandoned deep mines, can be compensated by having a fee attached to each ton of coal being mined presently, that fee going into an abandoned land reclamation fund. So as we get into a new era of vastly increased coal production, we can use some of the money generated by this production to go back and patch up the mistakes of the past.

Amendments to the Coal Mine Health and Safety Act, which hopefully will streamline and put some real teeth into the enforcement of that bill, will be adopted by this Congress. We hope that the enforcement of that act by the Carter administration will begin to bring down the increased fatal and nonfatal accident rates of deep mining.

Coincident with this legislation, ERDA has a real responsibility, looking at the coal development R&D program that it presently has underway, not in terms of one project versus another, but in terms of a comprehensive coal development program. By that I mean becoming aware of the fact that while we bring into the market new coal utilization technologies, we are not bringing into the market new, more efficient, safer ways of mining the coal. If we were to double the coal production in Appalachia, we would see that strip mining is a short-lived proposition and that eventually we will have to rely almost entirely upon deep-mined coal if we consider Appalachia to be a primary storage house of coal. We can't mine that coal with the pick-and-shovel methods that we are using now. The continuous miner is hardly a continuous miner. The room-and-pillar method of conventional deep mining leaves 40 to 50 percent of the coal reserves in place, whereas Germany is using a longwall coal-mining system and can recover up to 90 percent of that coal seam, offering greater safety and easier working conditions for the miners.

In addition to the coal mining R&D program, there has to be a mandatory manpower training program in this nation. We certify mechanics to repair our cars, but we don't certify miners to work in the hazardous conditions that they are faced with every day. We are seeing a number of experienced miners leaving the industry and going out of the region or into other jobs, and we are finding a great imbalance in the median age of the miners presently working. A great many returning Vietnam veterans are coming into the mining industry for the first time, having no previous skill. There may be opportunities for some on-the-job education by the experienced miners, but as their numbers decrease, we are finding more and more that the crews are made up of unskilled and untrained miners.

The federal government has opened several model deep mines throughout the coal fields in Appalachia and the Midwest, where on-the-job training of new miners and the implementation of new mining methods and new mining equipment could be done in a way in which we can insure that the fatal and nonfatal accident rates may eventually go down.

John F. O'Leary

Administrator, Federal Energy Administration

We have to start at the demand end of the line and then switch over to the supply end to see what, in fact, has to be done. Probably the best way to start on this business is to ask, who is going to burn the coal? It is not the householder; not the apartment dweller; not the school, to any great extent; not the hospital. It is going to be industry for the most part, and it is going to be utilities.

Then, what is the form of the coal that they are going to burn? I think probably what we are going to see is a lot of low-Btu gasification in a surprisingly short period of time. I would imagine that within the next ten years there will be a substantial spread of low-Btu gasifiers forced by the requirement of industry to move to coal, by attractive economic stimuli to industry to move to coal, and by constraints that are placed on the system by air quality requirements.

So we have then to work with the utility industry and the broad industrial category of processers and people who use coal for steam raising and what have you. Presently, they are not moving to coal. Last year, a very, very large chunk of industrial consumers of natural gas had to stop using that fuel, and they had to go to something else. The best evidence that we can obtain is that about 86 percent of them, at least by volume, didn't go to coal; they went to some oil product.

Now, the reasons for that are fairly straightforward. Coal is a messy material to handle. It has been widely regarded by industry, up until very recently, as obsolete. I can recall only three or four years ago, for example, hearing references, veiled references indeed, that coal is the obsolete fuel. Industrial perception of coal is that it causes all sorts of problems in the plant and even greater problems outside of the plant, from the standpoint of the dealings with the environmental community at large. So people don't really like coal, regardless of the economics.

Generally speaking, you find in an industrial setting that energy is a relatively small factor in an industrialist's budget, and, consequently, he is really quite content to pay a little bit more for oil or for natural gas if he can avoid the hassle of having strained feelings with the people in his community.

Now with regard to the utility sector, I think something quite different is at work. The utility sector at the moment is less likely to expend capital unnecessarily. First of all, the conventional wisdom in the utility industry--that the next ten years are going to be like the last ten years--has come into very serious question in the last two or three years. You recall now the exquisite difficulty of the utility industry in today's world in planning ahead. They must, in a heavily scrutinized arena, if they are going to go for a coal-fired plant, plan up to eight years ahead. If they are going to go for a nuclear-fired plant or a nuclear power plant, they must plan up to thirteen years ahead. And when they get there they had better be right. The utility that supplies power to this community was not right as a result of decisions made in the 1960s and in 1974 and 1975 and found itself with very substantial excess reserves. My understanding is that Potomac Electric Power Company found itself last year with reserves of about 40 percent, as opposed to the usual 20 percent.

The flattening in the demand curve for electrical energy after the Yom Kippur War, the 1973-74 experience, has really thrown projections into heavy question. Right now, as many of you know, there is a raging debate on whether or not the electric-generating industry will grow nationally at a rate of as low as 4 percent, as some people claim, or at a rate of as high as 8 percent, as others claim. There seems to be a consensus toward the low end of that side. But there still are those who say that, as a result of our overall problems in energy and our continued preoccupation with the environmental impacts, although total energy will grow below previous rates, electrical energy will grow more rapidly. That, of course, would kick it above the historical 7-percent into the 8- and perhaps even 8.5-percent range. This debate goes on. In the meantime the utilities are simply sitting there in a very confused state.

The second thing that has happened to the utility industry and that makes them somewhat dubious about this whole proposition of investment is that the regulatory lag that was so useful to them in the 1950s and 1960s has now turned around to bite them. Let me explain how regulatory lag can be useful. In my experience in the Federal Power Commission, we were seeing the end of an era, in the late 1960s, of declining costs for the utility industry, in that case, the gas utility industry. Our method of handling that was a periodic correspondence with the gas-serving utilities that were known as Dear John letters. They would start out something like this: "Dear John, We note that your costs have gone down; your rates have not gone down. Don't you think it is time that your rates went down?" There was always, of course, a lost-in-the-mail type of thing, and very long sequences between the question and the answer. The result was that there was a substantial regulatory lag. During that era, as a result of the sheer inability of the regulatory institutions to keep up with the declining cost structure of the industry, the utilities were doing very well, on the basis of regulatory lag.

In the late 1960s, the long downturn in real cost to the utility sector was halted. It has been going up ever since, and that same regulatory lag is still in the system. As a matter of fact, it might be worse now than it was before. This means that you fellows are finding yourselves now a couple of years behind the power curve instead of a year or two in front of the power curve, as was the case during those happy days. That has resulted in reduced earnings, and that has resulted in reduced selling prices for your common stock.

Consider what a utility has to do to put a billion dollars into a coal-fired plant. It has to borrow money, and everybody understands that. But, in addition to that, it has to put in an equity share, typically something between 30 and 50 percent of the total value of the plant. And how does it get that? By making an issue of its common or preferred stock or something akin to it. And when a utility does that in today's market, when the market price of the stock is at, near, or below its book value, it is watering its stock. And as Mr. Allen will tell you, utilities, in common with almost everybody else I know, are not really anxious to water their stock. Consequently, we are not finding that enormous surge toward investment.

Let us contrast that, just for a second, to the situation of ten or even five years ago. A utility, we must remember, makes money not on what it sells, but on what it uses to sell, that is, its rate base. The driver for this enormous expansion in the electric power industry and the gas industry in this country was, of course, the motivation to expand the rate base. This was one of the factors that made nuclear energy so very attractive; dollar for dollar, with a given cost of electrical energy at the busbar, you added a higher capital ingredient for nuclear energy than you did for coal-fired power plants, at least in the mid to late 1960s when they were being ordered so very heavily. So there was then strong motivation to make capital investments. Today, there is much less motivation because of the two factors that I have mentioned. I think we have to bear those in mind, then, when we think about what is going to happen to the coal industry.

There is another, somewhat more subtle, effect. At the moment, we--and by we I mean FEA--are attempting to persuade through legal mechanisms a fairly substantial chunk of the utility industry that can burn coal but is now burning oil to go back to coal. Every time we have looked at this, it appears to us that the change would be cost effective from the standpoint of the utility; that is, the utility would be trading investment dollars at the front end, to get their stacks and so on up to speed so that they can burn the coal within environmental constraints, against a lower-cost fuel delivered into their plants. And it appears to us that in at least many of the cases the life cycle costs are actually lower, and in some cases substantially lower, for the converter if he goes to coal rather than staying with his present fuel, generally oil.

There is enormous resistance to that. Why? It is simply this. The capital costs associated with upgrading that plant are as I have just described them, whereas the fuel cost passes through or flows through the rate mechanism, which is perfectly accommodated to taking a higher-cost fuel through directly to the customers without an awful lot of concern about who struck John from the regulatory commissions that control the rates of these companies.

So what I am describing then in the utility sector is an enormous amount of inertia in the system, and that, I think, we have to understand and compensate for before we can get to this great day of coal burning. It is not going to be easy. The current authorities and the FEA, that is, the ESECA and EPCA authorities whom I have just alluded to, are not enough. We are currently involved in an exercise with Senator Jackson and others on what is known as the Jackson Bill, which would confer additional authorities on FEA in order to provide a stronger momentum in the conversion of both industry and utilities to coal. I think it is going to be a fairly hard struggle to get sensible legislation through.

I might point out that there are two great weaknesses in the present legislation: First of all is the scope. It does confine us to those occasions when the utility does have the capacity to convert to coal now, and of course that won't solve our problem. Second, the present legislation puts the burden of proof where it never should be, from the standpoint of a bureaucrat, right smack on the government.

Those are the things that are not conventionally perceived to slow down the demand side of the equation. Those things that are conventionally perceived to affect the demand side are air quality rules and of course the uncertainties with regard to access to coal in terms of the long argument over the terms of strip mine reclamation and so on.

These are things that we should stabilize within the next year. We ought to have good legislation that the President can sign, although I am sure it will not please most of the people in the industry. It will incur additional costs in the industry in relation to strip mine reclamation. After another two or three years of litigating, we ought to have a stable regime with regard to strip mine restoration. Everybody ought to know what is involved, where you can go, what the timetable is, what the requirements are, and what the costs are. This is an item that we must get behind us because an awful lot of coal--if we are going to use coal--is going to be stripped not only in the West but in the East as well.

After we have looked at the demand side, we must see what we have on the supply side. There we have two warring contentions. The first one is that there is no problem. The second one is that there are enormous problems and you can't get there from here. I would suspect that we are going to find that the second one of these elements is going to be more true than the first. From the perspective of three years ago when Project Independent Blueprint came out, Harry Perry, Tom Hunter, and I, among others, were having a look at it. It really didn't seem that there was any single element in the system that you couldn't combat pretty effectively to double or triple or even quadruple coal production in a short term. And I think that that remains true today. But when you put all of the system together--when you ask yourself what you have to do in the way of institutional change, what you have to do in the way of convincing financiers to put their money into a chance like this, what you have to do to the utility sector to convince them to put their money in it--I think you are going to see an awful lot of the front-end money for the coal mines in the future coming from the utility sector.

There are some massive difficulties that we are going to experience. The transportation sector gives us a little bit of an insight into the institutional problems. There have been some estimates--they may be apocryphal--with regard to the loading of some selected railroads if we get up to the tripling and quadrupling of coal production. You hear scare stories that the line would actually be occupied 40 percent of the time; I suspect that in some bottleneck situations that would be true. During World War II, for example, I understand that line loading approaching that density was actually experienced in a few places like the Delaware water gap.

Now, if you do get queuing to that degree, any small breakdown of the system will completely distort it. That is one problem that we have to confront. Another problem is the long, rough history of the coal-mining industry with regard to labor relations. Here, it seems to me, you have a problem that must enter into every industrialist's decision whether to invest heavily in a facility that may or may not be disrupted over time. The experience over the last ten years has been a good deal more comfortable than the experience of the preceding twenty years. But, nonetheless, you do have a situation in coal mining that has been chronically less stable from the standpoint of labor-management relationships. If you look at the supply interruptions of oil and gas versus the supply interruptions of coal, you will see that this is one of the great constraints that operates throughout the system. It has to be faced squarely and taken into account.

Having accepted a great deal of the blame in a personal sense for the current reduction in productivity over the levels that prevailed, let us say, in the late 1960s in underground mining, particularly, I have asked myself whether indeed that blame was fairly placed. And in a self-serving way I will tell you that I think it was not.

I think that probably the significant downturn in productivity coincided not so much with my participation in the affairs of the Bureau of Mines, and in that context with the Health and Safety Act of 1969, but with the disappearance of the labor pools that had disciplined the labor force in Appalachia all during the 1950s and the 1960s. You will recall that during that long period of time from 1947 until the upturn, which occurred first in 1961--that was the first statistical upward movement in the coal industry in the postwar era--and indeed until the late 1960s, we had gone from a labor force of about 450,000 men to a low of 130,000, as I recall the numbers. Now, that meant that there were an awful lot of people standing in line for any job that occurred in the mines. All during the 1950s and 1960s this and the spread of the other single factor, the deployment of the continuous miner-conditioned productivity, were the conditioning elements behind productivity in the underground coal-mining industry during that period. When those labor pools dried up and as a result of the passage of the 1969 act, the industry was forced to avoid certain shortcuts that had crept in as a result of economic pressures on the system during the fifties and sixties. I think that those were the things that turned down the productivity line.

And what that suggests is that stable working relations, labor relations, will restore at least some degree of that productivity over time. They will never, however, I think, so long as we stand with our present system, return it to the pre-1969 levels. It was obvious that just too much of that was attributable, at least through hindsight, to short-cutting.

Now, another phenomenon with regard to the mining function itself that operates as a constraint is the incapacity of the mining industry historically to research and apply new technology. That remains true to the present time. We have two disassociated activities going on in this system: the Bureau of Mines and others conducting research and an industry that is much too busy to put it in place. That, given the lead times of deploying new technology through an industry as complicated as the coal mining industry, I think will condition the system for a good many years to come as well.

If you were to come up tomorrow morning with something as effective as, for example, the continuous miner, if history is any precedent, it would take fifteen to twenty years to deploy that piece of equipment and that new technology sufficiently throughout the system to make a massive change or at least a major change in productivity.

Indeed, we have one example, one sort of crude example, of a new technology that has the potential, at least, for a step change in individual labor productivity, and that is longwall mining. The spread, as those of you who have looked at it know, is very slow throughout this economy. There tends to be an enormous amount of inertia in this system and, I think, for good reason. That good reason is the preoccupation with today that has characterized this industry ever since I have been associated with it.

Well, those are some of the things that I think you ought to think about. The final one that I will burden you with is the issue of transportation. We discussed for just a moment the very heavy demands that will be placed on conventional systems if indeed production goes up. The institutional problem there of course prevents, or at least has prevented thus far, the spread of a much more rational transportation system, the coal slurry pipeline system. I think, if recollection serves, that the projected line that would come out of the coal belt into the mid-South would cross some forty-seven railroad rights of way. The owners despair of ever being able to negotiate agreements with each of those 47 railroads to cross with a competing transportation system. That brings it squarely back to the national government.

Does the government wish to provide for authority in one of its agencies to extend condemnation rights? This, has been openly debated for fourteen years. My recollection is that the first bill to extend the right of eminent domain to coal slurry pipelines went to Capitol Hill in 1963 as a result of a study that Harry Perry did. We are not at the end of the delay, largely because of the simple inertia in the system but also because of genuine concerns that the fate that occupied the eastern railroads with all of their financial difficulties will be visited upon the western railroads unless their large markets are retained. In this context, coal is looked upon as the salvation of some, at least, of the future of the western railroads.

So those, in brief, are some of the factors that are going to constrain the system. If there is a conclusion, it is simply this: We tend to think and say, to the point of its being cliché, that coal is demand constrained. If the worm should turn tomorrow--and we are going to do everything in our power to make it turn--I think we would find that the shoe would go rapidly on the other foot, that the supply constraints would become very real and very operative in a very, very short period of time.

ALLEN: I thought I heard very good news that the access to coal in the form of settled strip mine legislation would be in place in three years.

O'LEARY: Settled and litigated.

ALLEN: Right, settled down. You listed another uncertainty and didn't make any predictions. Where do we stand on clean air rules, our right to burn coal?

O'LEARY: I would think that if we are able to do our job with the massive capacity that any new administration brings to Washington in April of its first year in office, we ought to be able to meet that same type of a timetable.

ALLEN: Good luck.

O'LEARY: Yes, good luck. I think you are right.

HOUTHAKKER: I found Mr. O'Leary's remarks extremely helpful in covering a number of things that have not been covered so far, contrary to what he might have thought, especially what he said about labor. I thought that was quite interesting.

I just want to take up one point here that has to do with the conversion of power plants that presently burn oil toward coal. It seems to me that the essential point there is the control of oil prices. I believe that if oil were priced at thirteen or fourteen dollars a barrel, which it should be, the willingness of power plants to convert to coal would be considerably greater. I would certainly hope that an acceleration of realistic pricing of oil and gas is an important part of the forthcoming energy program of the President.

O'LEARY: No, I think in today's world that is not an accurate perception of what really happens universally in the utility area. Let me again make two points that I made earlier. To get from oil to coal requires some capital investment, and I have discussed the pain that is associated with that. To stay on a high-cost fuel, oil--even a higher-cost fuel, oil at the free market price--versus coal, is not painful from the standpoint of the utility. It may be to its customers but not from the standpoint of the utility because of the capacity of the utility to flow through on its purchase fuel adjustment clauses all of those costs. As a matter of fact, I can't think of a pricing regime alone--and Mr. Allen may wish to comment on this--that would offset the first feature and would force the industry to go into the conversion.

And there is one final point that has to be made in this context. The gas-serving utilities, and they may be more of a problem, are sitting down there in the Southwest kicking out energy at maybe six mills in many cases. The coal-fired replacement plant turning out electricity that is essentially indistinguishable from that six-mill power will be something like 35 mills, even though they go in with much lower cost energy. And these combinations, resistance to capital, the step change in the cost of new generating capacity, and, above all, the automatic flow-through mechanisms of the fuel cost really militate against the point that you made.

SPEAKER: I would like to ask Mr. O'Leary if the federal government's research and development with regard to coal will be facilitated toward cleaning up the use of coal and gasification to the same extent that we have channeled our resources into the breeder reactor program. And, second--a separate question altogether--what type of result do you foresee as coming out of a conflict between a federal policy toward slurry pipeline use, eminent domain powers, and use of federal lands versus the Montana and Wyoming statements about their use of their water rights?

O'LEARY: With regard to the first question, what we are seeing as a result of this administration's budgetary decisions is a significantly reduced breeder program. And, of course, the continued very strong funding for coal. My own judgment with regard to some of the coal investments that we are making is that they are not going to be demonstrated to be as cost effective as I would like. There are people here who can tell you some horror tales of the time when the program that we are now pursuing in 1977 was put in place in the early- to mid-1960s. There is a question as to whether if we had it to do all over again, we would be doing some of the things that we are currently embarked upon.

So what I am really saying is that it is not a matter of money, it is a matter of program content with regard to coal. I think most of us would agree that there is enough money in that program to get from here to there. I think we put too little attention, just generically, to low-Btu gasification. Had you asked me whether we were doing right in that five years ago, I would have said yes, along with many others. But in hindsight we put too little attention to that, and we put too little attention over time to atmospheric, fluidized bed research, going rather for somewhat more exotic approaches to this problem that may never pay off or may be a long way in the future. But money is not the issue, I think, here.

So, to recap a very long statement to a very simple question, we are downgrading the breeder program at a very rapid rate. Clinch River is essentially finished, and we will be upgrading through selectivity the content of the coal program and putting in whatever money is necessary to make it work.

The second part of the question is really much more difficult to deal with, I think, in that any time that you embark in these realms of government policy that we are finding ourselves in now, you get into inherent conflicts, among other things, between fuels. For example, your question runs into a psychological or philosophical conflict between nuclear energy on the one hand, or some aspects of nuclear energy, and fossil fuel on the other hand.

My own feeling is that the attitude of the administration toward conservation is going to smooth a lot of these problems over. We are going to find, I think, that the conventional projections with regard to demand for the whole basket of energy requirements in this country is going to go down below what any of us were thinking as being realistic even two or three years ago. The National Academy study that is now going forward is showing, for example, much lower projections of total energy demand today than anyone was really seriously considering three years ago and four years ago. And that element is probably the strategic element in our capacity to balance off these competing claims that exist within our system, and it is one that really requires, and I think will receive, the best attention we can give it. In that context the President has referred to nuclear energy as a last resort. Maybe what we should say is, in reality, if we can do all these things, energy is a last resort.

HOUTHAKKER: I think the question that we have argued about, the conversion of electric power plants, is too important to let go at this. I would like to point out in the first place that pass-through is not automatic. There are many states where pass-through of fuel costs does require a particular decision by the regulatory agencies. I think that is desirable, because the purpose of pass-through is not to put the utility industry on a cost-plus basis. It is, at least in some ultimate sense, to promote efficiency. Therefore, if the price of oil were to be increased to a market level, which would be an increase of, say, 30 to 40 percent, many utilities, although perhaps not all, would have to apply for rate increases, and at that point the utility regulatory commissions could quite appropriately raise the question of conversion to coal.

In the second place, I would point out that, earlier, you explained the attraction of nuclear power for the electric utility industry by the fact that they are regulated on their rate base, that they are allowed a certain rate of return on their rate base. Now, their conversion from oil to coal does also involve capital expenditures, which, by the same theory, the utilities would like because it also adds to their rate base. Therefore, I would maintain that a trend toward more realistic oil prices would be helpful in stimulating the demand for coal in the five utilities.

O'LEARY: I really think that I hear your point. Mr. Allen, who is the only person here whom I recognize as being from a utility--undoubtedly I do a disservice to many in the audience--can probably comment on it. The fact remains that most jurisdictions--and I would imagine that in terms of volume of burn, a reachable target--the vast majority of jurisdictions do in fact permit the pass-through. We find it throughout New England. Is that right, Mr. Allen?

ALLEN: That is generally right, yes. I would say that when the pass-through escalates by a factor of one or two or three, as it did after 1973, life is not exactly easy.

O'LEARY: What about 30 percent?

ALLEN: Thirty percent over thirty years would be just great.

O'LEARY: He, in addition to being in the utility business, is a lawyer.

Harry Perry

Consulting Economist, Resources for the Future, Inc.

We have been asked to consider what must be done. It is interesting to note that Frank Zarb, Jack O'Leary's predecessor, has said that he would no longer respond to that question. The only topic he would discuss is what is being done rather than what must be done. He gave as his reason for this his experiences with theoreticians at one end of this town and demagogues at the other.

This Forum, if it is to be useful, must at least take a shot at answering the question. What have we concluded in the last few days? There appears to be general agreement at the broadest levels that the nation will have to rely more heavily than it now is on its coal resources. But having said that we immediately have sharp differences over how much more and in what way these resources should be developed. However, it would be useful to take a minute or two to establish the underlying premises on which this conclusion, a greater reliance on coal, is based. Either implicitly or explicitly, it has been assumed that with the demographic pattern that will exist in the United States, a level of energy supply will be needed to meet economic and social goals for the nation that cannot be supplied by those fuels currently providing much of our energy demand.

Second, we cannot depend to any great extent on imports of energy because of national security questions and balance-of-payments difficulties.

Third, the United States has vast resources of coal and, with the operational breeder reactor, of nuclear fuels that can somehow be used to make the transition from an oil and gas economy to a coal-nuclear-electric economy.

Having identified what I think are the broad issues on which we can agree, the details, as they always do, become sticky. A whole host of issues arise. How much conservation can be achieved, and what kinds of changes, if any, must society make to reach projected demand levels for energy? It really makes a great deal of difference if we are trying to provide 20 quads of energy with coal in 2010 or 50 quads.

In what way can we provide energy from coal so that the best trade-offs are made so as to minimize adverse environmental effects on land, water, and air? How can the coal that will be needed be produced in the least socially disruptive way to the miners, to the public at large, and to local and regional governments? We have not answered these questions on how the coal option should be developed, and it is certain that two weeks, two months or the two days of this discussion we have had will still leave us with large differences of opinion on how to proceed.

Some suggestions on ways of resolving this problem have been offered, from resolving the major policy conflicts that have been identified as being barriers to coal development and use to providing a forum for reasoning together. I would like to examine the problem of what can be done in a somewhat different way, by considering what the limitations are on both supply and demand for being able to produce greater reliance on our domestic coal resource base. I would like to consider this for two different time frames, the five-to-ten-year period and the period beyond that, but not as far beyond that as Alvin Weinberg went in his presentation.

First, on the demand side for the next five to ten years, even with the decline in domestic production of oil and gas, there are only limited economic substitutions for potentials for coal in the short run. Coal is already being used to whatever extent that it can be in the metallurgical market, and it would be impossible, again in the short run and with existing technology, to have coal become a major factor in either the household or commercial sector.

In the electric utility sector the industry is already planning for their new capacity to be either coal or nuclear powered. Very little oil or gas is expected to be used. There is considerable controversy as to what can and should be done for the existing utility plants that are burning oil and gas. The problem is not a simple one, and efforts to resolve it to date have not been very successful, as was just mentioned by the previous speaker.

The reasons that it is not simple are, first, many of the plants that burn oil and gas would be unable to use coal without such major modifications in the boiler that it would be easier and cheaper to build an entirely new unit. Plants now burning oil and gas, but which were once able to burn coal, raise other kinds of problems. The economics of conversion depends on many factors: the age of the plant; the current existence of space for a coal storage pile; a way to deliver coal to the plant; the need for purchasing or refurbishing coal stockpiling, handling, crushing, and grinding equipment; the nature of the modifications needed to feed and burn the coal; the adequacy of ash collection, handling, and disposal facilities; and, finally, the ability of the converted facility to meet environmental standards. This latter point is extremely important, since older plants often have no space in which to put new control technology and equipment.

Obviously, the answers to these questions can only be determined on a case-by-case basis. But it appears to me, in spite of what you have just heard, that too much effort has already been devoted to this problem. FEA has estimated that a maximum of about sixty or seventy million tons of coal is involved, but when this has been reduced to those situations where the conversion is economic, the amount of coal will be much less.

In passing, it should be noted that if the conversion is not economic, it would be unfair to impose an economic penalty on the consumers of a particular utility so as to meet the broad national goals of reducing oil imports.

The final market for coal, again in the short term, is in the industrial sector. Coal still provides a sizable amount of coal to the industrial market, but as pollution standards, particularly those for air, become more stringent, this market could become much more difficult to even retain. On the other hand, a consensus appears to be evolving that burning oil and gas in industrial plants, many of which would be suitable for coal use, represents a waste of a scarce and very limited resource that could be used to better advantage elsewhere.

Converting existing industrial plants from oil and gas faces the same problems as the conversion of existing utility plants. In fact, because of their size and other factors, the economics for industrial conversions is generally less favorable than for an electric utility plant. Even for new plants the industrial sector can only use coal with greater difficulty than a utility plant. Unless clean fuels for coal can be supplied to the industrial user, or new combustion processes such as fluidized bed combustion can be brought to commercial realization, the industrial sector is faced with using control technologies that at least until now have been designed for the large consumer. Stack-scrubbing costs and disposal of waste sludge on relatively small industrial plants could be prohibitive.

Here then are three major technologic opportunities that need acceleration, not only for the industrial sector, but for other coal markets too: clean fuel from coal, cleaner combustion processes, and new control technology designed to be less costly in smaller plants.

However, it is in the longer run, beyond 1985, that the opportunities to increase the use of coal become much more attractive. The three technologic developments that I just mentioned could become commercial on a wide scale by then, and other longer-range opportunities to use coal more widely could exist. There are at least three other major areas where new developments would permit a greater use of coal. These are (1) advanced electric power cycles that can produce electricity at higher efficiency than that of present plants, thus using less fuel and producing the same number of kilowatt hours with less pollution; (2) the conversion of coal to low-sulfur boiler fuel or other petroleum products; and (3) the conversion of coal to substitute high-quality gas, to replace the dwindling supplies of natural gas.

While I have discussed these needs in terms of new technologies required if coal use is to be expanded, bringing them to commercialization will require a large number of nontechnologic developments and actions. I won't try to list them all, but I will mention a few. We would have to provide an adequately planned, financed, and administered R&D program; resolve emerging environmental concerns associated with the new technologies; find solutions to the political, social, legal, and regulatory issues that will emerge; raise the capital needed to finance the construction of very large plants; and, finally, assure that the private sector can be in a position to commercialize the new technologies as they become economic.

Until now, I have only looked at the demand side of the supply-and-demand equation, that is, what must be done if more coal is to be used. However, if coal demand could be expanded, we must also examine what must be done to assure that there are adequate supplies of coal to meet this demand.

As has been noted a number of times during the Forum, for the time period in which most people are interested, there are adequate coal reserves to meet any demand placed upon them. Whether we will be able to mine and deliver the coal to where it can be used is not as obvious. Because of the lead times required for constructing new mines, the amount of coal that could be produced by 1980 will fall within a relatively narrow range. Some slowdown in constructing new mines could occur, and a certain amount of acceleration would be possible if it were needed.

Because of the demand constraints for the short term for developing new markets for coal that I mentioned earlier, it appears that there will be adequate productive capacity to supply what is required. The adequacy of factors other than productive capacity is not as certain. Manpower will probably not be a limiting constraint in the short term, but as was mentioned a few moments ago, the railroads in the East may not be able to satisfy the transport demand for these requirements.

The critical period for planning, however, will be upon us shortly because the investment decisions must be made in the next year or two for the new mines that will be operational in 1981 and beyond. But the coal industry is faced with a host of uncertainties that will make these investment decisions difficult, and this could lead to significant delays in their being made.

On the production side these include a surface-mining legislation that was just mentioned, delays in the coal-leasing program, new mine safety legislation, coal transportation difficulties with the eastern railroads, changes in black lung benefits, horizontal divestiture questions, and a mining research program that may be needed.

By the way, these are very similar to the list of coal policy conflicts that Louise Dunlap indicated had to be resolved before coal would be used more widely.

Unless some of the uncertainties revolving around these issues are removed, the productive capacity that could be in place will not be. Other factors must also be explored that might limit the production and delivery of the needed quantities of coal in 1985 and beyond: manpower, mining equipment, water availability, and transportation systems. Where constraints seem to exist, new policies will have to be adopted to remove them; nothing has been done to start work on removing those constraints.

While the problems that must be resolved to get the coal resource to be used more widely have been identified, I confess I do not see clearly how the actions taken to resolve them will be initiated. We need leadership to see that the necessary measures are taken, but this leadership has not been evident in the executive branch, the Congress, in industry, or in the trade and professional societies. We need a catalyst to get these groups moving and to work together in order to start the actions that will make the transition to coal possible.

ALEXANDER GAKNER, Federal Power Commission: The information which resides in our agency shows that there are now plans for about 111,000 megawatts of coal-fired capacity to be built in a ten-year period. The consumption of natural gas has dropped by 25 percent in the last five or six years, from 4 to 3 trillion cubic feet, and in the area where most of the gas is now being used for electric power generation, there is a positive plan to reduce gas consumption by about 6 percent per year. So we are looking at gas phasing out naturally in a natural growth development process.

There are still some plans for about 20,000 megawatts for oil-fired capacity, the construction of which began prior to the embargo; it is too late to change. But even in the larger regional planning, simply the cost differential between oil and coal that you have with the oil/coal economy generation suggests that coal-fired plants are being used more intensively than oil-fired plants. We can see that probably in the early 1980s, there is going to be a reversal in the growth, in the use of oil, a reduction.

With that preamble, I heard you say that perhaps our effort to push that part of ESECA or EPCA, or perhaps this proposed S.977, the Jackson Bill, should focus on new plants rather than on the conversion of existing plants.

PERRY: I didn't say it in those words, but that is precisely what I meant.

GREGORY GOULD, Fuel Engineering Company: A question that has not really been given enough airing in this recent discussion is the tendency of energy costs to pretty much seek the same level. I think that we make a serious mistake and possibly are due for some very serious disillusionment if we anticipate coal as being cost effective from the standpoint of a real saving. I am not proposing or advocating that oil and gas prices should not rise; I think they should. But I think you would also find in the process, particularly over the life spans of the plant and equipment that would use the coal, that the apparent advantages pricewise for coal would disappear. I would be interested in hearing any comments you might have on that.

HOUTHAKKER: I agree with you that there is some tendency for the different kinds of fuel to be priced on a Btu basis. However, there are two major exceptions to this. In the first place the convenience of handling this is part of this. In fact, in some cases this calculation really becomes rather meaningless, as in hydro or nuclear energy. In the second place, cost of production is still a very important part of this. Now, at the moment, coal is priced well below oil on a real cost basis. If you take the price of oil to be the cost of imported oil, which it really should be, then domestic coal is much cheaper than imported oil. So there is, at the moment, a distortion that is due in large part to price controls.

In the long run it will still be true that there will be some tendency to this kind of modified Btu pricing. I would say that if we allow the markets to operate, then coal will be priced on its marginal cost of production, which by all accounts will rise much less than the marginal cost of producing oil, especially domestically, and even internationally if you take into account the additional profit that OPEC will try to raise from our increased demand for oil.

So this I think is the basic point, that the supply function of coal is basically flatter than the supply function of oil and that we can obtain large supplies of coal without having much higher prices. I don't go as far as some people say, and I have heard this from experts, that you can supply very large additional amounts of coal without raising the price. I don't believe that, and some of the things that have been said here by Mr. O'Leary and others also contradict that. It is clear that there is a rising supply price of coal. But I believe it rises much less rapidly than the supply price of oil and gas, and, therefore, if markets are allowed to operate, coal will indeed in the normal course of events replace some oil and gas.

Eric H. Reichl

President, Conoco Coal Development Company

Some of the things I am going to tell you will contradict what has been said before, and maybe that makes it a little more interesting.

In a nutshell, the first question on the future contribution of coal is an issue of the limited coal supply. I think Mr. O'Leary had a good point about the nearer term. Right at this moment, of course, we have really an excess supply. I think if you were to be successful in getting fairly good acceptance of this switching of oil- and gas-fired units to coal, you might then run temporarily into a shortage because there would be a little lag. We might say that it costs about seven or eight times as much per annual ton of coal to use it as it does to mine it. Once you have thought that through, you can see that, really, the mining is not going to be the constraint over, say, the twenty- or thirty-year term.

In addition, you will recall, it takes significantly less time to open and develop a coal property and put in a coal mine than it does to build a power plant, or a coke oven, or a gas plant, or a liquid plant. So I think the coal is going to be there. When you look at the future over the next twenty- or thirty-year period, supply is not going to be the constraint; everybody seemed to agree that manpower per se would not be. The other issue is capital. And I just don't think capital is going to be the issue because when a coal miner sees somebody invest a billion dollars in a power station, he will open a coal mine for it.

How are we going to use more coal? I don't think there is going to be a sudden change. There will be slower shifts, because the whole energy pattern cannot be thrown around and rapidly changed. We can slowly bend the trends. That is all. But the one thing you can count on fundamentally is that by the year 2000, we will still use our energy in roughly the same forms as we do now, namely, in electricity, something like a sixty-cycle alternating current; methane; and all kinds of liquids. These three will supply about 90 percent of all the energy we will use in the year 2000. So how does coal then contribute to that? Of course, we have discussed at great length the three areas, so let me comment briefly on "coal-to-power," "coal-to-gas," and "coal-to-liquids."

On coal-to-power I differ from what I think I have heard during the last few days. I believe that the bulk of coal used in the year 2000 is going to be in pulverized fuel boilers, exactly the same as we have today. I would guess that a large number of them will have scrubbers. There is a very simple reason; that is the best way to clean up. I think that is going to be recognized eventually. The thing that is very unfortunate is that we have such a discrepancy between this uniquely obvious problem and the R & D efforts that move in that direction. We spent $750 million of federal money alone on coal research this year, and out of that, $4 million, or about a half of a percent, is aimed at cleaning what is the largest single user and stinker in the coal business, namely, the stacks. That is a misalignment of our R&D effort with what is important, and I think this can be changed and will be changed. I just don't believe that these nostrums--and I use the term advisedly--of cogeneration, fluidized bed, MHD, combined cycle, low pressure, and high pressure are really going to affect this picture. Market penetration alone would keep you from doing it. Furthermore, there is a real question in my mind whether any of them really offers an advantage technically or economically.

Alvin Weinberg asked me to make a particular comment on fluidized bed. If you look at the fluidized bed problem, it doesn't get around the simple question of having to react 10 tons of air with a ton of coal and to take the flue gas product out of a stack. That is true for fluidized beds and pulverized coal. I have the cleanup problem, in both cases, certainly on the particulate side, because the fine particulates, which are the tough ones to recover, are just as much present in the off-gas from the fluidized bed as from an ordinary boiler. I have a lime problem, which in effect is probably, in terms of tonnage, greater than if I were to scrub with lime, which is the wrong way to scrub. I don't think lime scrubbing is really the answer.

In addition, you have a problem in terms of power requirement, which is not a minor issue in the operation of a fluidized bed. Consider the effort required to push the air through this bed; I would rather use that energy to clean up. So all in all it is my view, and this is a very personal one, that fluidized bed, although it is highly touted, is not going to change the picture.

Now, a comment on coal-to-gas. I have said the same thing often. I don't want to bore you with it. The key point is good technology is here, and the R & D effort, trying to make great improvements in it, is one of the mistakes. Mr. O'Leary just referred to that issue; he was quite right in what he said. I think we have created promises that are unfulfillable. You can make some improvement on this technology but nothing significant. I think the question is: Is it worthwhile to make synthetic gas or not with the technology we now have? I think it is. I think that gas is a very elegant form in which to distribute coal energy cleanly to the consumer. We have an existing distribution system that reaches just about everybody in the United States, and you cannot burden the small guy, the school, the hospital, the shopping center, and the home, of course, with anything in terms of cleanup. I think that if the consumer gets energy at all, it has to be cleanly usable, and gas is an elegant form for that.

We have somehow forgotten that there was a time not so very long ago when almost all the gas in the eastern part of the United States was made from coal by processes that were monstrous in comparison to what we can do today. It was worthwhile then to install gas distribution systems because it was a convenient form of energy. I think this is even more true now, and we have better ways of making the gas. I think this is one of the things that is likely to be a significant growth industry in spite of all that has been said.

I might specifically make a comment on low-Btu gas because there was quite a bit of discussion on that. I think of course it would be fine; it is a great idea, and I hope it will happen. I just don't believe it is quite as attractive a proposition as is claimed, for this reason: the problem here is that whenever you make a low-Btu gas, you must also invest in a distribution system and, in many cases, a new system--different burners, piping, and so forth. You cannot use the existing distribution system and you are not using the "roll-in" mechanism not only of the physical distribution but of the pricing that relates to it.

Finally, for the small industrial users, it is almost by definition that the plants will be very, very much smaller than, say, central methane-from-coal plants, which greatly eliminates the advantage of size. So when all is said and done, and you compare the cost of the methanation step, which is only the last finishing step in gas-from-coal, to make high-Btu gas from the low-Btu gas, versus installing a separate distribution system, it has turned out that the advantage was with making methane. There may be some instances where this will not be so, and then we will put in low-Btu gas.

I have a very brief comment on the liquid fuels, liquids from coal. First of all, there is no question that you can make them. They have been made commercially in other countries. The technology is available today if you would want to do it. The cost would be two and a half times what you can buy the oil for by import, so the issue is one of how to establish an incentive for anybody to do anything about it. The thing that worries me particularly was expressed by Dr. Houthakker and that is this issue of central decision making. I would almost like to ask him to repeat this. But the point he made, I think, was that you can bet your boots that if all the decisions on how to make liquid fuels from coal are made in one spot in this town, they are going to be the wrong decisions. We have already had examples of this. A better system would be to somewhere establish an incentive, or a force, or an order, a mandate for the oil industry, which now sells liquids, to assure that it will put synthetics in. It can be made from biomass or from corncobs or from coal, if you please, or shale. Let the industry decide what raw material to use and what process to use. Don't tell them how. Just tell them they have to do it. This has been done in other countries; this isn't by any means as difficult as it seems.

ALLEN: I have been unhappy until this afternoon, because nobody was looking at the prime customer of the coal industry, namely, the utilities. We have had a lot of attention now. I have been brought up by you and others to say, I think, that there is nothing that coal can't do in the way of expanding demand as long as there is an assured market, and that is usually followed in the fine print by saying that, by this, we mean a very tight, bankable, long-term contract from a utility. But you gave me the swerve a moment ago. You said all I have to do is start down the road building a 1,000-megawatt plant, and you will build a coal mine. Where are you going to get the money? From me?

REICHL: Not necessarily, no. I think that the assurance of the market is adequate. I am not negotiating a coal deal with you now, and everybody appreciates this, but I think on the whole that the real issue has been the assurance of use. This isn't just an issue between you and us. It is an issue of your being allowed to use this coal. This is one of the points that we ought to bring out. It isn't good enough for Don Allen to tell me, I am going to take this coal from you as long as the plant is here. We want to be sure and he wants to be sure that he has got a plant that can use this coal legally. He doesn't want someone to come in suddenly and shut it down and say, you can't use this coal any more. I think that is the problem.

ALLEN: This is the very problem, and that was the speech I wanted to make. I have done a certain amount of reading just because of my ignorance in the issue of coal. I delved into several financial essays by recognized authorities, and, independently, three of them said that the one real problem here on financing coal is that nobody wants to bear the risk of the nonusable coal. The banker doesn't want to. The coal producer can't because he has got to use my contract to bank. He wants my contract to pick up that risk. I can't do it with any assurance. I have to look at two people. I don't think I can push this off in all cases to my consumers, and I am not sure they should bear this. I don't think that the regulatory agencies feel that this is the risk of a long-term contract and a supervening change in the Clean Air Act, which is built into the Clean Air Act philosophy. That leaves only the utility investor, and he has already been somewhat burned by fortune and by the many things that we have been asked to do in the national interest and in responding to OPEC. He too is weary. I know of no one who will now take the open-ended risk expressed in the following question: Can you burn the coal under a new mine that you are going to open under a long-term contract?

There are obvious solutions to the problem, it seems to me. One is to stabilize the Clean Air Act, at least for new plants that are clean and right and meet new source performance standards when they are built. That is not yet in the fabric of the Clean Air Act, and perhaps it is what Jack O'Leary hopes to be able to do.

Another solution, it seems to me, is to find some means of having the government in some capacity share the risk with the utility family, both consumers and investors. That basically says that if we come along with some new demand in the name of public health or in the name of public security, the Project Independence argument, we will negotiate up from some baseline and share the cost of either back-fitting or shutting down.

I think it is time, if there is to be development of the coal industry, that we investigate this bottom line of who shares the ultimate risk of long-term investments from supervening and quite rational new demands that change the basic investment fabric with which you started out.

HIBBARD: Dr. Boyd, would you give us your impressions of the Forum in your role as a member of the Panel for Inquiry?

BOYD: The cancellation of the Kaiparowits coal-fired electricity-generating station illustrates the destructive power of the three new horsemen of the apocalypse: inflation, bureaucracy, and unstructured good intentions. Government, with or without a simply understood policy, can produce no coal and no electricity from coal. It can only delay and interfere with those who can.

The total energy-consuming and energy-producing system has become so enormous that individuals, even those in the seats of power, only nudge the system in the effective direction of change. They cannot change it markedly except to slow the whole process. The vast, diversified sources of coal are readily accessible. The large numbers, types, and locations of coal deposits available make this fuel the one that must assume the energy burden of our industrial economy, of our particularly high standard of living. As we are forced to evolve from an oil-and-gas economy to synthetics and shale and to nuclear energy and the more exotic sources of energy, coal can and must fill the gap.

Coal is relatively cheap. It requires a short but still significant time to convert deposits into producing mines. The transportation system can be modified, despite what we heard today, to meet these changing patterns of distribution more cheaply and more quickly than the plants can be modified to convert it to our daily needs.

This meeting warns of the dangers to the environment of the overindulgence in coal. This, to my mind, far outweighs the imagined dangers from the development of nuclear power. Both are, however, outweighed by the potential failure to provide sufficient energy for our industrialized socio-economic system. Of all the dangers, that of not having energy is by far the greatest. The dilemmas faced in the Ohio River basin and the Great Plains illustrate the urgency of that.

The President faces an awesome decision this month. If he should adopt a policy of interrupting the painstaking and costly evolution of the nuclear option, he can jeopardize the enormous manufacturing, construction, and operating capability that has been laboriously evolved with the development of the atom. It is simply impossible for the coal industry to replace the contribution that this nuclear industry must make to the economy. Coal and nuclear power must be developed simultaneously as complementary energy sources. The provision of energy from animal dung to fusion is a vast evolutionary process. It is one which has been significant in its changes and far reaching in its effects.

Although the contribution of oil and gas is declining now, those fuels are supported by significant resources. But this is the coming age of coal. It will be relatively brief in the course of human history, but we can only pray that mankind can use his inherent wisdom to adapt to it wisely.

MIGNON SMITH, *Washington-Alabama Report*: I wonder if Mr. Reichl or someone would comment or would assess the effect of the probable passage of the Udall Strip Mining Bill on the production and cost of coal, particularly as it applies to Appalachia.

HIBBARD: I think that there is someone who might be able to answer this question about the strip mine bill. Tom Hunter, you made a study of this. Do you still have a feel for what the strip mine bill might do in terms of available coal?

THOMAS W. HUNTER, energy consultant (coal specialist): I would hesitate to comment on that.

HIBBARD: It was once about 200 million tons, as I remember.

HUNTER: Oh, I think the 200 million tons you are referring to is the impact of the regulation of sulfur. Today, if the sulfur regulations that now exist were fully enforced, they would cut out about 200 million tons of usable coal. That is the problem. We really have to find a balance between these things. We all want a good environment and we also have to have energy. We must find a balance between these difficult problems.

PERRY: I can't give you a number on the cost, because it varies from location to location, but as the bill was written last year, it disqualified about 100 million tons of strip-mined coal if you do not include the alluvial valley floor provisions. That adds another 70 million tons of that curbed production.

COHN: I would like to add a comment on that point. The surface-mining legislation is not greatly at variance with a lot of other legislation that has been written recently. There are all kinds of concepts in it that are not susceptible, in my view, of evaluation, because they involve very substantial questions of interpretation and we don't know how they will finally be interpreted. I suspect that the estimates given by Mr. Perry are pretty good in terms of an analysis of what someone thinks the legislation means. We generally find out that it means a lot more than we thought at first. But in my view, again, that is only a part of the story. It is only a small part in some respects.

The way the legislation is set up, it calls for permits and licenses to be obtained. It involves hearings, opportunities for all kinds of delays to be injected into the proceedings. Having observed similar procedures under other legislation, I can tell you that nobody is going to get a permit under this legislation for a couple of years at least. And to that extent it will make coal unavailable for a very long time.

ALLEN: Overall, I am very much reassured to hear people who know how this problem can be managed, how to get the coal production, how to get the coal conversion, and how to use what we want under new circumstances.

I think the most chilling thing I have heard in the whole conference--and I agree with it--is the decision-making or conflict-resolution session under Milton Wessel, in which it was stated that we don't have any way of solving the divergent views of the public, all of which are legitimate. This relates to Herb Cohn's problem of just a moment ago, that it is the institutional problems that are going to slow us down and keep us from doing the things that we know we can do if we are given a clear right of way.

Finally, I am confused about the economics. I think Dr. Newcomb said that he could supply me with nice, clear coal ready to burn in a utility boiler for $1.50 an MBtu. I have been hearing about how I can get other kinds of clean fuel, high-Btu gas, low-Btu gas, possibly even liquids, ranging anywhere from $4.00 to $6.00. Finally I have heard the message that Eric Reichl has been trying to teach to me for six months: the way to go is to stick with conventional stuff and, if you have to, buy a scrubber.

SUMMARY OF FORUM

Lester Lave

The task of trying to summarize what went on in two and a half days is like trying to put shoes on a sockless octopus; but I'm grateful, since I could have been asked to teach the octopus to walk.

As I said two and a half days ago, I have got some good news and some bad news for you. The good news is that you and I (I presume) have learned a great deal from this Forum; a lot of progress has been made. The bad news is that we haven't come up with any solutions.

There are several themes that have been sounded and that I will try to summarize. I will try to keep my own opinions out, or at least keep them down to only 90 percent of what I say, and try accurately to report what has been said.

One of the first major issues that was raised was a general desire for a national energy policy. A national energy policy was desired, since we have not made noticeable progress in increasing coal production or, more generally, in resolving our nation's energy problems. We have papered over energy problems for the last several years by vastly increasing our imports of oil. We have created a climate of uncertainty that tends to freeze action, miring us more deeply in the problem. This was all summarized by David Rose on the first night: "Oh woe, oh woe!"

There were people who argued against having a national energy policy, and here I will inject my opinion. What we have gotten out of the Congress and the executive branch over the past decade are a series of pious declarations with little or no common sense involved in them. For example, in the 1970 amendments to the Clean Air Act was the definition of the "primary" air quality standard: a level of air quality that would not harm even susceptible people. In the water standard the Congress declared its desire to have no emissions, zero emissions, in the waterways. Finally, under NEPA, the Congress declared the policy of nondegradation. I submit that, from an ecological or any other scientific point of view, none of those make any sense. Would you buy a national energy policy (used car) from a group that gave us these pieces of legislation?

Personally, I don't believe that we could have a coherent, rational national energy policy. First, we are in an era of great change, as exemplified by the discussion of Kaiparowits. Since 1965 we have had a vast change in opinion regarding the environment, regarding the importance of the untoward consequences of burning coal or other fuels, and regarding what society should do about these. This era is not one where a particular leader changed our knowledge and opinions, but rather this era resulted from myriad influences, a general public movement. Leaders have had to look over their shoulder continually and, often, seeing no one behind them, had to run in front of the public so they could shout, "Follow me!"

A national energy policy is ruled out not only by shifting public perceptions but also by the pervasiveness of energy. Energy is embodied in every one of the goods and services that we consume. A national energy policy concerning supply would have to encompass a policy on land use, the uses of national parks, and the size of locks on the Ohio River. We must have a policy on deepwater ports for supertankers, drilling on the outer continental shelf, oil spills, restoration of strip-mined land, allocation of water rights, and air and water quality.

Formulating a supply policy toward energy is highly difficult, although possibly feasible. For demand it is simply impossible. A federal policy toward the demand for energy would have to look at insulation in our homes. It has already looked at auto fuel standards. Indeed, it might go to the extreme deciding whether children should be bused to achieve integration on the basis of how many gallons of gasoline it will consume.

The real problem, I submit, is that the public has not yet figured out what it is that we want. Our perceptions have changed, especially of the time horizon for viewing the problems. Consider the time horizon exemplified in the film *Black Coal, Red Power*: perhaps 1,000 years. Is mining coal compatible with these Indians living in harmony with their environment for a period of 1,000 years?

Our society seems to regard the long run as ten years. Nor do governmental commissions take a longer view. Finally, for most politicians or corporate leaders, the decision horizon seems to be one or two years.

Time horizon is important. If you calculate to two decimal places the average growth in real per capita income from the time when man first appeared on the earth up to the present, that rate of increase is zero. The increase in real per capita income over the last decade in the United States is something over 1 percent per year. It really depends on your time horizon. If you are optimistic, you might argue that 100 years ago we set off a course of history that was diametrically different from anything that had gone on before and that this will continue into the future. This 100-year horizon implies that our current problems are not very important; they will be solved very fast. As an economist, I caution you to realize that implicit in economists' analyses is that the future is going to continue to be better than the present.

If, instead, we take a much longer horizon, beginning one thousand or one million years ago, the answers are quite different. Instead, we are being short sighted by using up our current energy resources rapidly and, in so doing, creating long-term problems.

Thus, a further reason why we can't have an energy policy in the United States is that we don't know what we want. Nor do we know the energy alternatives open because there are new technologies being developed.

A second round of themes developed over the last two and a half days had to do with social costs. The first of these were changed lifestyles. Hendrik Houthakker expressed it best in asking, "Who gets the grandfather clause?" The Indians of the North Plains have a grandfather clause because they have been there for literally thousands of years. Do the current residents have an equal claim to protecting their life-styles?--to put it more neutrally than Houthakker. A further question involves which life-styles we intend to give special protection.

In exploiting western energy we are accelerating a process that is more than a century old. Before 1860, about 80 percent of the population lived in rural areas. By 1950, almost 90 percent of the population lived in urban areas, a massive change. Since World War II there has been a massive migration from the North Plains, from the Northeast to the South, and particularly to the West. By extracting western energy, we are accelerating the trend. I was impressed by Michael Enzi of Gillette and his description of the problems stemming from very rapid growth.

May I comment, as somebody from the Northeast, that declining population also presents social problems. In the Northeast the median age is higher than the national average, reaching the late fifties in some communities. There are communities where essentially all of the young people have moved out. Such communities have a large amount of social capital (roads, bridges, schools, et cetera) that will not be utilized in the future. The migration has meant fewer people to pay taxes to keep up the community. Problems are created by either rapid population growth in the West or declining population in the East.

You'll remember that David Rose issued a challenge during the first session of the Forum: he asked about the consequences of selling the entire United States for $250 an acre. I asked him later whether he would sell me his house for $250 an acre. He declined, but he offered to sell me MIT. (Come to think of it, it was Harvard he offered.)

A series of large issues have been raised, some of which haven't been given adequate attention. In dealing with these I feel rather like the fifth blind man approaching the elephant, who is given the elephant's residuals. It is known as getting one's hands dirty.

The first issue had to do with the demand. How will energy grow in the future? I tried to make evident that if energy grows at its historical rate of 4 percent per year, we will have a nonviable situation. What is the alternative? Three percent? One percent? Zero percent?

In the very long run, zero percent growth in energy use is the only equilibrium. But how do we get there? What does "the long run" mean? How many quads will we be using in the United States in 2010? Will it be the 300 quads that would come from a 4-percent growth rate, or 150 quads that would come from having a 2-percent growth rate due to conservation, or 100 quads due to lower GNP growth or a change in life-style, or 75 quads due to conservation, lower GNP growth, and changed life-style?

Implicit in the future is an array of social decisions. Important decisions include the fertility rate, desires about per capita income, and the work-leisure decision. Individual and social decisions on such issues are complicated, interdependent, and certain to change over time.

A second major issue has to do with other sources of fuel: oil, natural gas, uranium, and so on. Only brief reference has been made to the large uncertainties associated with the reserves of these fuels. We can only guess at how much oil will eventually be recovered from the outer continental shelf, how much natural gas would be forthcoming if we deregulate, or how much uranium will be found. Reserves are a central issue in evaluating the current Ford/MITRE report. Society's options and decision windows depend on whether we can be optimistic or pessimistic about the amount of uranium that will be available. While important, these questions go beyond this Forum and have been neglected. However, the answers to them condition the questions that we ask about coal.

I personally hope that we can proceed with immediate deregulation of oil and natural gas. In that glowing exchange between Houthakker and O'Leary, both sides had good points; but, if you don't allow prices to rise to what it costs to replace fuels, then you will be calling forth too little supply and increasing demand; if so, the market can't clear, and increasing regulation will be required to reconcile supply and demand.

The third of the major points concerns risk. We heard a lot about the nature of the risks that come from coal: health effects; accidents in the mine; coal workers' pneumoconiosis; transportation accidents; and, finally, damage due to air pollution stemming from the oxides of sulfur, the oxides of nitrogen, trace elements, particulates, and so on. There are also the environmental effects of subsidence, acid mine drainage, spoils piles, overburden, and acid rainfall. By the way, on the first evening I made what was much too conservative a statement. I said that the rainfall in New Hampshire had been measured below a pH of four. That statement is correct but incomplete. An Academy publication noted that readings as low as three had been derived from rainfall in New Hampshire. That is a reasonable quality acid, in case you need it for processing.

A series of ethical questions were raised by David Rose. How should we distribute income and energy between the rich and poor in the United

States? On a larger scale, how do we distribute income and energy between the rich and poor in the world? As usually happens, we are uncomfortable with these questions; we don't have good tools to handle them, and so we give them short shrift. These are important questions that cannot be neglected. In the extreme, one could envision a United States with a standard of living 100 times greater than the rest of the world. That would be a highly unstable situation leading to terrorism and war.

Another major equity issue concerns weighing the consumption of people today against that of 10, 100, or 1,000 years from now. Do we want to use up all the resources and let our grandchildren fare for themselves?

A possible long-term risk from burning fossil fuels is the increase in atmospheric CO_2. Coal gives rise to more than its share of visits, from extraction to burning. David Rose said that if we were to exercise prudent judgment, we might well ban all coal production. That is an extreme statement, but it does indicate that we will watch all energy production, particularly coal, more closely in the future.

I would like to sound a note of hope at this point. The lesson I got out of the discussion on Kaiparowits is that we have made a great deal of progress. Given what has been learned, a project of that sort would not be attempted. Instead, the utilities have discovered ways to reduce environmental effects, stay away from areas where effects would be egregious, and proceed in a way that leads to exchange of information and views. Similarly, the intervenors have learned to ask where plants can be located instead of opposing each case. While delays won't disappear or proposals all be approved, I think that a great deal has been learned.

Lest anyone believe that the environmentalists were being entirely unreasonable, the films showed us pictures of the smoke plume from the Four Corners plant. In many other cities, particularly in the Northeast, the levels of air pollution got markedly worse between 1969 and the mid-1970s and only now are beginning to improve. There is much room to improve.

The tumult of the 1960s and 1970s stems, first, from the public trying to decide what it wants and, second, from the lack of experience in dealing with environmental challenges.

Other problems associated with major expansion of coal production are labor relations, the boomtowns and bust towns, and changed lifestyle in the West.

During the debate between Houthakker and O'Leary, I jotted down some numbers. A 1,000-megawatt coal burning plant requires about 3 million tons of coal each year. At $20 a ton, that would be a fuel bill of about $60 million. Unregulated natural gas or OPEC petroleum would triple that fuel bill. That is, you would be paying approximately $120 million a year more for that fuel. That is a bit of an incentive to switch over to coal if it could be burned.

What are the problems for the future? The issues revolve around telling the public the nature of the alternatives we face and the associated problems and doing this in such a way as to be believed. The industry people complain that they have no credibility when they talk to the public, but the polls indicate that the general public doesn't believe the governmental officials either. Unfortunately, nobody has credibility on energy issues. It is absolutely essential that we get these facts out, that we begin to have some reasoned debate. Only then can we begin to get credibility.

There are a series of steps that can be done in the short term to alleviate the energy problem. The first is to build capacity where that is possible. There is little danger of overcapacity during the 1980s; delays in licensing and construction will see to that. The second is that even when licenses aren't granted, advanced planning is helpful. Someday we will need energy rapidly, and having detailed plans will be of immense value.

What about predictions? Frank Zarb gave one. He predicted increases in coal mining from the 660 million tons of coal mined in 1976 to about 1 billion tons in 1981. Without claiming to be an expert at all on this, I would be surprised if we reached 800 million tons per year (a 20 percent increase). I see a host of reasons, particularly on the demand side, to doubt Zarb's forecast. I believe the next five years will be like the last five, with wildcat strikes, uncertainty, and delays; we will continue to grope toward understanding our goals and possibilities rather than settling down to solve the problems. There will be more general understanding and debate over these issues, but I doubt there will be much resolution. That means that we will be in a worse situation in 1981 than now. My prediction is that we will be importing a great deal more OPEC oil, that we will be paying a lot more money for it, and that the problems will have become more evident. Thus, 1981 looks like a somewhat bleak year.

THE ACADEMY FORUM

Panel for Inquiry

Donald G. Allen, Vice-President, New England Electric System
James Boyd, President, Materials Associates
Hendrik Samuel Houthakker, Henry Lee Professor of Economics, Harvard University
David J. Rose, Professor of Nuclear Engineering, Massachusetts Institute of Technology

Program Committee

Walter R. Hibbard, Jr., University Distinguished Professor, College of Engineering, Virginia Polytechnic Institute and State University, *Cochairman*
Alvin M. Weinberg, Director, Institute for Energy Analysis, Oak Ridge, *Cochairman*

Peter L. Auer, Professor, Graduate School of Aerospace Engineering and Director, Laboratory for Plasma Studies, Cornell University
Louise C. Dunlap, President, Environmental Policy Institute
George R. Hill, Director, Fossil Fuel Plants Department, Electric Power Research Institute
Kenneth C. Hoffman, Associate Chairman for Energy Programs and Head of the National Center for Analysis of Energy Systems, Brookhaven National Laboratory

Kenneth A. Hub, Energy Systems Analyst, Energy and Environmental Systems Division, Argonne National Laboratory
John F. O'Leary, Administrator, Federal Energy Administration
Harry Perry, Consulting Economist
Raymond Zahradnik, Consultant and former Director of Coal Conversion and Utilization, Energy Research and Development Administration
Martin B. Zimmerman, Research Staff Economist, Energy Laboratory and Lecturer, Sloan School of Management, Massachusetts Institute of Technology

General Advisory Committee

Robert McC. Adams, Professor, The Oriental Institute
Kenneth J. Arrow, James Bryant Conant University Professor, Project on Efficiency of Decision Making in Economic Systems, Harvard University
David Baltimore, American Cancer Society Professor of Microbiology, Center for Cancer Research, Massachusetts Institute of Technology
Arthur M. Bueche, Vice-President, Corporate Research and Development, General Electric Company
Freeman J. Dyson, Professor of Physics, School of Natural Sciences, The Institute for Advanced Study
Donald S. Fredrickson, Director, National Institutes of Health
Gertrude S. Goldhaber, Senior Physicist, Brookhaven National Laboratory
Michael Kasha, Director, Institute of Molecular Biophysics, Florida State University
Daniel E. Koshland, Jr., Professor and Chairman, Department of Biochemistry, University of California, Berkeley, *Chairman*
Philip Morrison, Institute Professor, Department of Physics, Massachusetts Institute of Technology
John R. Pierce, Professor of Engineering, Steele Laboratory, California Institute of Technology
Alexander Rich, Sedgwick Professor of Biophysics, Massachusetts Institute of Technology
Frederick C. Robbins, Dean, School of Medicine, Case Western Reserve University
Lewis Thomas, President, Memorial Sloan-Kettering Cancer Center
Alvin M. Weinberg, Director, Institute for Energy Analysis, Oak Ridge
David A. Hamburg, President, Institute of Medicine, *ex officio*
Courtland D. Perkins, President, National Acadmey of Engineering, *ex officio*